Applications of Computational Intelligence in Concrete Technology

Smart and Intelligent Computing in Engineering

Series Editors:
Prasenjit Chatterjee, Morteza Yazdani, Dragan Pamucar, and Dilbagh Panchal

Artificial Intelligence Applications in a Pandemic
COVID-19
Salah-ddine Krit, Vrijendra Singh, Mohamed Elhoseny, and Yashbir Singh

Advanced AI Techniques and Applications in Bioinformatics
Loveleen Gaur, Arun Solanki, Samuel Fosso Wamba, and Noor Zaman Jhanjhi

IoT - Based Smart Waste Management for Environmental Sustainability
Biswaranjan Acharya, Satarupa Dey, and Mohammed Zidan

Applications of Computational Intelligence in Concrete Technology
Sakshi Gupta, Parveen Sihag, Mohindra Singh, and Utku Kose

For more information about this series, please visit: https://www.routledge.com/our-products/book-series

Applications of Computational Intelligence in Concrete Technology

Edited by
Sakshi Gupta, Parveen Sihag,
Mohindra Singh Thakur, and Utku Kose

CRC Press is an imprint of the
Taylor & Francis Group, an **informa** business

First edition published 2022
by CRC Press
6000 Broken Sound Parkway NW, Suite 300, Boca Raton, FL 33487-2742

and by CRC Press
4 Park Square, Milton Park, Abingdon, Oxon, OX14 4RN

CRC Press is an imprint of Taylor & Francis Group, LLC

Library of Congress Cataloging-in-Publication Data
Names: Gupta, Sakshi, 1989- editor.
Title: Applications of computational intelligence in concrete technology / edited by Sakshi Gupta, Parveen Sihag, Mohindra Singh, and Utku Kose.
Description: First edition. | Boca Raton, FL : CRC Press, 2022. |
Series: Smart and intelligent computing in engineering |
Includes bibliographical references and index.
Identifiers: LCCN 2021061018 (print) | LCCN 2021061019 (ebook) |
ISBN 9781032013022 (hbk) | ISBN 9781032026350 (pbk) | ISBN 9781003184331 (ebk)
Subjects: LCSH: Concrete–Testing–Data processing. | Computational intelligence.
Classification: LCC TA440 .A67 2022 (print) | LCC TA440 (ebook) |
DDC 620.1/36–dc23/eng/20220218
LC record available at https://lccn.loc.gov/2021061018
LC ebook record available at https://lccn.loc.gov/2021061019

ISBN: 9781032013022 (hbk)
ISBN: 9781032026350 (pbk)
ISBN: 9781003184331 (ebk)

DOI: 10.1201/9781003184331

Typeset in Times
by codeMantra

Contents

Preface

Concrete technology is a branch of civil engineering which is articulating concrete and technology together. Concrete which contains cement, aggregates, admixtures, etc. has numerous rheological properties. The behavior of concrete mix is quite complex due to heterogeneity in its constituents. Since there is a variety of input variables in the concrete mix and have complicated procedures to prepare it; challenges are there to control its strength. The various tests are conducted in the concrete to ascertain its compressive, flexural, split-up tensile, fatigue strengths, etc. Such tests exhibit the characteristics of the concrete likely to perform in service loads. On the basis of experimental inputs, the accurate prediction for the strengths of concrete can be done by deploying a computation intelligence mechanism. Various algorithms are used in machine learning techniques such as artificial neural network, fuzzy logic, genetic algorithm, adaptive network-based fuzzy inference system, support vector machine, random tree, random forests, and linear regression for the prediction of concrete strength. Various chapters on the prediction of the compressive, flexural, split-up tensile strengths and other properties of concrete are included in this edition. The modeling in the concrete with marble powder, recycled aggregate, fibers, and other materials are of prime focus in the various chapters of this edition. The readers would find it interesting to satisfy their quench in the prediction of various strength parameters in concrete by exploiting useful machine learning algorithms.

Fundamental concepts and essential analysis of various computational techniques are presented to offer a systematic and effective tool for better treatment of different applications related to concrete technology. It covers a wide range of applications. The various techniques will help in process modeling made for real-life complex processes in the area of concrete technology. This book is beneficial for those researchers who want to use computation intelligence techniques and their applications in the field of concrete technology.

Editors

Miss Sakshi Gupta is Assistant Professor in the Department of Civil Engineering, Amity School of Engineering & Technology, Amity University Haryana. She is pursuing Ph.D. from the Indian Institute of Technology, Delhi, India, and has completed M.Tech in Structural Engineering from National Institute of Technology, Kurukshetra, India in the year 2013 and B.Tech from MD University, Rohtak, Haryana, India in 2011. Her areas of interest include building information modeling, concrete technology, sustainable construction, construction technology & management, computation intelligence techniques, and soft computing techniques in civil engineering applications. She is a reviewer and editorial board member of many scientific journals. She has published more than 30 papers in reputed international and national journals and attended various conferences. She has also authored many books in Elsevier and Springer.

Dr. Parveen Sihag is presently working as an Assistant Professor at Chandigarh University, India. He has completed B.Tech in Civil Engineering, M.Tech in Water Resources Engineering, and Ph.D. in Civil Engineering Department, National Institute of Technology, Kurukshetra, India. His areas of interest are hydrology, infiltration, soft computing, and optimization. He has published more than 70 research papers in conferences and journals.

Dr. M.S. Thakur has been working as a Professor in the Department of Civil Engineering, Shoolini University, Solan HP India since 2015. He obtained his B.E. (Civil) from Punjab Engineering College Chandigarh (1980), M.Tech (Structures) from IIT Bombay (1988) and Ph.D. from Stevens Institute of Technology (2002), Hoboken, New Jersey USA. He has published papers in many international conferences and journals. He is not only an academician but is also an engineer having worked with the last posting as Engineer-in-Chief in HP Public Works Department in 2015.

Dr. Utku Kose received a B.S. degree in 2008 from computer education of Gazi University, Turkey as a faculty valedictorian. He received an M.S. degree in 2010 from Afyon Kocatepe University, Turkey in the field of computers and a D.S./Ph.D. degree in 2017 from Selcuk University, Turkey in the field of computer engineering. Currently, he is an Associate Professor at Suleyman Demirel University, Turkey. He has authored more than 100 publications including articles, edited books, proceedings, and reports. He is also a member of editorial boards of many scientific journals and serves as one of the editors of the Biomedical and Robotics Healthcare book series by CRC Press. His research interest includes artificial intelligence, machine ethics, artificial intelligence safety, biomedical applications, optimization, chaos theory, distance education, e-learning, computer education, and computer science.

Contributors

Anshul Aggarwal
Department of Civil Engineering
NIT Kurukshetra
Kurukshetra, India

Paratibha Aggarwal
Department of Civil Engineering
NIT Kurukshetra
Kurukshetra, India

Sejal Aggarwal
School of Architecture and Design
Manipal University
Jaipur, India

Yogesh Aggarwal
Department of Civil Engineering
NIT Kurukshetra
Kurukshetra, India

Hussain Alahmad
Department of Civil and Architectural Engineering
KTH Royal Institute of Technology
Stockholm Sweden

Fadi Hamzeh Almohammed
Department of Civil Engineering
Shoolini University
Solan, India

Ahmad Alyaseen
Department of Civil Engineering
Shoolini University
Solan, India

Sunita Bansal
Department of Civil Engineering
Manav Rachna University
Faridabad, India

Ahmed Elbeltagi
Agricultural Engineering Department, Faculty of Agriculture
Mansoura University
Mansoura, Egypt

Deepika Garg
Department of Mathematics
G.D Goenka University
Gurugram, India

Sakshi Gupta
Department of Civil Engineering
ASET, Amity University
Gurugram, India

Karam Hammadeh
Department of Computer Science and Engineering
Vel Tech Rangarajan Dr. Sagunthala R&D Institute of Science and Technology
Chennai, India

Veena Kashyap
Department of Civil Engineering
Shoolini University
Solan, India

Utku Kose
Department of Computer Engineering
Suleyman Demirel University
Isparta, Turkey

Bhupender Kumar
Department of Civil Engineering
Shoolini University
Solan, India

Navsal Kumar
Department of Civil Engineering
Shoolini University
Solan, India

Pranjal Kumar Pandey
Department of Civil Engineering
NIT Kurukshetra
Kurukshetra, India

Arunava Poddar
Department of Civil Engineering
Shoolini University
Solan, India

Rabee Rustum
School of Energy, Geoscience, Infrastructure and Society
Heriot-Watt University
Dubai, United Arab Emirates

Khyati Saggu
Department of Civil Engineering
Panipat Institute of Engineering and Technology
Samalkha, India

Nitisha Sharma
Department of Civil Engineering
Shoolini University
Solan, India

Parveen Sihag
Department of Civil Engineering
Chandigarh University
Ajitgarh, India

Balraj Singh
Department of Civil Engineering
Panipat Institute of Engineering and Technology
Samalkha, India

Tanvi Singh
Department of Civil Engineering
Panipat Institute of Engineering and Technology
Samalkha, India

Salwan Tajjour
Department of Centre of Excellence, Energy Science and Technology
Shoolini University
Solan, India

Mohindra Singh Thakur
Department of Civil Engineering
Shoolini University
Solan, India

Ankita Upadhya
Department of Civil Engineering
Shoolini University
Solan, India

1 Usage of Computational Intelligence Techniques in Concrete Technology

Utku Kose
Suleyman Demirel University

CONTENTS

1.1 INTRODUCTION

Today's modern world is approaching rapid advancements thanks to many digitally transformed tools. However, such tools are highly supported with computational aspects forming analytical and algorithmic solutions. Here, the most important contribution is by the field of artificial intelligence (Dean et al., 1995; Neapolitan & Jiang, 2018). Artificial intelligence is known as the most innovative research area for not only the present world but also for the future. When artificial intelligence is examined, it is possible to observe that there is a wide variety of algorithmic architectures effectively used for developing different techniques. These techniques are generally used to ensure the following solution methods (Garnham, 2017; Lan, 2020; Neapolitan & Jiang, 2018):

- Iteratively finding the most optimum values for the target mathematical model of the problem,

DOI: 10.1201/9781003184331-1

- Iteratively processing data samples to find a hidden pattern, which could not be detected by using traditional analytical approaches.

The first solution method is generally associated with optimization-oriented efforts, which are examined under classical optimization by Mathematics. On the other hand, the second solution is related to statistically empowered iterative algorithms, which can optimize some parameters so that the used algorithm may be sensitive for further data flows. Briefly, these solutions are used to form different soft computing techniques under the umbrella of artificial intelligence. Since these techniques are associated with computational runs, a more formal name, computational intelligence, has been widely used to define them (Azar & Vaidyanathan, 2015; Eberhart & Shi, 2011). For today's advanced problems, which are requiring more contributions by alternative data-processing algorithms, computational intelligence is often used for deriving alternative findings and shaping the way of the scientific literature.

When it is deeply examined, it is possible to see that computational intelligence techniques often use iterative data-processing so that simulating a learning mechanism, which seems to be a digitally transformed way of human-side learning capabilities. In the context of artificial intelligence, these techniques are collected under also machine learning. Machine learning is known as the most effective sub-area of artificial intelligence as machine learning techniques can learn from past data samples to ensure predictive or descriptive findings for newly encountered new data flows (Dean et al., 1995; Eberhart & Shi, 2011; Neapolitan & Jiang, 2018). Although descriptive solutions tend to be associated with instant connections among the observed data, predictive solutions are mostly for deriving new data, which are used for determining the future state of events, problems, and so on. When the usage of computational intelligence techniques is observed, it is possible to see that there is a great interest for predictive solutions since determining future state is among popular focuses, and even descriptive outcomes are highly connected with predictive further steps.

Early periods of computational intelligence techniques are associated with not daily life before. However, with more developments within the last quarter of the 20th century and the 20-year period of the 21st century so far, foundations by computational intelligence have been widely seen in our daily life. It is remarkable that for even today's daily life solutions, there has always been an effort in terms of solving engineering problems. That situation is similar when different tools and objects, which have a critical role in human life, are discussed. From healthcare to education, or defensive technologies to communication, Computational intelligence is highly associated with all different fields of life (Andina & Pham, 2007; Kose & Koc, 2014; Yu et al., 2018). Here, construction technologies have been often supported by predictive outcomes of Computational Intelligence.

Concrete technology is a remarkable component where computational intelligence techniques have been widely employed. As associated with that, the literature needs reference works for the outcomes reported for the combinations of different computational intelligence techniques to deal with different applications of concrete technology. As an introductory approach, this chapter briefly gives pre-information about this book. The following sections discuss the use of different computational intelligence/soft computing solutions to find effective outcomes. However, this chapter uses original information by the authors and enables readers to have the necessary knowledge for understanding this book.

As associated with the aim of this chapter, the next section provides information about the models used in this book. After that, the third section is devoted to brief information about this book and the outcomes. This chapter ends with conclusions and some ideas about the future.

1.2 COMPUTATIONAL INTELLIGENCE MODELS FOR CONCRETE TECHNOLOGY

Computational Intelligence is a wide family of soft computing techniques to deal with advanced problems. By focusing on the field of artificial intelligence, it is possible to examine hundreds of different techniques with even variations so that there is a great scope for the researchers to use alternative solutions for deriving findings. Figure 1.1 provides a general scheme for widely used Computational Intelligence models.

In this book, some remarkable techniques were used. However, the majority of the techniques were with the use of neural network-based models. That's because of the advanced accuracy capabilities of neural networks and perceptron foundations. However, there are also some other strong competitive techniques and additional components to ensure hybrid formations for improved outcomes. The following sections include general information on these techniques as combined with the original content by the authors.

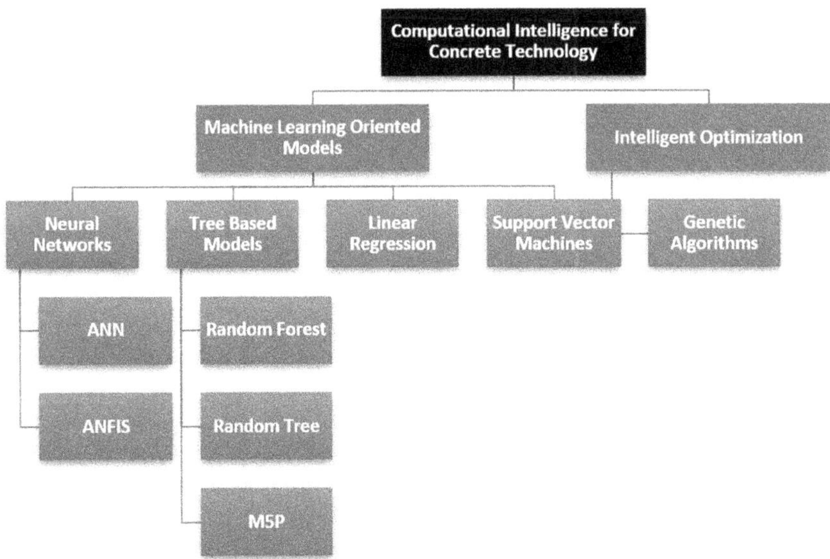

FIGURE 1.1 Computational intelligence models for concrete technology applications.

1.2.1 Artificial Neural Networks (ANN)

Artificial neural network (ANN) is the strongest and long-time used technique coming from the machine learning side of artificial intelligence. With its computational capabilities, it is among the widely used computational intelligence models to deal with advanced problems. Briefly, a typical ANN architecture had set of nodes which divided into layers (input layer, hidden layer can be multi-layers or zero, output layer), biases, weights between these nodes, and the ways of flowing the information on the neural network as feed-forward networks, the signal in this model transfer in one way from the input layer to the output layer. Furthermore, in feedback networks, the signal can travel in both directions by the loops between the hidden layers, and this type of connection has a memory for an internal state to process a sequence of inputs (Abraham, 2005). According to Benardos and Vosniakos (2007), there are four elements to clarify optimal ANN's architecture: number of layers (especially hidden layers), number of nodes (neurons) in each layer, type of active function in each node/layer, and the algorithm used in training plays an essential role in determining the value of weight and biases of nodes. The backpropagation (BP) algorithm is a supervised learning algorithm; it is used to train a multi-layer neural network, which is done by iteratively updating the weights of nodes to minimize the value of the error function in weight space, and this technique is called as the gradient descent or delta rule (Ampazis et al., 1999; Zweiri et al., 2003). Figure 1.2 represents a general scheme for a typical ANN (Bre et al., 2018).

1.2.2 Adaptive Neuro-Fuzzy Inference System (ANFIS)

Adaptive neuro-fuzzy inference system (ANFIS) is a technique in which the number of hybrid combinations of FIS and ANN is used to find out the various aspects and,

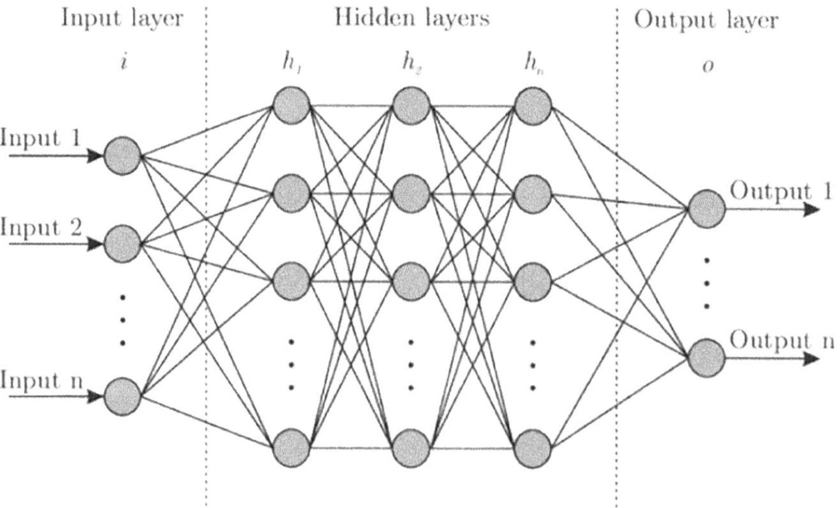

FIGURE 1.2 Artificial neural networks (Bre et al., 2018).

as a result, used to prepare the model forecast for the future responses. This inference system can combine the benefits of fiber length with neural network principles in a single framework to solve problems. ANFIS is a hybrid machine learning technique that transforms a given input into a target output by using a fuzzy interference system model. It also enables the analysis easier. The training and test data set is utilized to create the ANFIS model, and evaluate the built model's prediction ability during the model building process, respectively. The concept behind using a test data set for ANFIS model validation is that the model starts overfitting the training data set after a certain point in the training. The parameters decide the performances of a fuzzy system which describes the membership function and the rule-based use in the system specifies how often a fuzzy system performs. The various parameters can be modified and can be combined with a fuzzy system and ANNs to build a neuro-fuzzy system (Ahmadi-Nedushan, 2012; Poddar et al., 2018; Kumar et al., 2020; Poddar et al., 2020). The benefits of both approaches are combined in neuro-fuzzy systems, which merge the natural language description of fuzzy systems with the learning properties of ANNs. ANFIS is a composite in which the parameters of the fuzzy system are fixed using an adaptive BP learning algorithm. By using the fuzzy sets, ANFIS combines the human-like reasoning structure of fuzzy systems. ANFIS is a multilayer feed-forward network where each node performs a specific role on incoming signals (node function). In this process two inputs x and y can be considered and one output z. During the ANFIS operation mainly five processing stages take place (Figure 1.3). The main objective of the ANFIS is to incorporate the best features of fuzzy systems and neural networks. ANFIS approach mainly uses the membership function parameters which can be adjusted using either alone or in combination with various types of methods. One of the common methods for computing the output is the Sugeno-type fuzzy model in which the fuzzy preposition worked upon the 'If' rule condition. Sugeno-type model is one of the simplest methods to detect because

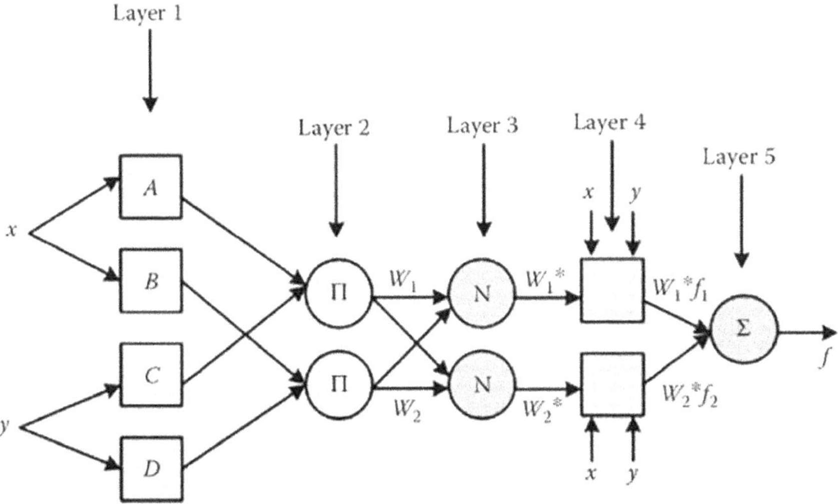

FIGURE 1.3 General flow of the genetic algorithm (Shukur et al., 2015).

it has fewer rules and its parameters can be estimated from numerical data using optimization approaches.

1.2.3 Genetic Algorithm (GA)

Genetic algorithm (GA), which is one of the most popular and global search techniques, is used to optimize a dynamic problem with several difficult-to-obtain parameters or functions (Ahmad et al., 2010; Gopan et al., 2018). This evolutionary optimization algorithm is comprised of four steps: population, selection, crossover, and mutation. The way the population is presented (chromosome), the health mechanism in the selection, and the mutation rate play a significant role in obtaining an optimum response (Sastry et al., 2005). Chromosome reproduction efficiency is measured by the number of chromosomes that reproduce. GA can perform a wide range of stochastic chromosomal operations, including genetic operations such as crossover and mutation and evolutionary operations such as selection (Figure 1.4). The sorting operator separates the fitter chromosomes from the fewer fit chromosomes, which improves population fitness across generations. When two or more chromosomes

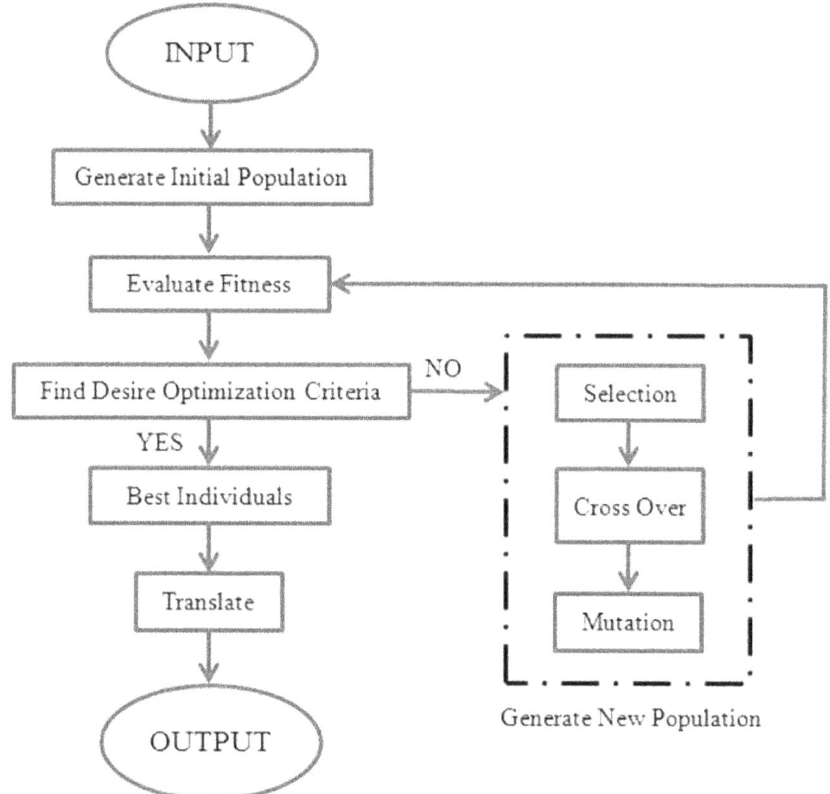

FIGURE 1.4 General model of the ANFIS (Aghbashlo et al., 2018).

combine to form a new generation, this is known as crossover recombination; by combining the strengths of the two or more parent chromosomes, crossover recombination aids in the development of more competent individuals in the population. The cross-site between the two chromosomes is selected randomly, and the genes are exchanged between the parent and offspring chromosomes to give rise to better offspring; it is possible to replicate healthy chromosomes. By spontaneously changing the foundation blocks, mutation incorporates genetic variation into the new population. It encourages the algorithm to explore the whole solution space and keeps it stuck in local minima. The next wave of the population is generated by the crossover and mutation operators working together. The method is repeated until the number of generations or a specific value is achieved or the health characteristics plateau at a particular value has been reached. Like many other optimization algorithms, GA is helpful to determine essential parameters of machine learning techniques such as ANN. Here, the primary goal of using a hybrid GA-ANN model is to minimize the objective function and increase the power and speed of ANN by adjusting a set of weights and biases (Gopan et al., 2018; Momeni et al., 2014). As long as it is possible to catch some parameters from a Computational Intelligence/Machine Learning technique, GA can be used for different types of techniques eventually.

1.2.4 Random Forest (RF)

The random forest (RF), first described by Breiman (1996), is an ensemble learning method for regression and classification that has been chosen for the arrangement of many nonlinear or difficult engineering problems (Thakur et al., 2020). RF is made up of a number of tree predictors, each of which is created from the random vector, i.e., using input vector sampled independently and built-in different ways. It was used to assess the accuracy of concrete compressive strength prediction. To generate training, a data set is divided randomly which is known as bagging (Singh et al., 2019). The out-of-bag error testing validation or using the rest of the data is computed by each tree in the forest. The total out-of-bag error rate is then calculated by adding the mistakes from all the trees (Chopra et al., 2018). The number of trees to be produced (k) and the number of input variables (m) utilized at each node to generate a tree are two user-defined parameters in RF regression. Only a few factors are looked at each node to find the appropriate split. As a result, the RF regression is made up of a number of (k) trees (Singh et al., 2019). Furthermore, each tree is expanded to reduce classification error, although the result is accentuated by random selection. The RF model's drawback is determining how much the prediction error grows as the data output for exact variables is permuted (Lu et al., 2020). The general flow of the RF is shown in Figure 1.5 (Wu et al., 2019).

1.2.5 Random Tree (RT)

The random tree (RT) model is an alternative approach, which is a machine learning algorithm that combines two existing algorithms: (1) RF ideas are used to connect single model trees. (2) RTs are decision trees with a linear model in each leaf that is optimized for the constrained subspace defined by that leaf (Pfahringer, 2010).

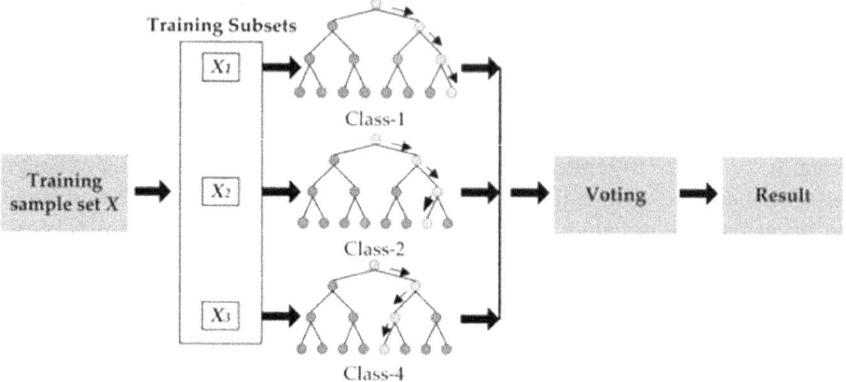

FIGURE 1.5 General flow of the RF model (Wu et al., 2019).

In this modeling technique, data are divided randomly into training and testing data sets and construct a tree with *k* number of random features at each node. It builds a highly specific classifier for a large number of data sets and generates an internal impartial estimate of the generalization error as the forest construction continues. It also works with a large number of input variables and evaluates the relevance of factors in defining classification, resulting in an internal unbiased evaluation of the generalization error as the forest building progresses (Deepa et al., 2010). Meanwhile, the training error is calculated internally, therefore cross-validation or bootstraps aren't required to assess the training stage's precision. It's worth noting that the result for regression issues is calculated by taking the average of all the forest individuals' responses (Lu et al., 2020). averaging a suitable number of trees is important for successful prediction performance since the trees are semi-arbitrary and hence not perfect in separation. Because of the randomization process, adding additional trees to an ensemble will never significantly degrade performance, but as with most ensemble algorithms, any enhancements will ultimately degrade performance (Pfahringer, 2010).

1.2.6 Linear Regression (LR)

Linear regression (LR) is one of the most broadly utilized statistical tools because gives a basic technique to building up a useful relationship among factors. It is a set of informative components that removes non-significant variables one by one until the remaining variables are all significant. Any progression variable with the lowest absolute t-statistic, which is defined as the percentage of the coefficient to its standard error, is removed (Dunlop & Smith, 2015). LR is the mathematical equivalent defining the best line through the data set. Parameter values are found that limit the amount of-the-squares deviation of the trial from the determined qualities. It is assumed that these errors are randomly scattered and that systematic errors are not present. Systematic errors can only be removed by correct experimental design (Leathetbatrow, 1990). It is also helpful in finding the correlation coefficient which

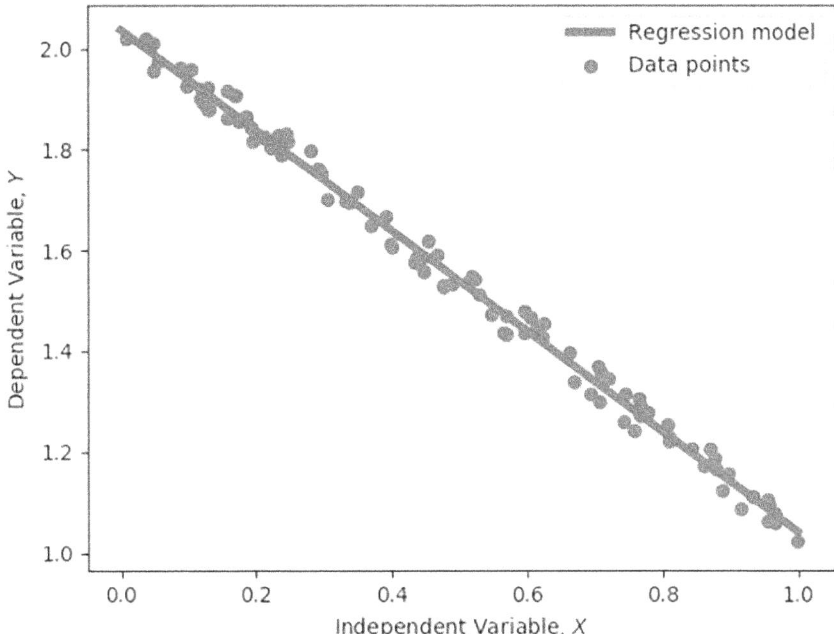

FIGURE 1.6 General outcome when the LR model is used (Date et al., 2021).

tends to describe the variations in the data set. The more the value is closer to 1, the more the data are reliable (Deepa et al., 2010). Figure 1.6 shows a general outcome when the LR is used (Date et al., 2021).

1.2.7 M5P Model

The M5P decision tree algorithm is employed for qualitative and continuous variables, as well as missing values, in high-dimensional tasks (Ali et al., 2020). A model tree is one of the regression trees, which use the linear function in their leaves. They are similar to a linear equation in piece-wise form. An M5P model tree is capable of estimating continuous mathematical features (Thakur et al., 2021; Mohammed et al., 2020; Almasi et al., 2017). For reducing the threat of over-fitting, pruning is applied in this algorithm. To obtain better information with less deviation in the values of the intra-subset class in each branch, a separation approach is used. Tree growth, trimming, and smoothing are the three primary processes in the M5P preparation process (Thakur et al., 2021). The data are studied using a decision tree, which is a tree with roots, branches, and leaves. In general, recursion is used to identify the optimum outcome for the anticipated class, and it entails selecting an attribute in the root node, adding a branch for each of the conceivable values, and repeating the operation for each branch using only the attributes that reach the branch (Ali et al., 2020). There are two stages to the growth of such a tree. The construction of the tree is the initial phase. A recursive process of splitting the data space into multiple pure subspaces in terms of classes is activated based on a learning sample. The forecast criterion is used to make the estimate. The tree will be pruned in the second phase. To obtain strong predictive performance, the final step removes the unrepresented branches. This operation necessitates the use of a criterion to identify which branches should be pruned. The new leaves are labeled based on the distribution of observations utilized in the learning process after they have been pruned (Ali et al., 2020; Blaifi et al., 2018). The M5 tree method can handle very high dimensionality and deals with continuous problems rather than discrete ones. It shows the piecewise details of each linear model that was built to approximate the data set's nonlinear relationship. Information about the partition norm of the M5 model tree has been obtained based on the error calculations of each node. Calculate the standard deviation of the class value that reaches the node to determine the error. To break at this node, choose the attribute that maximizes the reduction of expected errors caused by the test of each attribute.

1.2.8 Support Vector Machine (SVM)

Support vector machines (SVM) were developed by (Muller et al., 1999; Smola and Scholkopf, 1998) and gained popularity due to high classification and prediction performance (Campbell et al., 2000; Yang et al., 2002). SVM creates a decision boundary or the best fit line that can segregate n-dimensional spaces into classes so that a new data point can be applied easily in the correct category. This best decision boundary is called a hyperplane. The hyperplane is derived by selecting training data points using selected data points called support vectors. The distance of vector from the hyperplane is called margin and optimal hyperplane maximized the margin. Unlike other logistics regression in which output lies in the range of [0, 1], the threshold values of SVM lie between [−1, 1] which act as a margin (Figure 1.7).

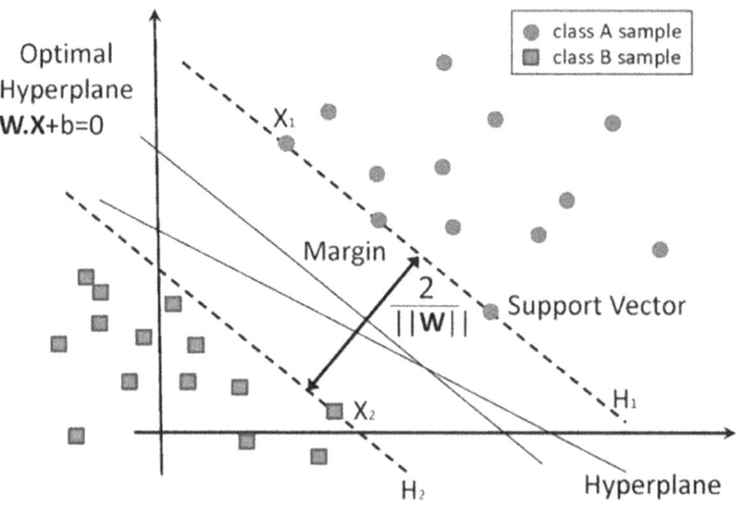

FIGURE 1.7 Representation of the SVM approach (García-Gonzalo et al., 2016).

SVM algorithms use a set of mathematical functions that are defined as the kernel. The kernel function calculates dot product in high-dimensional variables space using low-dimensional space data input without making any assumptions about data features.

1.3 PREDICTIVE COMPUTATIONAL INTELLIGENCE IN CONCRETE TECHNOLOGY

This book is related to predictive efforts for mostly finding the compressive strength of the concrete. Different parameters and components in the concrete must be important to forward the way of predictive efforts. On the other hand, there is also a chapter for predictive outcomes regarding the ultrasonic pulse velocity (UPV) of the concrete. By examining these under two subsections, the following sections provide brief information about the findings.

1.3.1 Prediction of Compressive Strength of the Concrete

As predictive applications have a critical place in effective outcomes in dealing with the nature of the concrete and its role within different application areas, computational intelligence techniques have been widely used and compared for several problem issues. As Alyaseen et al. point out in this book, concrete compressive strength prediction is important to understand the material behavior associated with nonlinear, complicated, or unknown interactions among the components. They indicate the issue as follows: Modeling concrete compressive strength based on design mix proportions is a highly nonlinear issue that is difficult to solve using traditional mathematical methods. Because of that, the ANN with BP training method has historically been employed to simulate material behavior, and the initial weights and biases heavily affect its convergence probability since it relies on the gradient descent methodology. However, that method has a significant disadvantage that adversely affects the performance of the ANN. It is feasible to hide the BP algorithm's tendency to converge at suboptimal locations by using ANN and GA. While combining two different methods increases the global search capacity of GA while simultaneously improving the searchability of the BP algorithm, it does it more effectively. So, to solve the issue, they apply a hybrid solution for the initial determination of the neural network weights and biases, and run the GA algorithm for getting an optimal initial set of weights and biases, which were then fine-tuned using the Levenberg-Marquardt back- propagation (LMBP) technique. As compared to the commonly used hybrid ANN-GA model, this chapter finds that the backpropagation neural network (BPNN) method comes with consistent predictions throughout the training and testing stages, demonstrating the resilience of the hybrid modeling approach. Furthermore, the BPNN model ensures the required performance level faster than the ANN-GA model. So, the suggested model may be utilized as a decision support tool, assisting technical personnel in predicting the compressive strength of a particular concrete design mix. It will further significantly reduce the work and time required to create a concrete mix for a specific compressive strength without the need for repeated trials.

As associated with the related compressive strength prediction solutions, the other chapter by Upadhya et al. comes with a widely comparative work to evaluate performance outcomes by RF, RT, and LR models, focusing on the glass fiber reinforced concrete this time. However, they used a data set containing 156 samples from the literature, and they randomly divided those data sets into two sets: with 70% of the data being used for training and 30% being used for testing. An assessment of the performance during the testing and validation stages validated the efficacy of the better models. In the research by Upadhya et al., cement, fine aggregate, coarse aggregate, water, superplasticizer, fiber length, glass fiber, and curing days are the variables in the data set as input parameters to obtain the compressive strength at the output side. Comparison of the predicted values of testing results indicated that the RF model worked better than other soft computing-based models. The minimum error values indicate that the predicted compressive strength values are quite close to the actual compressive strength values. The performance assessment of the RF model gives better results among other applied models with the higher correlation coefficient (CC) values as 0.989325 with the lower error RMSE 3.632436 and MAE 2.821574 for estimating the compressive strength of glass fiber reinforced concrete and also NSE 0.977148, WI 0.977925, LMI 0.859511 are much closer to the actual values similarly, the values of SI and BIAS are 0.081217 and −0.86702 respectively which is minimum in RF as compare to RT and LR model. As a final touch, the sensitivity analysis realized in that research concludes that the most significant parameter affecting the compressive strength of glass fiber reinforced concrete is the curing days.

In the other chapter by Almohammed, the predictive analyses were done for the compressive strength of concrete with basalt fiber reinforced polymer (BFRP). He briefly applied RF, M5P, M5P-Stochastic, ANN, and RT-based models to predict the values for defining the best model. However, the research consisted of a data set, which was collected from various research papers. As done by Upadhya et al., the total data set is divided into 70% of the training data set and 30% of the testing data set. Cement, fine aggregate/crushed sand, coarse aggregate, water, superplasticizer, fly ash, BFRP, diameter of BFRP, length of BFRP, curing time were the input variables whereas compressive strength of the concrete with BFRP was the output. According to the evaluation indices of root mean square error (RMSE), mean absolute error (MAE), and CC, the evaluation of Almohammed recommend the RF as the best model to predict the compressive strength for the concrete with BFRP with a value of CC (0.9968, 0.9813), RMSE (1.0876, 2.027), and MAE (0.8193, 1.5336) for training and testing data sets, respectively.

Another work by Sharma et al. points out the use of steel fiber. In the chapter associated with that work, analysis over the effect of input variables such as cement, fine aggregate, coarse aggregate, water, superplasticizer/high range water reducer, fly ash, steel fiber, and curing days was done on the compressive strength of the concrete mix. The models in this research were ANN, ANN-cross-validation in ten folds and LR. Evaluation of developed models was with CC, MAE, and RMSE. However, cement, fine aggregate, coarse aggregate, water, superplasticizer/high range water reducer, steel fiber, and curing days are considered as input parameters in ANN-M7. Results obtained from the ANN-M7 were found to be most suitable to predict the concrete compressive strength. Among all the applied models, the ANN-M7 predicts better results followed by ANN-cross-validation (ten folds) with the highest CC values of 0.9788 and 0.9059 with lower error values such as MAE 2.9744 and 4.2074 and RMSE values of 3.9719 and 6.7505 respectively in the testing data set. Also, the ANN-M7 model has minimum error bandwidth, and it is a good fit for predicting the output.

Considering the ANN and ANFIS models, Kashyap et al. used a total of 65 random data rows for training and a total of 28 data rows for the testing phase. As the input parameters, aspect ratio, percent of fiber, and the number of days were used in this chapter. The prediction of compressive strength was for the concrete with coconut fiber. This chapter concludes that ANFIS Triangular-based model performs well for the determination of compressive strength of concrete with a CC, RMSE, and MAE values of 0.97, 1.56 and 1.01 and 0.84, 3.87, and 2.70 for the training data set and testing stage respectively as compared to other membership functions. The results also suggest the improved performance of the ANFIS model as compared to the ANN model for determining the compressive strength of the concrete.

In the associated literature, self-compacting concrete containing silica is another research component to apply computational intelligence for predictive purposes. In the chapter by Pandey et al., ANN, linear progression, SVM, RF, and hybrid bagging technique were all used to predict the compressive strength of self-compacting concrete containing silica as supplementary cementitious materials. In this sense, they used an experimental data set combined from several published works and it was applied to train and test for the models. The data set used in models consists of input and output parameters with cement, sand, coarse aggregate, silica, superplasticizer, and water binder ratio as input and 28 days compressive strength as the output. The evaluation was done through correlation coefficient, MAE, and RMSE calculations. According to the findings, the ANN-based bagged model outperforms every other model with $R2=0.9289$, MAE = 4.2069 and RMSE = 5.9004 in the prediction of compressive strength. Also, ANN-based model performs better than techniques of LR, SVM, and RF with $R2=0.9249$, MAE = 4.3658, and RMSE = 6.0064. Bagging-based RF model perform better than RF model with $R2=0.899$, MAE = 5.4527, RMSE = 7.5371. As this chapter includes sensitivity analysis, it was observed that the cementitious materials were the most influencing parameter in comparison to other input parameters for the prediction. As the authors indicate, the results coincide well with the experimental results in all the phases of validation so as to clarify the accuracy of the proposed B-ANN model. In that model, an improvement in compressive strength of self-compacting concrete is shown with the use of silica as a partial replacement of cement.

1.3.2 Prediction of Ultrasonic Pulse Velocity of the Concrete

Another predictive problem application way for the concrete is related to UPV. As Singh et al. indicate in their chapters, the UPV is one of the main and popular non-destructive tests of concrete which examine the cracks, quality, homogeneity, and defects within the concrete. In order to deal with that issue, they provided a research work, in which the values of UPV were predicted using the three soft computing models, such as M5 rule (M5PR), M5P tree (M5PT), and RF. The research area was the old hostel of Panipat Institute of Engineering and Technology, Samalkha Panipat, which was found to have some signs of distress and surface cracks. Here, the tests carried out were UPV, core test, half-cell potential test, carbonation test, and chloride test. The findings showed that all the soft computing models can predict the UPV. But RF is the model which can predict the UPV accurately. The values of CC and RMSE are seen as 0.9541 and 0.2565 respectively for the RF, which was much more ideal than the values of M5PR (0.8492 and 0.4197) and M5PT (0.8246 and 0.4694). So, the authors point out that the RF is the model which can be used in the prediction of UPV. Their chapters also show that although the experimentation of the UPV is not a simple task, the values of the UPV can be predicted by using different soft computing models.

1.4 CONCLUSIONS

This book shows that the use of computational intelligence models have a great role in predictive outcomes. When compared, it is possible to see that especially neural-network-oriented models show better results, since they have more complicated solution mechanisms. That's also because more parameters to optimize may be giving better accuracy capabilities for machine learning-based models. Of course, alternative models can give predictive values in the context of both compressive strength and UPV prediction.

When the future possibilities in terms of using computational intelligence for concrete technology are considered, the following points may be expressed:

- Predictive analytics is among key research efforts for dealing with problems in the concrete technology field.
- More complicated models can give better results when compared with alternative models from the literature.
- The literature on computational intelligence and a high number of solution mechanisms by the artificial intelligence field show that there will be always a remarkable research opportunity in the intersection of computational intelligence and concrete technology.
- It is remarkable that different concrete materials may be forwarding research to deal with different input–output balances in terms of data flow used for specific applications such as compressive strength prediction.
- Further research for computational intelligence and concrete technology may include intense use of the Internet of Things and robotic systems as well as smart environments to catch more data for improved outcomes.
- In order to support predictive results, descriptive methods and more data pre-processing hybrid formations may be included in further research.

REFERENCES

Abraham, A., 2005. Artificial neural networks. In: *Handbook of Measuring System Design*. Peter Syden-ham and Richard Thorn (Eds.), John Wiley & Sons, London, pp. 901–908.

Aghbashlo, M., Hosseinpour, S., and Mujumdar, A. S., 2018. Artificial neural network-based modeling and controlling of drying systems: A review. *Intelligent Control in Drying*, pp. 155–172.

Ahmad, F., Mat-Isa, N., Hussain, Z., Boudville, R. and Osman, M., 2010. Genetic algorithm-artificial neural network (GA-ANN) hybrid intelligence for cancer diagnosis. *2010 2nd International Conference on Computational Intelligence, Communication Systems and Networks*, Liverpool, United Kingdom, 28-30 July 2010.

Ahmadi-Nedushan, B., 2012. Prediction of elastic modulus of normal and high strength concrete using ANFIS and optimal nonlinear regression models. *Construction and Building Materials*, 36, pp. 665–673.

Ali, M., Talha, A. and Berkouk, E. M., 2020. New M5P model tree-based control for doubly fed induction generator in wind energy conversion system. *Wind Energy*, 23(9), pp. 1831–1845.

Almasi, S. N., Bagherpour, R., Mikaeil, R., Ozcelik, Y. and Kalhori, H., 2017. Predicting the building stone cutting rate based on rock properties and device pullback amperage in quarries using M5P model tree. *Geotechnical and Geological Engineering*, 35(4), pp. 1311–1326.

Ampazis, N., Perantonis, S. and Taylor, J., 1999. Dynamics of multilayer networks in the vicinity of temporary minima. *Neural Networks*, 12(1), pp. 43–58.

Andina, D. and Pham, D. T., (Eds.). 2007. *Computational Intelligence: For Engineering and Manufacturing*. Springer Science & Business Media, The Netherlands.

Azar, A. T. and Vaidyanathan, S., (Eds.). 2015. *Computational Intelligence Applications in Modeling and Control*. Switzerland: Springer International Publishing.

Benardos, P. and Vosniakos, G., 2007. Optimizing feedforward artificial neural network architecture. *Engineering Applications of Artificial Intelligence*, 20(3), pp. 365–382.

Blaifi, S.A., Moulahoum, S., Benkercha, R., Taghezouit, B. and Saim, A., 2018. M5P model tree based fast fuzzy maximum power point tracker. *Solar Energy*, 163, pp. 405–424.

Bre, F., Gimenez, J. M., and Fachinotti, V. D., 2018. Prediction of wind pressure coefficients on building surfaces using artificial neural networks. *Energy and Buildings*, 158, 1429–1441.

Breiman, L., 1996. Bagging predictors. *Machine Learning*, 24(2), pp. 123–140.

Campbell, C., Cristianini, N., and Smola, A. J., 2000. Query learning with large margin classifiers. *Proceedings of the Seventeenth International Conference on Machine Learning*, Morgan Kufmann Publishers Inc., San Francisco, CA, USA, pp. 111–118.

Chopra, P., Sharma, R. K., Kumar, M. and Chopra, T., 2018. Comparison of machine learning techniques for the prediction of compressive strength of concrete. *Advances in Civil Engineering*, 2018, pp. 1–9.

Date, P., Arthur, D., and Pusey-Nazzaro, L., 2021. QUBO formulations for training machine learning models. *Scientific Reports*, 11(1), pp. 1–10.

Dean, T., Allen, J., and Aloimonos, Y., 1995. *Artificial Intelligence: Theory and Practice*. Benjamin-Cummings Publishing Co., Inc, San Francisco, CA.

Deepa, C., SathiyaKumari, K., and Sudha, V. P., 2010. Prediction of the compressive strength of high performance concrete mix using tree based modeling. *International Journal of Computer Applications*, 6(5), pp. 18–24.

Dunlop, P. and Smith, S., 2003. Estimating key characteristics of the concrete delivery and placement process using linear regression analysis. *Civil Engineering and Environmental Systems*, 20(4), pp. 273–290.

Eberhart, R. C. and Shi, Y., 2011. *Computational Intelligence: Concepts to Implementations*. Elsevier. ISBN:978-1-55860-759-0.

García-Gonzalo, E., Fernández-Muñiz, Z., García Nieto, P. J., Bernardo Sánchez, A., and Menéndez Fernández, M., 2016. Hard-rock stability analysis for span design in entry-type excavations with learning classifiers. *Materials*, 9(7), p. 531.

Garnham, A., 2017. *Artificial Intelligence: An Introduction*. Routledge, London. https://doi.org/10.4324/9780203704394.

Gopan, V., Wins, K., and Surendran, A., 2018. Integrated ANN-GA approach for predictive modeling and optimization of grinding parameters with surface roughness as the response. *Materials Today: Proceedings*, 5(5), pp. 12133–12141.

Kose, U. and Koc, D., (Ed.). 2014. *Artificial Intelligence Applications in Distance Education*. IGI Global, Hershey, PA.

Kumar, N., Shankar, V., and Poddar, A., 2020. Agro-hydrologic modelling for simulating soil moisture dynamics in the root zone of Potato based on crop coefficient approach under limited climatic data. *ISH Journal of Hydraulic Engineering*, 28(1), pp. 310–326.

Lan, G., 2020. *First-order and Stochastic Optimization Methods for Machine Learning*. Springer Nature, Switzerland AG.

Leathetbatrow, R. J., 1990. Using linear and nonlinear-regression to fit biochemical data. *Trends in Biochemical Sciences*, 15, pp. 455–458, ISSN: 0968-0004.

Lu, S., Koopialipoor, M., Asteris, P.G., Bahri, M., and Armaghani, D.J., 2020. A novel feature selection approach based on tree models for evaluating the punching shear capacity of steel fiber-reinforced concrete flat slabs. *Materials*, 13(17), p. 3902.

Mohammed, A., Rafiq, S., Sihag, P., Kurda, R., Mahmood, W., Ghafor, K., and Sarwar, W., 2020. ANN, M5P-tree and nonlinear regression approaches with statistical evaluations to predict the compressive strength of cement-based mortar modified with fly ash. *Journal of Materials Research and Technology*, *9*(6), pp. 12416–12427.

Momeni, E., Nazir, R., Jahed Armaghani, D., and Maizir, H., 2014. Prediction of pile bearing capacity using a hybrid genetic algorithm-based ANN. *Measurement*, 57, pp. 122–131.

Muller K., Smola, A., Ratsch, G., Scholkopf, B., Kohlmorgen, J., and Vapnik, V., 1999. Using support vector machines for time series prediction, in: B. Scholkopf, J. Burges, A. Smola, ed., *Advances in Kernel Methods: Support Vector Machine*, MIT Press, Cambridge, MA.

Neapolitan, R. E. and Jiang, X., 2018. *Artificial Intelligence: With an Introduction to Machine Learning*. CRC Press, Boca Raton, FL.

Pfahringer, B., 2010. *Random Model Trees: An Effective and Scalable Regression Method*. University of Waikato, Department of Computer Science, Hamilton, New Zealand.

Poddar, A., Kumar, N., and Shankar, V., 2018. Effect of capillary rise on irrigation requirements for wheat. *Proceedings of International Conference on Sustainable Technologies for Intelligent Water Management (STIWM-2018)*, IIT Roorkee, India.

Poddar, A., Preeti, N. K., and Shankar, V., 2020. *Artificial Ground Water Recharge Planning Using Geospatial Techniques in Hamirpur Himachal Pradesh, India*. Roorkee Water Conclave. Indian Institute of Technology Roorkee and National Institute of Hydrology, Roorkee during February 26-28, 2020.

Sastry, K., Goldberg, D., and Kendall, G., n.d. Genetic algorithms. In *Search Methodologies*, Springer, Boston, MA, pp. 97–125.

Shukur, A. F., Chin, N. S., Norhayati, S., and Taib, B. N., 2015. Design and optimization of valveless micropumps by using genetic algorithms approach. *Journal of Engineering Science and Technology*, 10(10), pp. 1293–1309.

Singh, B., Sihag, P., Tomar, A. and Sehgal, A., 2019. Estimation of compressive strength of high-strength concrete by random forest and M5P model tree approaches. *Journal of Materials and Engineering Structures*, 6(4), pp. 583–592.

Smola, A.J. and Scholkopf, B., 1998. A tutorial on support vector regression. Tech. rep., NeuroCOLT2 Technical Report NC2-TR-1998-030.

Thakur, M.S., Pandhiani, S.M., Kashyap, V., Upadhya, A., and Sihag, P., 2021. Predicting bond strength of FRP bars in concrete using soft computing techniques. *Arabian Journal for Science and Engineering*, 46(5), pp. 4951–4969.

Wu, X., Gao, Y., and Jiao, D. 2019. Multi-label classification based on random forest algorithm for non-intrusive load monitoring system. *Processes*, 7(6), p. 337.

Yang H., Chan, L., and King, I., 2002. Support vector machine regression for volatile stock market prediction, IDEAL 2002. *Lecture Notes in Computer Science*, 24412, pp. 391–396.

Yu, K. H., Beam, A. L., and Kohane, I. S., 2018. Artificial intelligence in healthcare. *Nature Biomedical Engineering*, 2(10), pp. 719–731.

Zweiri, Y., Whidborne, J. and Seneviratne, L., 2003. A three-term backpropagation algorithm. *Neurocomputing*, 50, pp. 305–318.

2 Developing Random Forest, Random Tree, and Linear Regression Models to Predict Compressive Strength of Concrete Using Glass Fiber

Ankita Upadhya, Mohindra Singh Thakur, and Nitisha Sharma
Shoolini University

CONTENTS

2.1 INTRODUCTION

In the construction industry, concrete is a material that is broadly used as a building material and is also utilized in retrofitting and rehabilitation of masonry and concrete structures. Concrete has several desirable features, including high compressive strength, rigidity, and durability under a variety of climatic conditions, but it is also

DOI: 10.1201/9781003184331-2

weak under tension (Chandramouli et al. 2010). Normally, continuously deformed steel bars or high tensional wires are used to support cement concrete (Kumar et al. 2020). Although steel bars are a common type of concrete reinforcement, constructions subjected to very aggressive conditions, such as seawater and deicing salts, Steel reinforcements have been shown to speed up the degradation of concrete due to expanding rust (Lee et al. 2007). Researchers have come up with the addition of fibers such as polymers like basalt fiber, steel fibers, glass fiber, carbon fibers, and other fibers into concrete to eliminate the effect of corrosion and to increase strength and durability. Fiber reinforced concrete (FRC) has been strengthened by the inclusion of various fibers in concrete to eliminate the effect of corrosion and to increase strength and durability (George et al. 2019). Controlling breakage, considerably increasing elasticity, flexural strength, and improving deformity characteristics are all important functions of fiber in a concrete composite mix (Yadav et al. 2020).

Glass fiber is one of the most extensively used fibers, with a higher surface area to weight ratio and higher tensile qualities than other varieties (Hilles and Ziara 2019). Glass fibers can be stranded or chopped, but they're commonly coated with epoxy resin to improve ductility, matrix bonding, and load transmission through fibers (Faleschinia et al. 2020).Glass fiber in concrete has been studied in both non-modified and glass fiber modified concrete. The results reveal that adding glass fiber to mortar enhances its qualities while adding 1800 g/m^3 of glass fibers to concrete increases its mechanical properties. Compressive strength increased by 31.5% (Małek et al. 2021). To investigate the behavior of basalt fiber reinforced concrete and glass fiber reinforced concrete, it was discovered that adding glass fiber reinforced concrete to the mix improves the coefficient of elasticity or elastic modulus and compressive strength of concrete (Arslan 2016). An endurance of fiber reinforced concrete was determined using rapid chloride and migration penetration tests. The findings show that the performance of glass fiber reinforced concrete has a stronger impact on the concrete's mechanical and durability properties (Liu et al. 2019). The outcome of glass fiber on moisture absorption reduction was considerably increased when the water content was 0.30% and 1.35%. The moisture absorption of the specimen was reduced from 3.49% to 1.99% using glass fiber, with a maximum reduction of 43.0% (Yuan and Jia 2021). According to the findings, the addition of glass fibers in the range of 0.3%–0.6% boosts the compressive strength and improves the engineering qualities of the concrete (Akbari and Abed 2020). To investigate the impact of the type of fiber, i.e., steel and glass fiber. On hybrid-reinforced self-consolidating concrete, the appropriate dose of steel and glass fibers is 1.0% and 0.05%, according to the results, which outperforms other combinations in terms of mechanical and durability (Ganta et al. 2020).

The amount of glass fiber in cement mortar boosts the mortar's compressive and flexural strengths. Glass fiber-reinforced mortar specimens, on the other hand, performed better than non-fibrous ones (Kelestemur et al. 2014). For an American concrete institute mix design with lengths of 0.60 and 2.54 cm, glass fibers of 0.5%, 1.0%, and 1.5% were employed to determine a water-binder ratio of 0.5 and 18 cm slump. It was found that increasing the number of glass fibers in concrete increases the concrete strength both compressive and flexural, in addition to lowering the thermal conductivity (Wang et al. 2020). Glass fibers containing 1.25% increasing fiber concentration resulted in decreased workability but enhanced fly ash-based geopolymer concrete (GPC) which has a high density, compressive, and flexural strength (Nematollahi et al. 2014). To build specific technologies, scientists and mathematicians are applying scientific and mathematical principles. Engineered tools, such as machines, structures, and goods, are developed using these technologies to assist engineers in solving problems (Emiroglu et al. 2012).

In recent years, Artificial intelligence has been widely applied in civil engineering fields such as building materials, geotechnical engineering construction management, and highway and transportation engineering to analyze the data obtained and developed models and overcome data challenges to give accurate and robust vulnerability prediction of models (Nhu et al. 2020). Artificial neural networks (ANN), fuzzy logic, tress, and gene expression programming are examples of AI approaches and many more techniques have been applied to determine the material behavior and characteristic of concrete specimen analyzed impacts of various constraints on the 28-day strength (Baykasoglu et al. 2004). To forecast the behavior of large-scale columns, under axial loading, it is explored the strain behavior of substantial fiber-reinforced columns. Only a few small-scale specimen-based models were found to agree with trial outcomes to an acceptable result (Naderpoura et al. 2019).

When glass fiber reinforced concrete with different combinations of admixtures is compared to control specimen, it increases by 12% and 90%. The experimental data was compared to the projected strength of glass fiber reinforced concrete which included compressive and impact strength. The results reveal that the strength predicted by ANN is quite similar to the experimental values, with only a small margin of error (Sangeetha and Shanmugapriya 2020a). The performance of several models, such as the M5P tree and linear regression (LR), was investigated using multilayer perceptron algorithms to determine the optimal model performance. The results indicate that the performance of tree-based models were way better than multilayer perceptron and LR because correlation coefficient (CC) is higher with lower RMSE values (Deepa et al. 2010). For the advancement of fiber-reinforced concrete, different techniques such as neuro-fuzzy inference system, and neural networks, genetic programming neural networks and stepwise regression/LR were utilized. Ten models were compared with experimental findings from the literature. The results show that the accuracy machine learning models have been shown to be more satisfactory by using soft computing formulation (Cevik 2011). The performance of a random forest (RF) model based on the beetle antennae search method is evaluated in order to estimate the uniaxial compressive strength of lightweight self-compacting. It was found that the model beetle antennae search-random forest (BAS-RF) achieved higher prediction of exactness with correlation of 0.97 and with the low RMSE esteems on preparing and testing data set (Zhang et al. 2019). Under axial compression, the compressive strength of concrete columns covered with GFRP was studied using an ANN model to predict the experimental results with R-squared value of 0.992 for training, 0.999 for testing and 0.999 for the validation indices, the ANN model and experimental results exhibited a high correlation (Sangeetha and Shanmugapriya 2020b).

2.2 MODELING TECHNIQUES

2.2.1 Random Forest (RF)

It is an ensemble learning method for regression and classification, and this approach that was initially introduced by Breiman (1996) has been chosen for the arrangement of various nonlinear or complex engineering problems (Thakur et al. 2020). RF consists of a combination of several tree predictors, where each tree is generated from the input vector using a random vector sampled independently and built with different combinations which is used in the present study to determine the prediction accuracy of compressive strength of concrete. To generate a training data set, data set is divided randomly which is known as bagging (Singh et al. 2019). Each tree in the forest computes the out-of-bag error testing validation or uses the rest of the data. Then, errors from all the trees are accumulated to find the total out-of-bag error rate (Chopra et al. 2018). Two user-defined parameters are required for RF regression, i.e., and the number of trees to be grown (k) number of input variables (m) used at each node to generate a tree. At each node, only selected variables are searched through for the best split. Therefore, the RF regression comprises of number of (k) trees (Singh et al. 2019). Moreover, each individual tree is extended in order to minimize the classification error; however, the result is exaggerated by the random selection. The limitation of RF model is to decide to what extent the prediction error increases as the data output for precise variables is permutated (Lu et al. 2020). As seen in Figure 2.1, the basic structure of the RF algorithm.

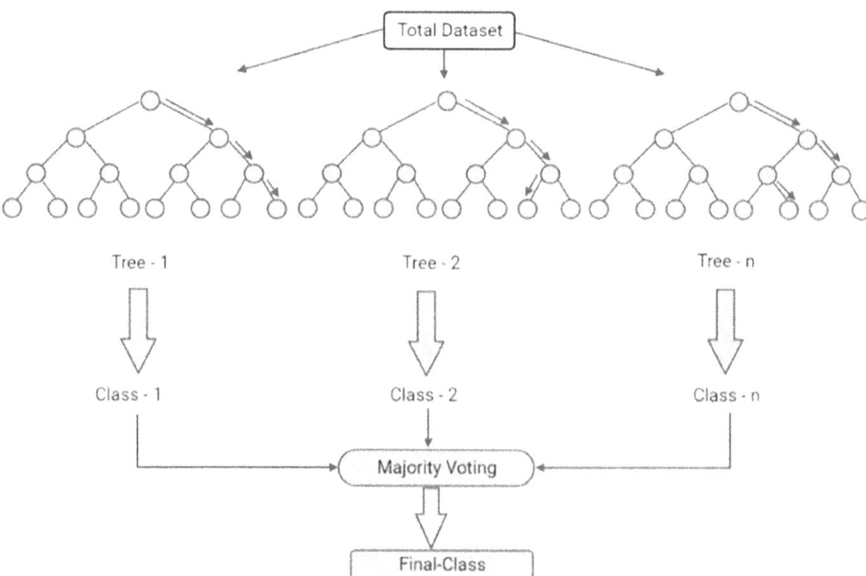

FIGURE 2.1 Structure of random forest.

2.2.2 Random Tree (RT)

Random tree model is an alternative approach, basically the combination of two existing algorithms in machine learning: single model trees are joined with RF ideas. RT are decision trees where each and every leaf holds a linear model which is optimized for the limited subspace defined by this leaf (Pfahringer 2010). In this modeling technique data is divided randomly between training and testing data set and constructs a tree with *k* number of random features at each node. For numerous data sets, it produces a highly specific classifier and handles a very huge number of input variables and estimates the importance of variables in defining classification that generates an internal unbiased evaluation of the generalization error as the forest building progresses (Deepa et al. 2010). Meanwhile, the training error is computed internally, cross-validation or bootstraps are not essential for assessing the precision of the training stage. It is worth mentioning that the output for the regression problems is considered by taking the average of the responses of all the forest individuals (Lu et al. 2020). As the trees are semi-arbitrary and accordingly not optimal in separation, averaging an appropriate number of such trees is essential for the performance of good prediction. Due to the randomization process, adding more trees to an ensemble will never significantly degrade performance, but as for most ensemble methods any enhancements reduce ultimately Pfahringer (2010).

2.2.3 Linear Regression (LR)

It is one of the most broadly utilized statistical tools because gives a basic technique to building up a useful relationship among factors. It is set of informative factors in the model and takes out non-significant variables one at a time until all the remaining factors are significant. At any progression variable with the smallest absolute *t*-statistic, that is the proportion of the coefficient to its standard error, will be eliminated (Dunlop and Smith 2015). LR is the mathematical equivalent defining the best line through the data set. Parameter values are found that limit the amount of-the-squares deviation of the trial from the determined qualities. It is assumed that these errors are randomly scattered, and that systematic errors are not present. Systematic errors can only be removed by correct experimental design (Leathetbatrow 1990). It is also helpful in finding CC which tends to describe the variations in the data set. More the value is closer to 1, more the data is reliable (Deepa et al. 2010).

2.2.3.1 Methodology and Data Description

A total number of data sets containing 156 readings from the past literature which is shown in Table 2.1. The total data set is divided randomly into two sets, i.e., 70% of data is used for training and 30% of data is used for testing. An assessment of the performance during the testing and validation stages validated the efficacy of the better models. The variables in the data set as an input parameter are cement, fine aggregate, coarse aggregate, water, super plasticizer, fiber length, glass fiber, and curing days to achieve the compressive strength which is an output parameter. As seen in Table 2.2, statistics features of the training, testing and validation to achieve the output, i.e., compressive strength. As seen in Figure 2.2, the flow chart is plotted to understand the methodology in detail.

TABLE 2.1
Data Set Details

S. No.	Name of Author	Publication Year	Total Data Set
1	Kelestemur et al.	2014	16
2	Akbari and Abed	2020	16
3	Arslan	2016	4
5	George et al.	2018	4
6	Liu et al.	2019	16
7	Yuan and Jia	2021	18
8	Hilles and Ziara	2019	10
9	Ganta	2020	24
10	Wang et al.	2020	48
Total			156

2.2.3.2 Performance Evaluation Criteria

To determine the preciseness of the estimated value eight statistical indices are used to determine the exactness of estimated model for both training and testing stages and quantified by CC in which the value of CC ranges from −1 to 1, higher the result of the correlation the results is found to be more accurate. RMSE and mean absolute error (MAE) vary from 0 to infinity which are helpful in evaluating the errors. Minimum errors show better prediction, whereas scattering index (SI), BIAS (average difference among actual and predicted values), Nash–Sutcliffe efficiency (NSE), Willmott's index (WI), and Legates and McCabe's index (LMI) range from 0 to 1 which can be expressed as:

$$CC = \frac{N\left(\sum_{i=1}^{N} GH\right) - \left(\sum_{i=1}^{m} G\right)\left(\sum_{i=1}^{a} H\right)}{\sqrt{\left[N\sum_{i=1}^{N} G - \left(\sum_{i=1}^{N} G\right)^2\right]}\sqrt{\left[N\sum_{i=1}^{N} H^2 - \left(\sum_{i=1}^{N} H\right)^2\right]}} \tag{2.1}$$

$$RMSE = \sqrt{\frac{1}{n}\left(\sum_{i=1}^{n}(H-G)^2\right)} \tag{2.2}$$

$$MAE = \frac{1}{N}\left(\sum_{i=1}^{N}|G-H|\right) \tag{2.3}$$

$$NSE = 1 - \left[\frac{\sum_{i=1}^{N}(G_i - H_i)^2}{\sum_{i=1}^{N}\left(G-\bar{H}\right)^2}\right] \tag{2.4}$$

TABLE 2.2
Features of Data Set

Statistics/ Paramters[1]	C (kg/m^3)	FA (kg/m^3)	CA (kg/m^3)	(W) (kg/m^3)	SP (%)	FL (%)	GF (%)	CD	CS (MPa)
					Training Data Set				
Minimum	240	203.6	561	0.37	0	0.6	0	3	18.01
Maximum	600	1100	1253.8	222.8	9.1	24	3	56	99.1
Mean	393.7614	765.8855	887.2195	168.6919	2.7265	7.98862	0.60532	25.4495	43.9795
Standard deviation	85.95337	1.9932333	232.20697	63.53950	3.189187	5.512669	0.581698	12.57973	23.52397
Kurtosis	0.020536	1.347862	−1.452536	1.4022652	−0.79433	−0.02245	1.25317	0.53696	−0.02533
Skewness	0.140288	−1.313713	0.3397079	−1.357919	0.850135	0.253943	0.956078	0.417444	1.119403
					Testing Data Set				
Minimum	240	203.6	561	0.37	0	0.6	0	3	18.7
Maximum	600	1100	1253.8	222.8	9.1	24	1.5	56	95.3
Mean	393.1702	747.1285	909.2874	161.1506	3.104894	8.531064	0.558723	21.51064	44.72511
Standard deviation	96.08084	211.0238	236.3653	67.9377	3.265617	5.150076	0.496311	13.93755	24.2889
Kurtosis	−0.17224	0.71304	−1.56436	0.914159	−1.08329	0.240633	−0.91858	0.50148	−0.46215
Skewness	0.361457	−1.14107	0.129497	−1.22221	0.675216	−0.0312	0.518592	0.80274	0.948738

[1] C = cement, FA = fine aggregate, CA = coarse aggregate, W = water, SP = super plasticizer, FL = fiber length, GF = glass fiber, and CD = curing days.

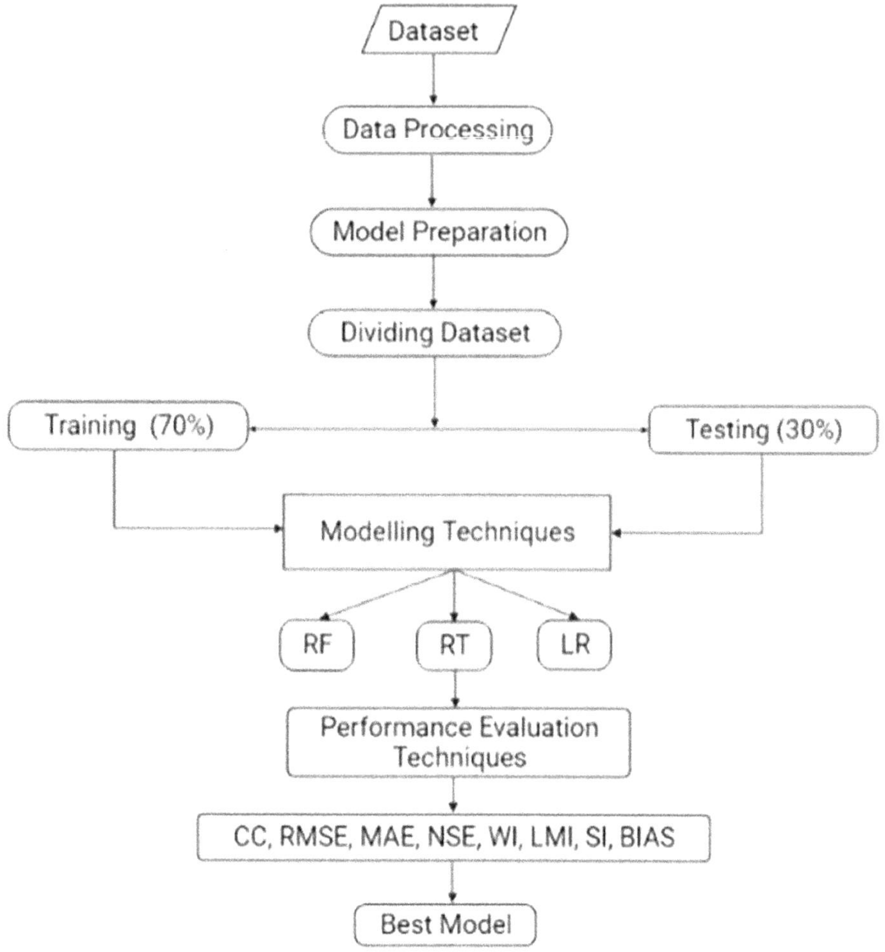

FIGURE 2.2 Methodology of data set.

$$\mathrm{WI} = 1 - \left[\frac{\sum_{i=1}^{N} (H_i - G_i)^2}{\sum_{i=1}^{N} \left(\left| H_i - \bar{G} \right| + \left| G_i - \bar{G} \right| \right)^2} \right] \tag{2.5}$$

$$\mathrm{LMI} = 1 - \left[\frac{\sum_{i=1}^{N} \left| H_i - G_i \right|}{\sum_{i=1}^{N} \left| G_i - \bar{G} \right|} \right] \tag{2.6}$$

$$\mathrm{BIAS} = \frac{\sum_{i=1}^{N} (G_i - H_i)}{\sum_{i=1}^{N} G_i} \tag{2.7}$$

$$SI = \sqrt{\frac{\sum_{i=1}^{N}\left[(H_i - \bar{H}) - (G_i - \bar{G}\right)^2}{\sum_{i=1}^{N} G_i^2}} \tag{2.8}$$

Where,

G = Observed values
H = Average of observation
$\bar{H}$ = Predicted value
N = Number of observations

2.3 RESULT ANALYSIS AND DISCUSSION

2.3.1 Result of RF and RT Model

Developing the RF model is trial and error process. Both of these models are based on decision tree. Eight different statistical parameters shown in Table 2.3 were used for the development of models for both training and testing stages. Table 2.3 suggests that the performance indices of RF model is better than RT model with the values for CC is 0.989325, RMSE 3.632436, MAE 2.821574, NSE 0.977148, WI 0.977925 LMI 0.859511, SI 0.081217, and BIAS −0.86702 with the numfeatures (K) = 2 and numiterations = 100. An agreement graph is plot among actual and predicted values as display in Figures 2.3 and 2.4 using RF and RT model for both the stages. Figure 2.3 indicates that the overall performance of RF is satisfactory in predicting the compressive strength of concrete mixed with glass fiber because their values lie close to the line of perfect agreement.

TABLE 2.3
Performance of RF, RT, and LR Models

Modeling Technique	CC	RMSE (MPa)	MAE (MPa)	NSE (MPa)	WI (MPa)	LMI (MPa)	SI	BIAS
				Training Data Set				
RF	0.994953	2.369599	1.522872	0.989759	0.990099	0.919834	0.05388	−0.041
RT	0.996423	1.978844	0.777972	0.992858	0.993097	0.959046	0.04499	−9.17431
LR	0.969511	5.738013	4.55089	0.939951	0.941991	0.760435	0.13106	−2.75229
				Testing Data Set				
RF	0.989325	3.632436	2.821574	0.977148	0.977925	0.859511	0.081217	−0.86702
RT	0.984728	4.769937	3.226766	0.960595	0.962055	0.839336	0.10665	0.099404
LR	0.974579	5.404799	4.425553	0.949408	0.951171	0.779647	0.120845	−0.24036

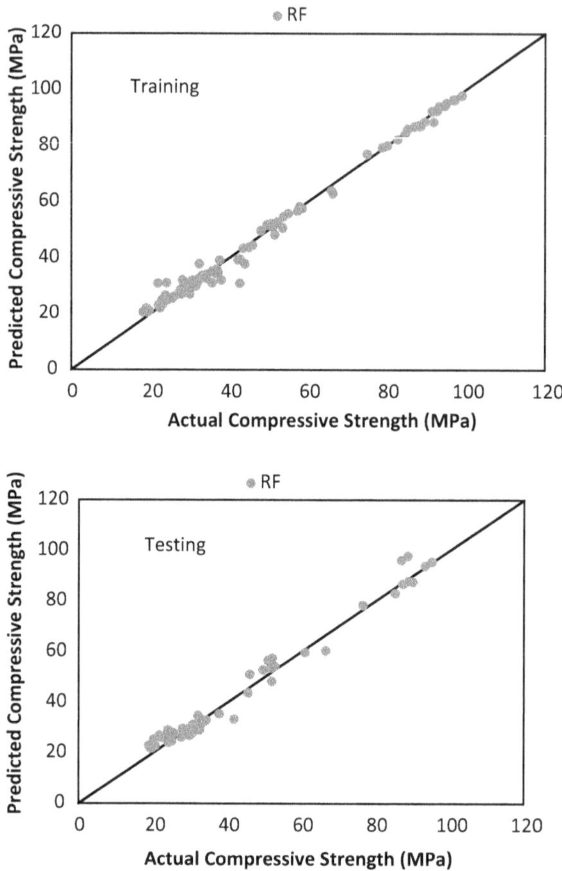

FIGURE 2.3 Scatter graph shows actual versus predicted values of compressive strength by using RF model for both data sets.

2.3.2 Result of LR Model

Developing the LR model the prediction is based on the relationship between two variables. Eight different statistical parameters shown in Table 2.3 were used for the development of models for both training and testing stages for predicting the compressive strength of concrete mixed with glass fiber. Table 2.3 suggests that the performance indices of LR model with the values for CC 0.974579, RMSE 5.404799, MAE 4.425553, NSE 0.949408, WI 0.951171, LMI 0.779647, SI 0.120845, and BIAS −0.24036. An agreement graph is plot among actual and predicted values as display in Figure 2.5 using LR model for both the stages. The equation is generated by LR technique as written below:

$$C.S = 0.2847 * C + 0.0133 * FA + 0.0478 * CA + 0.2673 * W + 2.6113 * SP$$
$$+ 4.0401 * FL + -3.2052 * GF + 0.245 * CD + -209.482 \quad (2.9)$$

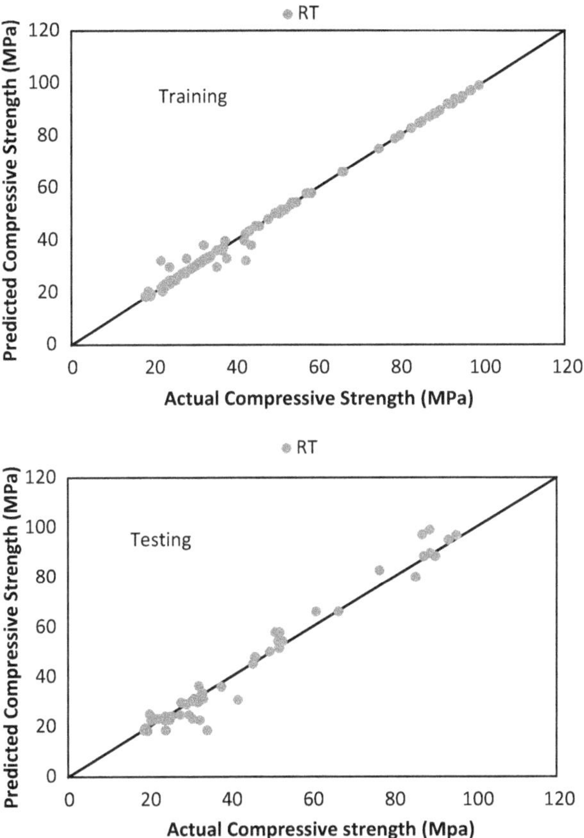

FIGURE 2.4 Scatter graph shows actual versus predicted values of compressive strength by using RT model for both data sets.

2.4 INTERCOMPARISON AMONG COMPUTING MODELS

In this study, comparison of soft computing techniques has been done to predict the compressive strength of concrete using glass fiber. Performance of all eight statistical parameters were evaluated for both training and testing stages as listed in Table 2.3. The comparison of the predicted values of testing results with best model is shown in Table 2.4, which indicates that the RF model works better than another soft computing-based model. Figure 2.6 presents the results of training and testing stages for RF, RT, and LR with minimum error which indicates that the predicted values of compressive strength lie very close to actual values of compressive strength. For estimating the compressive strength of concrete mixed with glass fiber, the performance assessment of the RF model gives better results than other applied models with higher CC values of 0.989325 and lower error RMSE 3.632436 and MAE 2.821574 and also NSE 0.977148, WI 0.977925, LMI 0.859511 are much closer to the actual values. Similarly, the SI and BIAS values in the RF model are 0.081217 and 0.86702, respectively, which are the lowest in comparison to the RT and LR models.

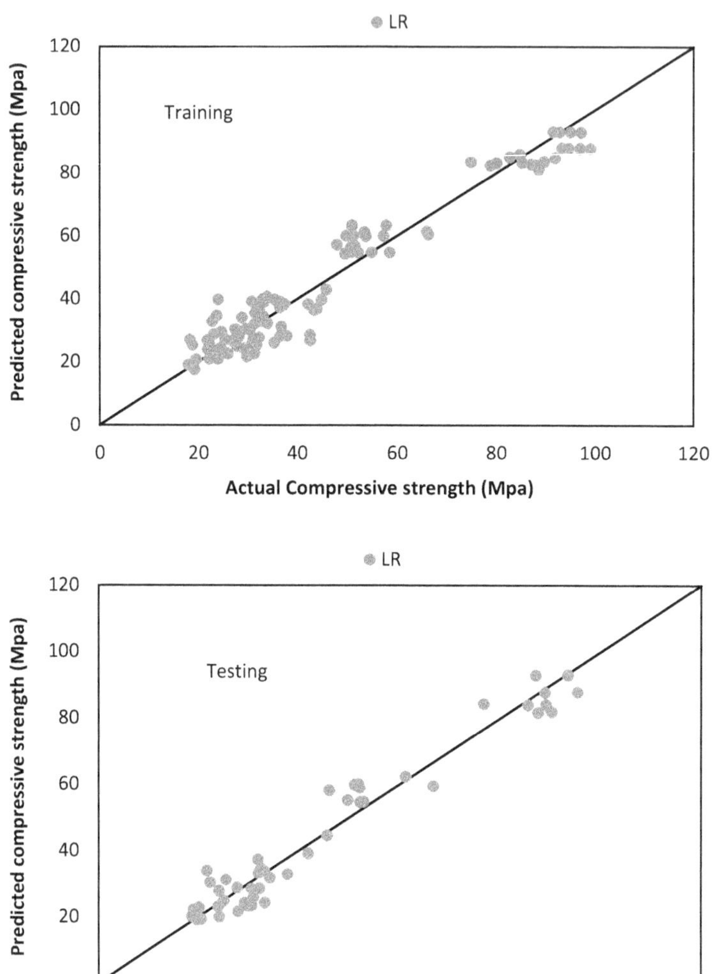

FIGURE 2.5 Scatter graph shows actual versus predicted values of compressive strength by using LR model for both data sets.

TABLE 2.4
Comparison of RF, RT, and LR Models

Models	CC	RMSE	MAE	NSE	WI	LMI	SI	BIAS
RF	0.989325	3.632436	2.821574	0.977148	0.977925	0.859511	0.081217	−0.86702
RT	0.984728	4.769937	3.226766	0.960595	0.962055	0.839336	0.10665	0.099404
LR	0.974579	5.404799	4.425553	0.949408	0.951171	0.779647	0.120845	−0.24036

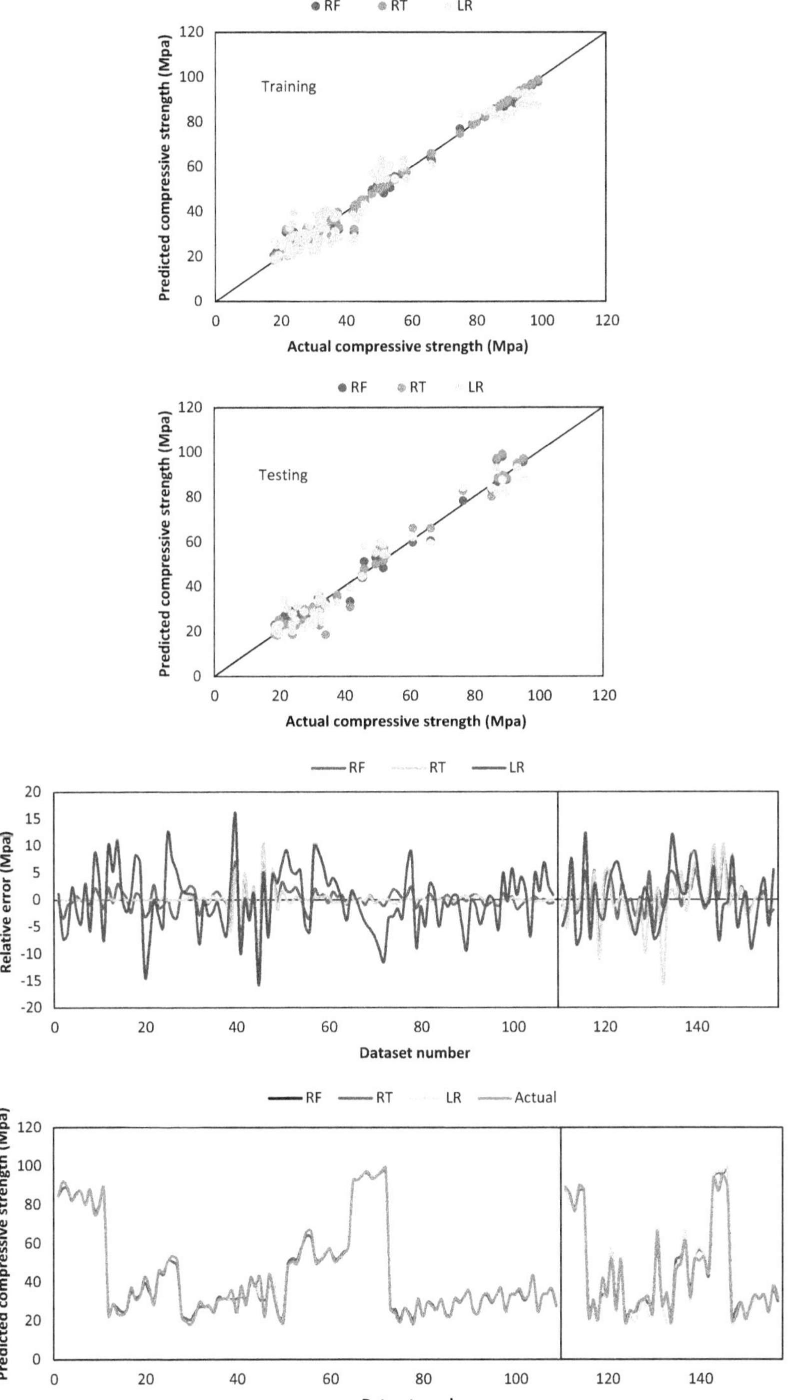

FIGURE 2.6 Comparison between actual versus predicted values of RF, RT, and LR models.

2.5 SENSITIVITY ANALYSIS

In order to assess the accuracy, the sensitivity was carried out to determine the most influencing input parameter for the prediction of compressive strength of concrete mixed with glass fiber. In this best performing model was selected for sensitivity analysis and training data is organized and done by eliminating single input parameter and outcomes is listed in the form of coefficient of correlation, root mean square error (RMSE) and absolute error as shown in Table 2.5 is considered. As per the results of Table 2.5 indicates that CD has major influence in predicting the compressive strength of concrete missed with glass fiber with RF-based model in comparison with other input parameters.

2.6 CONCLUSION

This paper explores the capability of machine learning techniques for the prediction of compressive strength of concrete mixed with glass fiber. Modeling techniques such as RF, RT, and LR has been applied to assess the most significant parameter. Eight statistical attributes are used to evaluate the performance of developed models, i.e., CC, RMSE, MAE, NSE, WI, LMI, SI, and BIAS. The results of performance evaluation reveals that the working of RF model is achieved with the value for CC 0.989325, RMSE 3.632436, MAE 2.821574, NSE 0.977148, WI 0.859511, LMI 0.859511, SI 0.081217, and BIAS −0.86702, which is better in predicting the compressive strength of concrete. The results obtained from sensitivity analysis conclude that CD is the most significant parameter which affects the compressive strength of concrete mixed with glass fiber for this data set.

TABLE 2.5
Sensitivity Analysis Using RF-Based Models

Input Combination								Output		RF-Based Model		
Cement (C) kg/m³	Fine Aggregate (FA) kg/m³	Coarse Aggregate (CA) kg/m³	Water	Super Plasticizer (SP) %	Fiber Length (FL)mm	Glass Fiber (GF) %	Curing Days (CD)	Compressive Strength (Mpa)	Parameter Removed	CC	RMSE	MAE
									-	0.9871	4.0536	3.1192
									C	0.9879	3.8376	2.9975
									FA	0.9878	3.9551	3.0509
									CA	0.9872	4.0351	3.1435
									W	0.9876	3.97	3.0602
									SP	0.9869	4.0745	3.1644
									FL	0.9842	4.4442	3.6039
									GF	0.9892	3.603	2.8112
									CD	**0.9715**	**5.8786**	**4.6926**

REFERENCES

Akbari J. and Abed A. Experimental evaluation of effects of steel and glass fibers on engineering properties of concrete. *Focussed on Structural Integrity and Safety: Experimental and Numerical Perspectives*, 54 (2020): 116–127. Doi: 10.3221/IGF-ESIS.54.08.

Arslan M.E. Effects of basalt and glass chopped fibers addition on fracture energy and mechanical properties of ordinary concrete: CMOD measurement. 114 (2016): 383–391. Doi: 10.1016/j.conbuildmat.2016.03.176.

Baykasoglu A., Dereli T., and Tanis S. Prediction of cement strength using soft computing techniques. *Cement and Concrete Research*, 34(11) (2004): 2083–2090. Doi:10.1016/j.cemconres.2004.03.028.

Breiman, L. Random forests. *Machine Learning*. 45(0) (2001): 5–32.

Cevik A. Modelling strength enhancement of FRP confined concrete cylinders using soft computing. *Expert Systems with Applications*, 38(5) (2011): 5662–5673. Doi: 10.1016/J. Eswa.2010.10.069.

Chandramouli K., Srinivasa R.P., Pannirselvam N., and Sekhar T. Strength properties of glass fibre concrete. *ARPN Journal of Engineering and Applied Sciences*, 5(4) (April 2010): 1–6.

Chopra, P., Sharma, R.K., Kumar, M., and Chopra, T. Comparison of machine learning techniques for the prediction of compressive strength of concrete. *Advances in Civil Engineering* 2018(4) (2018). Doi: 10.1155/2018/5481705.2.

Deepa C., SathiyaKumari K., and Sudha V.P. Prediction of the 339 compressive strength of high performance concrete mix using tree based 340 modelling. *International Journal of Computer Application*, 6(5) (2010): 18–24.

Dunlop P. and Smith S. Estimating key characteristics of the concrete delivery and placement process using linear regression analysis. *Civil Engineering and Environmental Systems*, 20(4), (December 2003): 273–290. Doi: 10.1080/1028660031000091599.

Emiroglu M.E., Bilhan, O., and Kisi, O. Neural networks for estimation of discharge capacity of triangular labyrinth side-weir located on a straight channel. *Expert Systems with Applications*, 38(1) (2012): 867–874. Doi: 10.1016/j.eswa.2010.07.058.

Faleschinia F., Zaninia M.A., Hofera L., Toskaa K., Domenicoc D.D., and Pellegrinoa C. Confinement of reinforced concrete columns with glass fiber reinforced cementitious matrix jackets. *Engineering Structures*, 218 (2020): 110847. Doi: 10.1016/J. Engstruct.2020.110847.

George R.S., Das B.B., and Goudar S.K., *Durability Studies on Glass Fiber Reinforced Concrete*. Springer Nature Singapore Pte Ltd. (2019). Doi: 10.1007/978-981-13-3317-0_67.

Hilles M.M. and Ziara M.M. Mechanical behaviour of high strength concrete reinforced with glass fiber engineering science and technology. *An International Journal*, 22 (2019): 920–928. Doi: 10.1016/J.Jestch.2019.01.003.

Kelestemur O., Yildiz S., Gokcer B., and Arici E. Statistical analysis for freeze–thaw resistance of cement mortars containing marble dust and glass fiber. *Materials and Design*, 60 (2014): 548–555. Doi: 10.1016/J.Matdes.2014.04.013.

Kumar D., Rex L.K., Sethuraman V.S., and Saravanan V.G.B. High performance glass fiber reinforced concrete. *Materials Today: Proceedings*, 33 (2020): 784–788.

Leathetbatrow R.J. Imperial College of Science, London.

Lee J.Y., Kim T.Y., Kim T.J., Yi C.K., Park J.S., You Y.C., and Park Y.H. Interfacial bond strength of glass fiber reinforced polymer bars in high-strength concrete. *Composites: Part B*, 39 (2008): 258–270.

Liu J., Jia Y., and Wang J. Experimental study on mechanical and durability properties of glass and polypropylene fiber reinforced concrete. *Fibers And Polymers*, 20(9), (2019): 1900–1908. Doi: 10.1007/S12221-019-1028-9.

Lu A.B., Dereli T.R., and Tanis S. Prediction of cement strength using soft computing techniques. *Cement and Concrete Research*, 34 (2004): 2083–2090. Doi: 10.1016/j.cemconres.2004.03.028.

Lu M.E., Lu A.B., and Yildiz S. ANFIS and statistical based approach to prediction the peak pressure load of concrete pipes including glass fiber. *Expert Systems with Applications*, 39(2012): 2877–2883. Doi: 10.1016/J.Eswa.2011.08.149.

Lu S., Koopialipoor M., G.P Asteris., Bahri M., and Armaghani DJ. A novel feature selection approach based on tree models for evaluating the punching shear capacity of steel fiber-reinforced concrete flat slabs. *Materials*, 13 (2020): 3902; Doi:10.3390/Ma13173902.

Małek M., Jackowski M., Łasica W., Kadela M., and Wachowski M. Mechanical and material properties of mortar reinforced with glass fiber: An experimental study. *Materials*, 14 (2021): 698. Doi: 10.3390/Ma14030698.

Naderpoura H., Nagaib K., Fakhariana P., and Hajia M. Innovative models for prediction of compressive strength of FRP-confined circular reinforced concrete columns using soft computing methods. *Composite Structures*, 215(2019): 69–84. Doi: 10.1016/J.Compstruct.2019.02.048.

Nematollahi B., Sanjayan J., Chai J.X.H., and Lu T.M. Properties of fresh and hardened glass fiber reinforced fly ash based geopolymer concrete. *Key Engineering Materials*, 594–595 (2014): 629–633 Trans Tech Publications, Switzerland Doi: 10.4028/Www.Scientific.Net/KEM.594-595.629.

Nhu VH., Shirzadi A., Shahabi H., Singh SK., Ansari AN., Clague JJ., Jaafari A., Chen W., Miraki S., Dou J., Luu C., Gorski K., Thai Pham B., Nguyen HD., and Ahmad BB. Shallow landslide susceptibility mapping: A comparison between logistic model tree, logistic regression, naïve bayes tree, artificial neural network, and support vector machine algorithms. *International Journal of Environmental Research and Public Health*, 17(8) (2020): 2749. Doi: 10.3390/ijerph17082749.

Pfahringer B. *Random Model Trees: An Effective and Scalable Regression Method*. The University of Waikato, Hamilton Working Paper: 03/2010 (June 2010).

Sangeetha P. and Shanmugapriya M. Artificial neural network applications in fiber reinforced concrete. *First International Conference on Advances in Physical Sciences and Materials Journal of Physics: Conference Series*, 1706 (2020a): 012113. Doi:10.1088/1742-6596/1706/1/012113.

Sangeetha P. and Shanmugapriya, M. GFRP wrapped concrete column compressive strength prediction through neural network. *SN Applied Sciences*, 2 (2020b): 2036 Doi: 10.1007/S42452-020-03753-4.

Singh B., Sihag P., Tomar A., and Sehgad A. Estimation of compressive strength of high-strength concrete by random forest and M5p model tree approaches. *Journal of Materials and Engineering Structures*, 6 (2019): 583–592.

Thakur, M.S., Pandhiani, S.M., Kashyap, V. Upadhya, A. Sihag, P. Predicting bond strength of FRP bars in concrete using soft computing techniques. *The Arabian Journal for Science and Engineering*, 46 (2021): 4951–4969. Doi: 10.1007/s13369-020-05314-8.

Wang W.C., Wang H.Y., Chang K.H., and Wang S.Y. Effect of high temperature on the strength and thermal conductivity of glass fiber concrete. *Construction and Building Materials*, 245 (2020): 118387. Doi: 10.1016/J.Conbuildmat.2020.118387.

Yadav G.S., Prabhanjan N., Sahithi G., Sangeetha G., Srinivas A., and Krishna A.S. Strength investigation of fly ash based concrete waste steel fibre and polypropylene fibre as reinforcing materials. *International Conference on Emerging Trends in Engineering*, (2020): 155–161 Doi: 10.1007/978-3-030-24314-2_21.

Yuan Z. and Jia Y. Mechanical properties and microstructure of glass fiber and polypropylene fiber reinforced concrete: An experimental study. *Construction and Building Materials,* 266 (2021): 121048. Doi: 10.1016/J.Conbuildmat.2020.121048.

Zhang J., Ma G., Huang Y., Sun J., Aslani F., and Nener B. Modelling uniaxial compressive strength of lightweight self-compacting concrete using random forest regression. *Construction and Building Materials*, 210 (2019): 713–719. Doi: 10.1016/J.Conbuildmat.2019.03.189.

3 Prediction of Compressive Strength at Elevated Temperatures Using Machine Learning Methods

Anshul Aggarwal, Yogesh Aggarwal, and Paratibha Aggarwal
NIT Kurukshetra (Haryana)

CONTENTS

3.1 INTRODUCTION

Concrete being a non-combustible material does not add to the fire load. But during the fire, concrete undergoes some structural changes and gets physically transformed. Such transformations are irreversible and affect their mechanical properties. Thus, it is imperative to study the physical changes that concrete undergoes during a fire. But due to a lack of any standardised test/method, researchers have been using different methods and techniques to study such situations.

Currently, researchers have been using controlled temperature simulations of fire and have produced temperature–time curves that are popularly known as fire curves. The various parameters in such tests include a change in the material composition. Generally, the basic composition is kept constant and an additive or strength enhancing material is added to vary the mechanical properties, such as compressive strength.

DOI: 10.1201/9781003184331-3

The present study has chosen predictive modelling to predict the compressive strength of concrete samples with diverse compositions, which include steel fibres, desert sand, superplasticizers, etc. Such a dataset provides a comprehensive view of the nature of behaviour and the trends of various concrete mixes under elevated temperatures.

Through this chosen method, models have been built on the data already available with the authors obtained from previously conducted experiments from the research works of other researchers. By utilizing techniques such as machine learning and data mining, the study aims to find the best predictive method that can be used to predict the compressive strength of concrete at elevated temperature

3.2 DATASET AND METHODOLOGY

The dataset includes observations referenced from the research of various published works. All the observations include the material composition of concrete along with the temporal conditions that were kept while casting and testing concrete.

Overall, a set of 22 parameters has been incorporated that has been used to predict the compressive strength at various temperatures. Each of the parameters has been assigned a particular identifier and has been checked for its dependency on the final results of strength. Parameters with their identifying codes have been given in Table 3.1.

The data has been summarised through visualisations made through RStudio software in the form of frequency distributions (Figures 3.1 and 3.2).

In the figures, the *Y*-axis signifies the number of occurrences for the component in particular categories labelled on *X*-axis. For details of the full dataset, Table 3.9 may be referred to.

3.3 PREDICTIVE MODELLING

Predictive models have been built on R language that includes machine learning algorithms, and have been used to build statistical and graphical interpretations. Data modelling with R has been operated through software called RStudio, which is an integrated development environment that allows interacting with R more readily.

Data analysis has been done in the following methods:

- Linear regression method
- Regression tree method
- Boosting method
- Neural network method

3.4 ANALYSIS OF RESULTS AND DISCUSSION

3.4.1 Linear Regression Method

This method assumes that a linear relationship exists between the parameters and the variable. Since the study relates to 22 such parameters, the multiple linear regression

TABLE 3.1
Parameters and Their Identifying Codes

Sr. No.	Parameter	Identifying Code
1.	Cement (kg/m^3)	Cement
2.	Water (kg/m^3)	Water
3.	Sand (kg/m^3)	Sand
4.	Gravel (kg/m^3)	Gravel
5.	Fly ash (kg/m^3)	Fly_Ash
6.	Silica fumes (kg/m^3)	Silica
7.	Desert sand (kg/m^3)	Desert_sand
8.	Water/binder ratio	WB_ratio
9.	High range water reducer (L/m^3)	HRWR_lpm3
10.	Coir (percentage of total volume)	Coir_PTV
11.	Polyvinyl alcohol (percentage of total volume)	PVA_PTV
12.	Poly propelene (kg/m^3)	PPE_kgpm3
13.	Steel fibres (% of T.Vol)	Steel_PTV
14.	Admixtures	Admix_kgpm3
15.	Superplasticizers	Suppl_kgpm3
16.	Normal curing	Curing_nrml
17.	Mechanical curing	Curing_Mech
18.	Open to air cooling	Cooling_OTA
19.	Mechanical cooling	Cooling_Mech
20.	Heating rate (°C/min)	Heatrate_Cpm
21.	Heat exposure at desired temp (h)	Exposure_Hrs
22.	No. of specimen per temp point	Spec_num
23.	Compressive strength	Comp_Str

is used, and the prediction results have been calculated by analysing the dependence of variables on each other and their influence on the compressive strength.

The results thus calculated are in the form of coefficients for each variable which can be put into a linear equation as given below, to find the predicted value.

$$\text{Compressive Strength} = c + \sum_{i}^{n} x_i * a_i \tag{3.1}$$

where c = intercept; x_i = parameter, a_i = coefficient and n = number of parameters.

Coefficient Prediction: The values thus predicted are given in Table 3.2.

Scatter Plot: The dependence compressive strength on each parameter has been summarised in the scatter plot (Figures 3.3 and 3.4). Each figure has black dots that signify the real values and a line that depicts the predicted values. Also, a band has been used to show a 95% confidence interval.

Result Parameters: To check whether a particular method fits the dataset, a number of accuracy parameters have been calculated as given in Table 3.3.

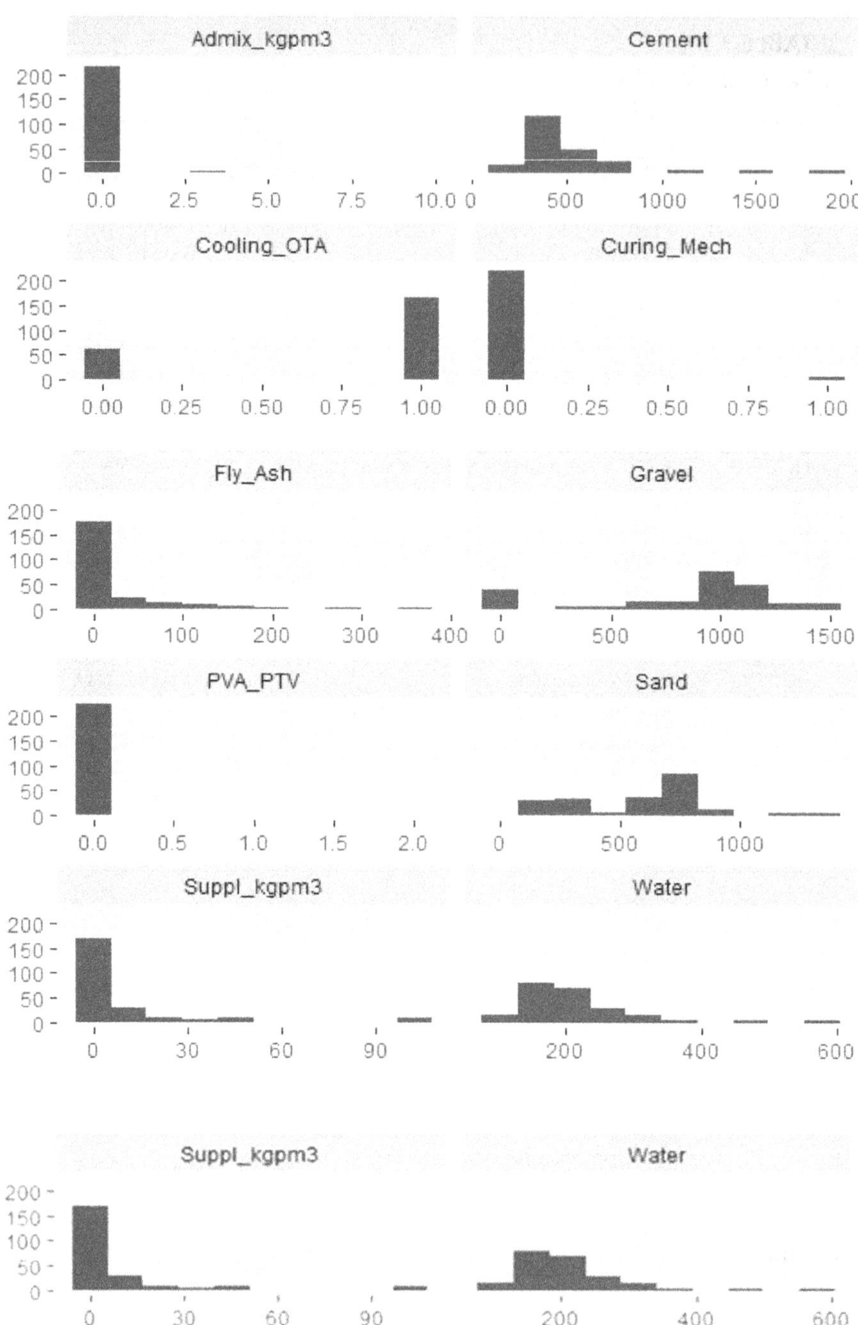

FIGURE 3.1 Frequency distribution of cement materials.

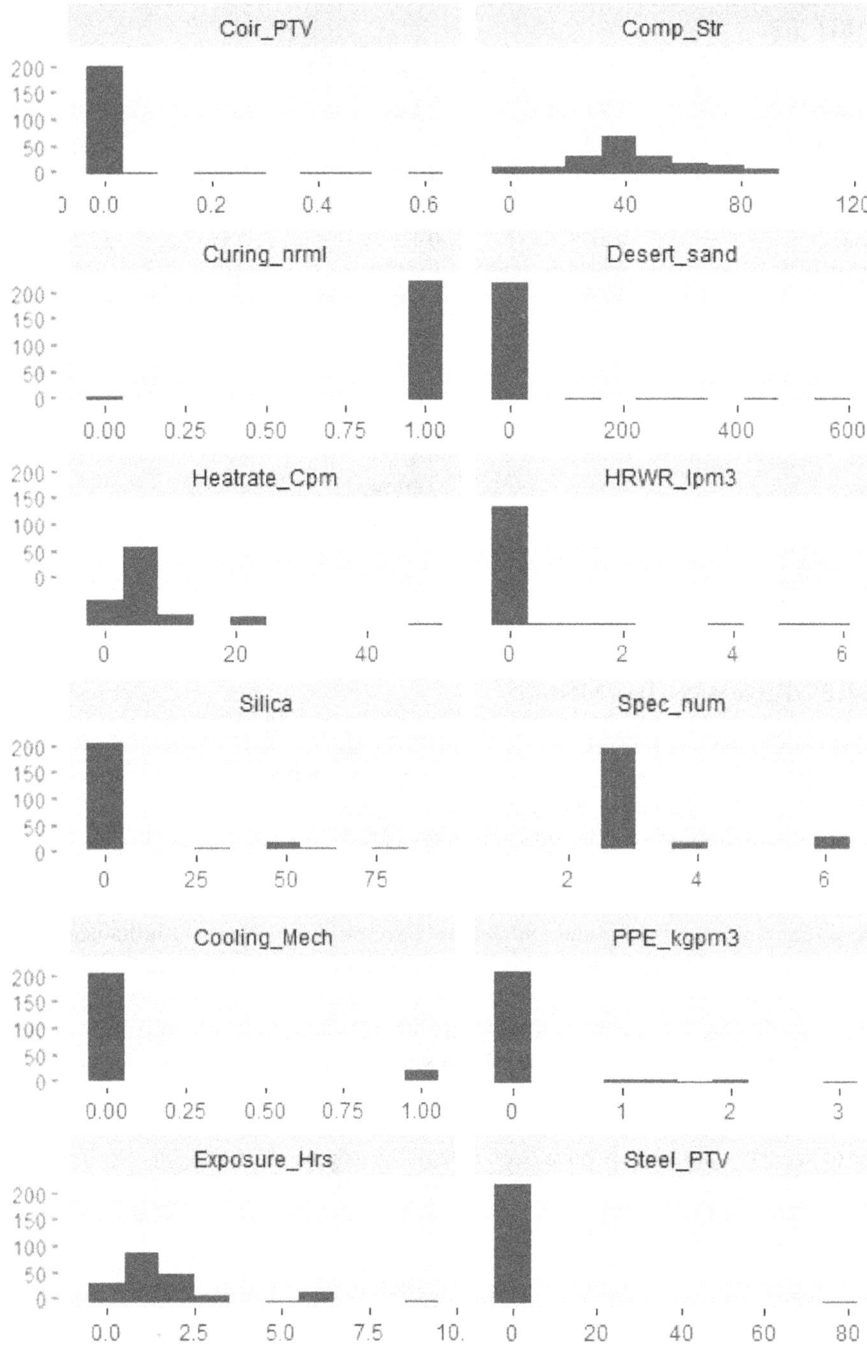

FIGURE 3.2 Frequency distribution of cement materials.

TABLE 3.2
Coefficients of Parameters Built through Linear Regression

Parameters	Coefficient	Std. Error	*t* value	Pr(>|*t*|)
Intercept	23.481	6.833136	3.436	0.000852
Cement	−0.0235	0.00882	−2.666	0.008911
Water	0.0657	0.026305	2.499	0.014026
Sand	0.0118	0.003759	3.142	0.002193
Desert_sand	0.0278	0.009005	3.083	0.002627
WB_ratio	−68.326	11.07374	−6.17	1.36E−08
Coir_PTV	−11.584	7.409219	−1.563	0.121015
PVA_PTV	−12.212	3.724522	−3.279	0.001422
PPE_kgpm3	7.0865	1.65568	4.28	4.20E−05
Steel_PTV	0.1085	0.071875	1.51	0.134084
Admix_kgpm3	1.7412	0.801456	2.173	0.032108
Curing_nrml	6.3938	3.619799	1.766	0.080301
Cooling_OTA	7.9395	3.516847	2.258	0.026084
Cooling_Mech	15.998	3.782207	4.23	5.08E−05
Heatrate_Cpm	−0.2112	0.126528	−1.669	0.098191

3.4.2 Regression Tree Method

A regression tree is a type of supervised learning algorithm in which trees are developed after iterations for accuracy. At each level, a number of variables are checked and splits are made to grow a tree into different branches.

The tree thus developed is shown in Figure 3.5.

Percentage MSE Increase: This signifies the extent to which any parameter influences the decision of putting a particular parameter at any node in the tree. In the given analysis, gravel, sand and cement have produced the most influence in the forming of the tree. A percentage increase in (MSE) is given in Figure 3.6.

OOB Error: The Out-of-bag (OOB) versus the number of tree graphs (Figure 3.7) helps in selecting the number of iterations (of tree) needed to produce the smallest error possible. The given dataset has been set to iterate to at least 100 trees to minimise the error in predictions.

Result Parameters: To check whether a particular method fits the dataset, a number of accuracy parameters have been calculated as given in Table 3.4.

3.4.3 Boosting Method

Boosting works as an addition to the regression trees or random forest methods. It refines the regression tree results by attaching an influencing factor that it calculates by iterations. It is a general approach that can be applied to many statistical learning methods for regression. The values thus are given in Table 3.5.

Test MSE: To calculate the number of Iterations that are needed to run a boosting analysis, a graph has been prepared between test MSE and the number of trees

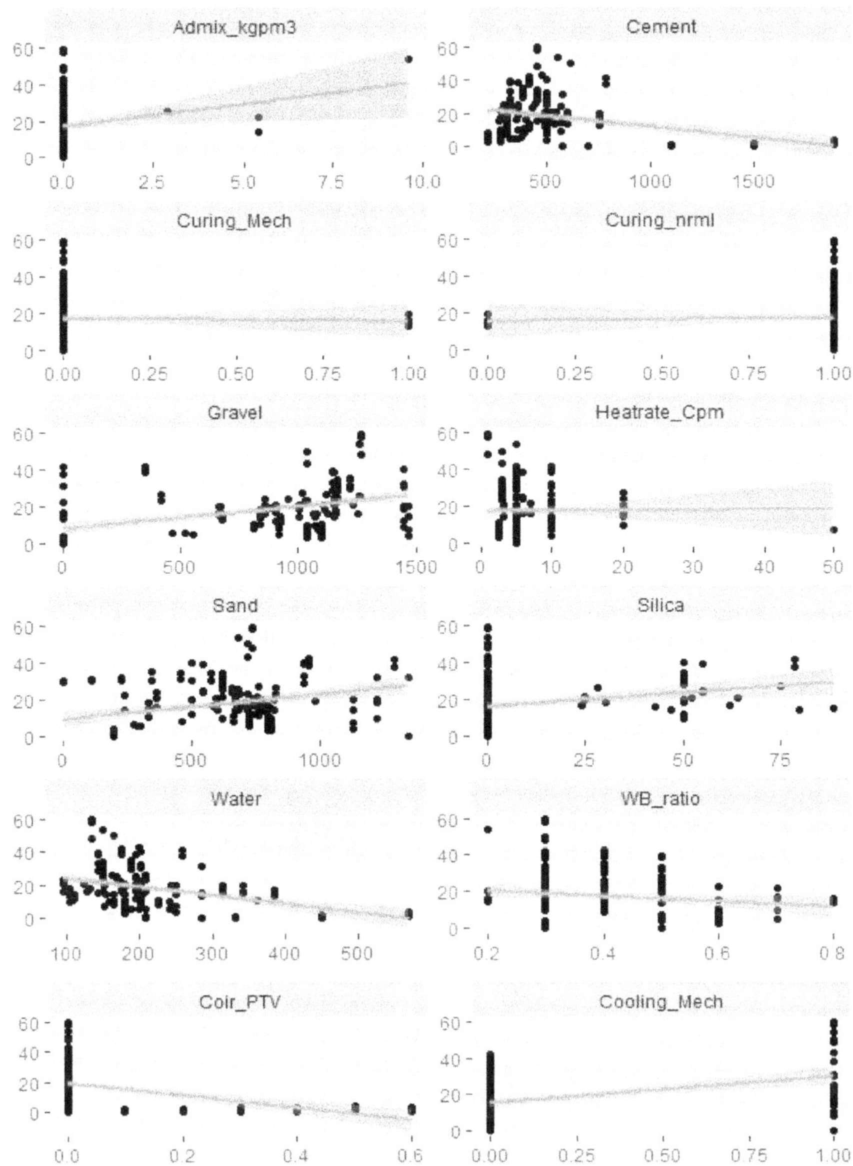

FIGURE 3.3 Scatter plot of parameters with compressive strength.

(Figure 3.8). Through this graph, it has become evident that the minimum number of trees (iterations) needed to produce the least MSE is around 500. Thus, for the dataset in the present study, the method has been run for producing 500 iterations of the trees.

Result Parameters: To check whether a particular method fits the dataset, a number of accuracy parameters have been calculated as given in Table 3.6.

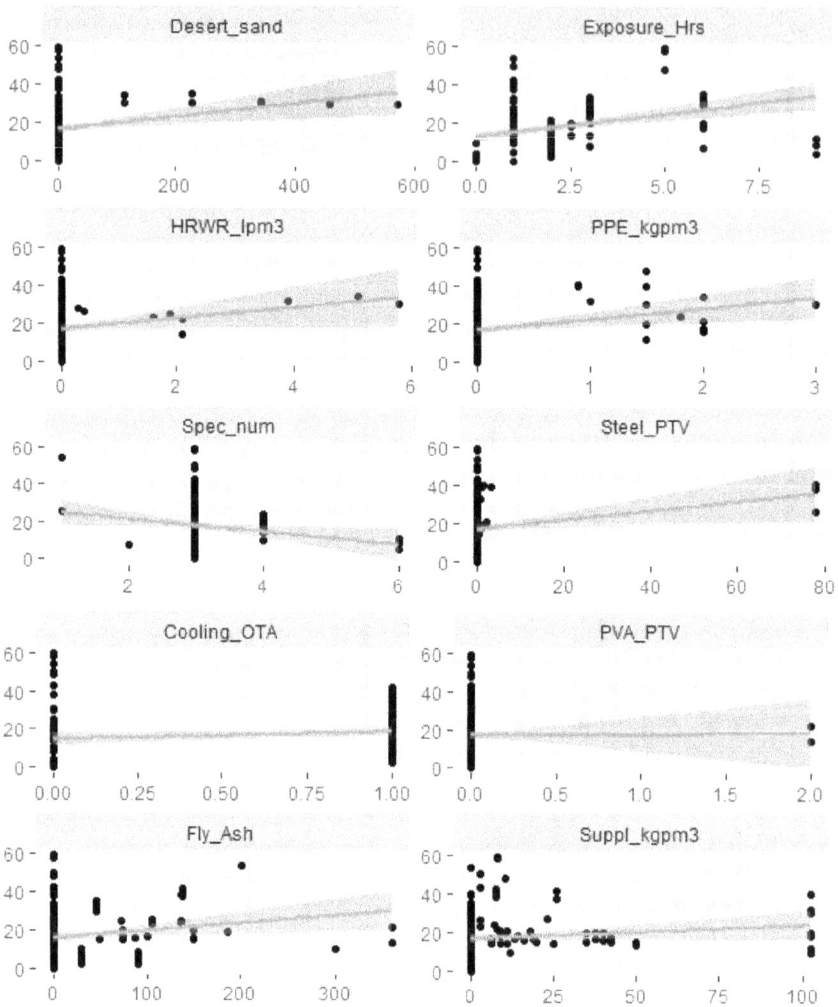

FIGURE 3.4 Scatter plot of parameters with compressive strength.

TABLE 3.3
Result Parameters for Linear Regression

Sr. No.	Result Parameter	Value
1.	Residual standard error	7.412
2.	Multiple R^2	0.6854
3.	Adjusted R^2	0.6427
4.	F-statistic	16.03
5.	p-value	2.2e-16

yes
Exposure < 0.5
no
1.095
n=19
WB_ratio >= 0.55
Heatrate < 2.75
Fly_Ash < 23
6.878
n=12
13.13
n=10
Heatrate >= 4.15
Water >= 200.1
30.05
n=8
18.2
n=10
33.51
n=18
Silica < 48.5
24.92
n=11
Exposure >= 1.5
10.51
n=9
Water < 130.2
14.83
n=10
21.1
n=11

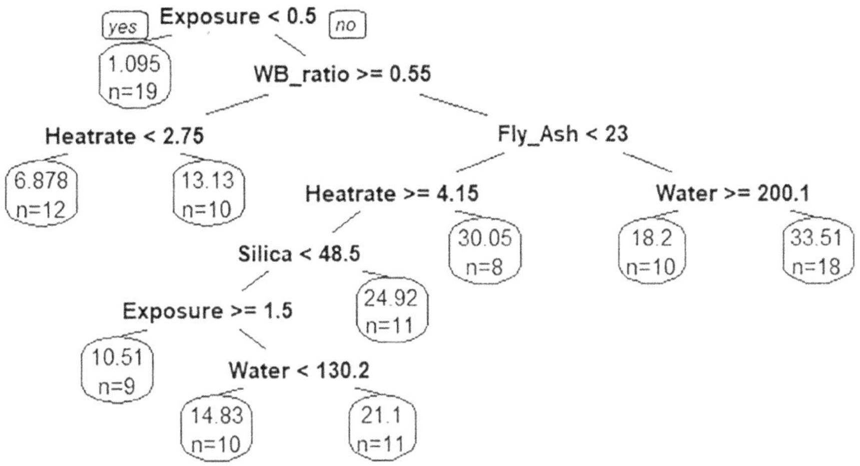

FIGURE 3.5 Regression tree

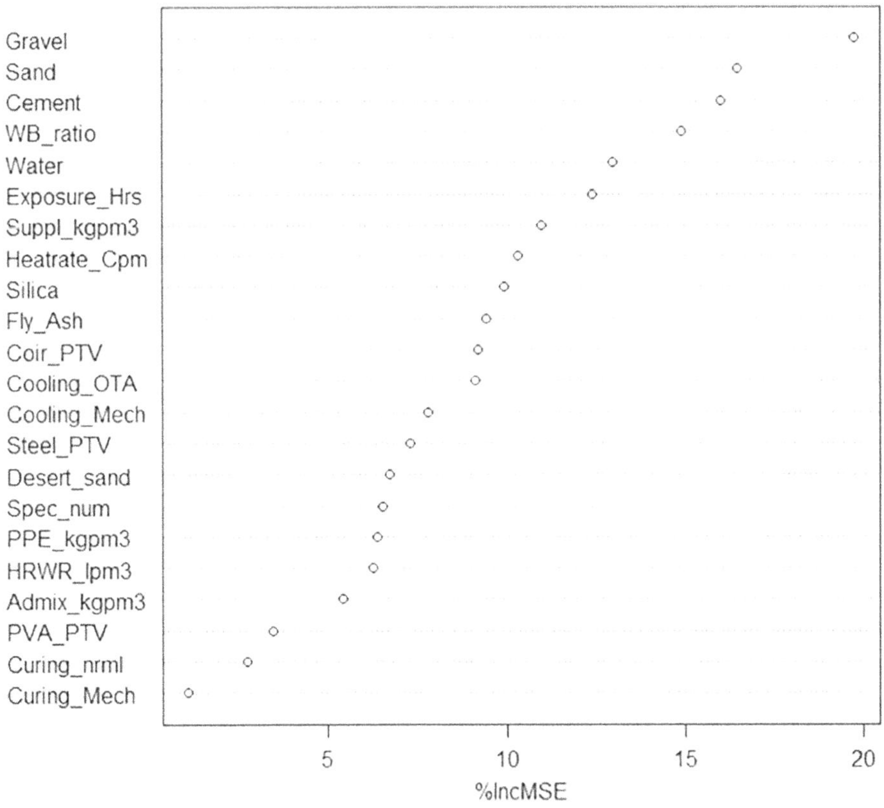

Gravel
Sand
Cement
WB_ratio
Water
Exposure_Hrs
Suppl_kgpm3
Heatrate_Cpm
Silica
Fly_Ash
Coir_PTV
Cooling_OTA
Cooling_Mech
Steel_PTV
Desert_sand
Spec_num
PPE_kgpm3
HRWR_lpm3
Admix_kgpm3
PVA_PTV
Curing_nrml
Curing_Mech
5
10
15
20
%IncMSE

FIGURE 3.6 Percentage increase in MSE.

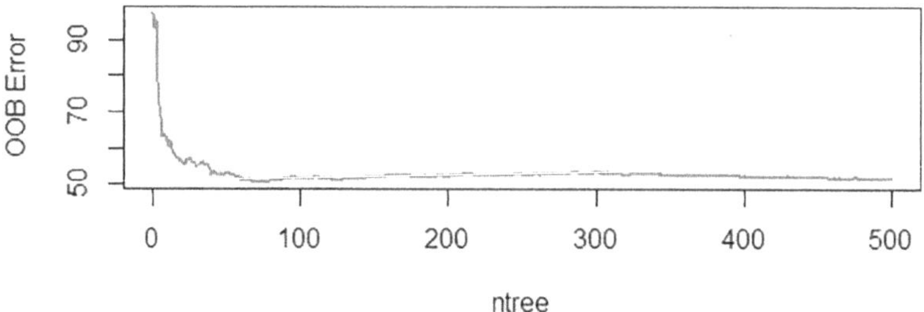

FIGURE 3.7 OOB error in a regression tree.

TABLE 3.4
Result Parameters for Regression Tree Method

Sr. No.	Result Parameter	Value
1.	Number of trees	100
2.	No. of variables tried at each split	13
3.	Mean of squared residuals	42.844
4.	Percentage variation explained	71.89

TABLE 3.5
Relative Influence Factors Determined by Boosting

Parameter	Relative Influence Factor	Parameter	Relative Influence Factor
Sand	1.74E+01	Cooling_OTA	9.53E−01
Gravel	1.71E+01	Coir_PTV	5.24E−01
Cement	1.29E+01	Steel_PTV	3.79E−01
Water	1.25E+01	Spec_num	6.25E−05
Exposure_Hrs	1.10E+01	Desert_sand	0.00E+00
Suppl_kgpm3	7.24E+00	HRWR_lpm3	0.00E+00
Fly_Ash	6.25E+00	PVA_PTV	0.00E+00
WB_ratio	5.84E+00	PPE_kgpm3	0.00E+00
Cooling_Mech	3.18E+00	Admix_kgpm3	0.00E+00
Heatrate_Cpm	3.08E+00	Curing_nrml	0.00E+00
Silica	1.74E+00	Curing_Mech	0.00E+00

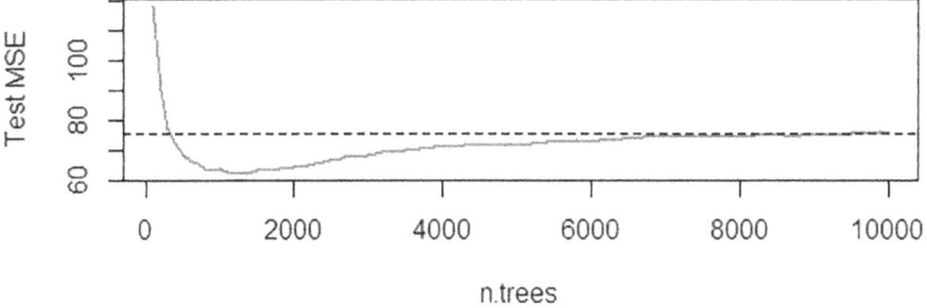

FIGURE 3.8 Test MSE in boosting method.

TABLE 3.6
Result Parameters of Boosting Method

Sr. No.	Result Parameter	Value
1.	Number of trees	500
2.	No. of variables tried at each split	13
3.	Mean of squared residuals	45.844
4.	Percentage variation explained	69.92
5.	Error	62.08

3.4.4 Neural Network Method

The artificial neural network method processes information in parallel throughout the nodes. It is a complex adaptive system that has the ability to change its internal structure by adjusting the weights of inputs.

The values predicted versus the real values have been shown in Table 3.7.

The plot for the above table has been shown in Figure 3.9. The figure comprises black dots that signify the real values and a blue line that shows the graph for the predicted dataset through a neural network.

Also, a grey envelope has been used in the figure to signify the 95% confidence interval, which has shown that most of the real values fall inside the envelope. Thus, it can be inferred that Neural Network may be used to predict the compressive strength.

Below is the comparison of the results of prediction by all the four given methods.

In Table 3.8, it can be seen that the mean square error and the mean square prediction error are the smallest for the boosting method.

TABLE 3.7
Testing versus Predicted Values from Neural Network Method

Sr. No.	Testing Value	Predicted Value	Sr. No.	Testing Value	Predicted Value
1	43.5	38.8249048	35	50	53.2831532
2	48.9	38.821768	36	37	32.8015773
3	48.8	54.2703938	37	33	32.3436917
4	42	40.832212	38	39.7	−6.4531324
5	42	40.832212	39	1.5	−0.8726439
6	74.7	78.4276454	40	2.4	3.8911869
7	97.8	75.0557965	41	3.1	2.6434277
8	32	51.4037261	42	3.9	3.5745312
9	9.9	17.4203588	43	3.4	5.4611222
10	9	14.2331137	44	2.7	6.4181217
11	97.3	112.6799869	45	7.1	10.6959959
12	40	38.283311	46	12	9.8526236
13	28	29.7973551	47	12.5	10.568272
14	40	38.8547166	48	15.5	12.5894056
15	40	39.245793	49	42.4	33.6491537
16	40	38.8547166	50	31.3	33.6491537
17	68.6	56.0914594	51	31.5	33.6491537
18	85.1	64.7707188	52	27.4	33.6491537
19	59.4	37.9335818	53	60	61.4557835
20	85.1	73.3459877	54	29.9	33.4233715
21	63.4	53.51399	55	41.2	33.4233715
22	75	71.424867	56	39.1	32.8993118
23	85	80.5696706	57	76.7	73.8421534
24	45.5	44.8363466	58	80.3	84.1500003
25	49.5	46.2988329	59	82.3	80.1790706
26	44	39.5715669	60	76.2	83.2748044
27	45.5	39.5802548	61	48	39.7684015
28	42	48.9218856	62	25	36.6770377
29	42	48.9228563	63	48	36.6770377
30	33	45.0581304	64	35	38.7201288
31	37	50.4675191	65	66	72.5703943
32	41	50.5557031	66	66	70.5138791
33	51	51.9433258	67	49.4	61.2076049
34	49.5	53.2441435	68	15.6	36.2703148

3.5 CONCLUSION

In this work, several studies have been referenced for their data on compressive strength. Afterwards, a comparison has been drawn between the data analytics methods to find the best method that can be used to predict the compressive strength behaviour of concrete. Based on the present study, the following conclusions can be drawn:

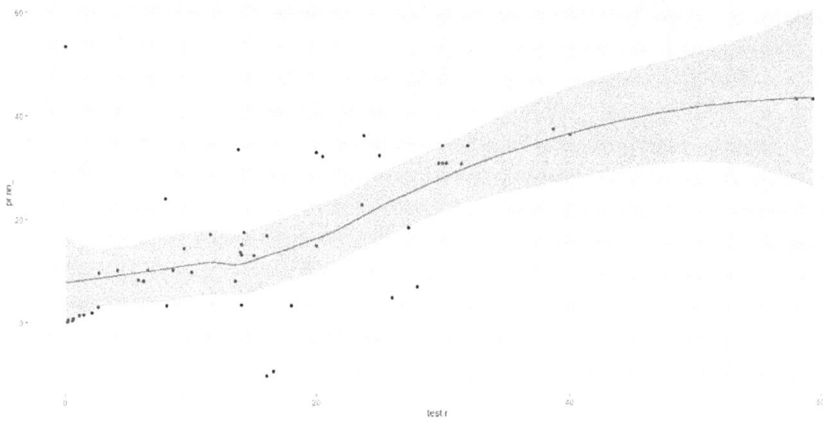

FIGURE 3.9 Neural network plot.

TABLE 3.8
Comparison of Performance of all the Methods

Sr. No.	Methods	Mean Squared Error (MSE)	Mean Squared Prediction Error (MSPE)
1.	Linear regression method	64.20	113.54
2.	Regression tree method	35.75	85.72
3.	Boosting method	2.583	7.57
4.	Neural network method	5.42	131.74

1. Boosting is the best method among all the used methods for predicting data based on compressive strengths as follows:
   a. Boosting has the lowest mean square error, which means the majority of the real values lie quite close to the predicted ones.
   b. Boosting has the lowest mean square prediction error, which means the probability to predict a particular value from a dataset is highest in the boosting method.
2. A neural network is the second best method that can be used to predict the data related to concrete. As most of the values of real data lie inside the 95% confidence interval band, it can be safely assumed that neural network prediction is true 95% of the times.

TABLE 3.9
Data Set Used for Modelling[a]

											28-Day Compressive Strength (MPa) at Temperatures in °C																
R. No	C	W	S	G	FA	SF	WBR	AC	HR	HE	20	23–28	100	150	200	300	350	400	450	550	600	650	700	750	775	800	850
								AC3 0																			
Abaeian et al., 2018	320	144	938	1146	0	0	0.45	0.55	-	1		41.8		40.2			39.4			38.3				28.5			
	320	144	938	1146	0	0	0.45	0.95	-	1		43.5		41.3			40.8			39.2				32.9			
	320	144	938	1146	0	0	0.45	1.35	-	1		44.5		42.8			41.7			40.9				38.7			
	320	144	938	1146	0	0	0.45		-	1		48.9		45.1			42.9			41.8				39.5			
								AC1 AC4																			
Chan et al., 2000	300	165	660	810	0	0	0.55	0 0	10	1	35		31.5		24.5			23.45			15.75					5.3	
	300	165	660	810	0	0	0.55	0 0	5	3	35		25.2		29.75			23.45			10.5					5.3	
	300	205	240	0	360	0	0.31	2.1 5.4	10	1	40		41.2		36			40			28					16	
	300	205	240	0	360	0	0.31	2.1 5.4	5	3	40		32		38			26.8			20					8	
								AC2																			
Chan et al., 2000	270	176	750	1156	0	0	0.65	2	7	2	47						36.3			24.5			16.2		9.3		3.3
	270	176	750	1156	0	0	0.65	2	7	2	34						32.4			23.1			16.6		13.2		7.5
	270	176	750	1156	0	0	0.65	2	7	2	42						36.6			23.3			17.2		10.7		3.6
	270	176	750	1156	0	0	0.65	2	7	2	42						40.9			30.4			21.6		16.8		7.6
								AC2 AC6																			
Kumar et al., 2019	450	135	732	1263	0	0	0.30	0 7.875	-	-	77					85			82.7		58						
	450	135	732	1263	0	0	0.30	1.5 10.485	-	-	75					78.4			74.4		48.2						
	450	135	732	1263	0	0	0.30	0 7.875	-	-	73					88.5			81.8		59.3						
								AC5																			
Drzymała et al., 2018	290	170	680	1261	72	0	0.47	2.9	5	1	46				41.15			36.5			25.16						
	548	150	680	1261	202	0	0.20	9.59	5	1	98				87.98			78.2			53.76						

(*Continued*)

TABLE 3.9 (*Continued*)
Data Set Used for Modelling[a]

R. No	C	W	S	G	FA	SF	WBR	AC		HR	HE	28-Day Compressive Strength (MPa) at Temperatures in °C																
												20	23–28	100	150	200	300	350	400	450	550	600	650	700	750	775	800	850
Ibrahim, 2017	370	200	280	465	0	0	0.54			-	2		9.89							7.607			5.79					
	340	190	300	520	0	0	0.56			-	2		8.95							7.267			5.177					
	320	210	290	550	0	0	0.66			-	2		8.5							7.547			5.033					
								AC2	AC3																			
Jessie & Santhi, 2019	262.5	210	588.8	1251	87.5	0	0.60	0	0	6	1		34.9														15.8	
	357.5	196	542.7	1153	137.5	55	0.36	0	0	6	1		97.3														24.9	
	357.5	196	542.7	1153	137.5	55	0.36	0	3.12	6	1		113.5														38.7	
	357.5	196	542.7	1153	137.5	55	0.36	1.82	0	6	1		99.1														23.8	
								AC2	AC3																			
Li et al., 2019	462	185	671	1024	0	0	0.40	0	0	-	-		25	24		23	21.5											
	572	200	646	898	0	0	0.35	0	7	-	-		69	64		59	50											
	572	200	646	898	0	0	0.35	1	7	-	-		68.6	65		63	59											
	572	200	646	898	0	0	0.35	2	7	-	-		66.1	60.7		59.8	55											
	572	200	646	898	0	0	0.35	3	7	-	-		60.2	56.5		51.3	50											
								AC6																				
Patel et al., 2020	500	385	820	0	0	0	0.77	6		-	1	38				37.22			35.19								13.3	
	450	385	805	0	50	0	0.77	6		-	1	42				37.37			33.02								14.99	
	400	385	795	0	100	0	0.77	6		-	1	46				35.63			32.21								16.58	
	350	385	780	0	150	0	0.77	6		-	1	50				34.61			30.49								16.02	
								AC3	AC6																			
Tanyildizi & Coskun, 2008	600	180	668	1002	0	0	0.30	0	9	3.3	2.5		33			31			29			20					8	
	600	180	668	1002	0	0	0.30	1	9	3.3	2.5		40			39			48			29					8	
	600	180	668	1002	0	0	0.30	0	0	3.3	2.5		37			36			40			34					7	

[a] C = Cement, W = water, S = sand, G = gravel.

REFERENCES

Abaeian, R., Behbahani, H. P., & Moslem, S. J. Effects of high temperatures on mechanical behavior of high strength concrete reinforced with high performance synthetic macro polypropylene (HPP) fibres. *Construction and Building Materials*, 2018, 165, 631–638. Doi: 10.1016/j.conbuildmat.2018.01.064.

Anita Jessie, J., & Santhi, A. S. Effect of temperature on compressive strength of steel fibre reinforced concrete. *Journal of Applied Science and Engineering*, 2019, 22(2), 233–238. Doi: 10.6180/jase.201906_22(2).0004.

Chan, Y. N., Luo, X., & Sun, W. Compressive strength and pore structure of high-performance concrete after exposure to high temperature up to 800°C. *Cement and Concrete Research - CEM CONCR RES*, 2000, 30, 247–251. Doi: 10.1016/S0008-8846(99)00240-9

Drzymała, T., Jackiewicz-rek, W., Gałaj, J., & Sukys, R. Assessment of mechanical properties of high strength concrete (HSC) after exposure to high temperature. *Journal of Civil Engineering and Management*, 2018, 24(2), 138–144. Doi: 10.3846/jcem.2018.457.

Ibrahim, R. K. The effect of elevated temperature on the lightweight aggregate concrete. *Kurdistan Journal of Applied Research*, 2017, 2(3), 193–196. Doi: 10.24017/science.2017.3.38.

Kumar, D., Ranade, R., & Whittaker, A. *Advanced concretes for high temperature applications*. 2019. https://www.researchgate.net/publication/335728521.

Li, L., Zhang, R., Jin, L., Du, X., Wu, J., & Duan, W. Experimental study on dynamic compressive behavior of steel fiber reinforced concrete at elevated temperatures. *Construction and Building Materials*, 2019, 210, 673–684. Doi: 10.1016/j.conbuildmat.2019.03.138.

Patel, V., Singh, B., Ojha, P., & Mohapatra, B. Effect on mechanical properties and stress strain characteristics of normal and high strength concrete at elevated temperature. *Journal of Building Materials and Structures*, 2020, 7, 199–209. Doi: 10.5281/zenodo.4077425.

Tanyildizi, H., & Coskun, A. The effect of high temperature on compressive strength and splitting tensile strength of structural lightweight concrete containing fly ash. *Construction and Building Materials*, 2008, 22, 2269–2275. Doi: 10.1016/j.conbuildmat.2007.07.033.

4 Implementation of Machine Learning Approaches to Evaluate Flexural Strength of Concrete with Glass Fiber

Ankita Upadhya, Mohindra Singh Thakur, Nitisha Sharma, and Fadi Hamzeh Almohammed
Shoolini University

CONTENTS

4.1 INTRODUCTION

There has been significant growth in the improvement of the latest age category of concretes in recent decades. The building industry is rapidly evolving, and the materials used in construction are emerging (Kumar et al., 2020). Concrete has significant advantages over other sustainable building materials, such as durability,

DOI: 10.1201/9781003184331-4

formability, and required mechanical strength. However, it does have certain disadvantages, such as poor yield stress and energy absorption capacity (Kizilkanat et al., 2015). Furthermore, because concrete has high compressive strength but a low-tension strength, fibers can help to improve structural characteristics. The majority of the problem comes when there is observable cracking on the concrete surface, which results in an uncomfortable concrete surface. These fractures are often apertures for penetration of water and salt into the concrete, which increases the probability of reinforcement corrosion (Akbari and Abed, 2020). The presence of fiber increases the fracture toughness, modulus of elasticity, and other technical properties of concrete, mortars, and cement matrix may all be improved (Nataraja et al., 2005). According to studies, glass fiber, which with its impact strength and elasticity modulus, absence of combustion, and cheap cost, has the potential to substantially improve the deformability, toughness, and impact resistance of cement-based materials. When chopped into them. Fibers incorporated into concrete may prevent the spread of micro-fractures and the emergence of new cracks in the matrix and substantially improve flexural strength and wear resistance (Xiaochun et al., 2017, Upadhya et al., 2021). The use of high-performance fiber-reinforced concrete improves the beam's seismic performance. Fibers can help with shrinkage cracking, water absorption, flexural and tensile strength, and impact resistance (Wagh et al., 2020).

Tassew and Lubell (2014), when glass fiber was added to ceramic concrete, the compression, flexure, and shear toughness all rose, while the workability dropped. Overall, the findings show that the durability and mechanical performance of glass fiber reinforced ceramic concretes can be achieved and appropriate for use in construction components. Sivakumar et al., (2017) investigate that there was a decrease in workability as the fiber level increased. Glass fibers were discovered to have no beneficial influence on concrete's compressive strength. With increasing fiber dose, glass fibered self-compacting concrete improved its split compressive and flexural strength. Nematollahi et al., (2014) found that the inclusion of glass fibers to heat-cured fly ash-based geopolymer concrete decreased workability but enhanced density, compressive, and flexural strengths. The geo-polymer concrete (GPC) mix with 1.25% glass fibers increased flexural and compressive strengths by 8.5% and 34%, respectively.

Ghugal and Deshmukh (2006) in this research study the workability of wet glass fiber reinforced concrete diminishes as the fiber content increases, increases in fiber content enhance the moist density of concrete, although only slightly. At 28 days, the percentage rise in cube strength attributes, flexural strength, split tensile strength, and bond strength when compared to regular concrete was 4.5, 4.0, 4.5, and 3.0, respectively. Atewi et al., (2019) performed a test utilizing nano-silica in the range of 0%–4%, with substitution up to 0.7% glass fiber, which was shown to improve self-compacting concrete's mechanical and permeability characteristics. The mortar samples with distributed glass fiber reinforcing were studied by Maek et al., (2021). It has been discovered that adding glass fibers to mortar increases its characteristics. In comparison to the base sample, the concrete containing 1800 g/m^3 of glass fibers had the greatest mechanical characteristics (31.5 maximum compressive strength by 29.9%, flexural strength increased by 29.9%, and split tensile strength increased by 97.6%).

According to the literature review, the usual test for assessing the characteristics of concrete is laborious and time-consuming, which has an impact on the project's cost. The empirical formula technique demonstrates the nonlinear relationship between the dependent and independent variables. Researchers were employing several machine learning methods to anticipate the outcome, according to the literature. Various machine learning techniques, such as artificial neural networks (ANN), M5P, support vector machines, Gaussian process, random forest (RF), and random tree, have been used on difficult engineering issues for the prediction of the strength characteristics of concrete mixes throughout the last two decades (Thakur et al., 2021, Chopra et al., 2018, Singh et al., 2019, Upadhya et al., 2021). Yıldızel and Öztürk (2016) investigate the artificial bee colony optimization and traditional backpropagation methods were used to analyze 143 distinct outcomes of four-point bending tests on glass fiber reinforced concrete mixes with varied temperature, slump values and fiber content. It has been determined that ANN models can be used to evaluate the outcome of a decision and the characteristics of fiber-reinforced concrete specimens. Emiroğlu et al., (2012) concluded that adaptive-network-based fuzzy inference system (ANFIS) and logistic regression are efficient approaches for forecasting peak pressure load value of concrete pipes incorporating glass fibers, and peak pressure load values may be anticipated using ANFIS and Logistic regression without completing any experiments in a relatively short amount of time with extremely low margins of error. Furthermore, the ANFIS model outperforms the multi-linear regression (MLR) model in terms of prediction. Yildizel and Öztürk (2016) study the correlation between the bending strength and the quantities of the mixture. According to the findings, the ANN model approved a greater relationship between the bending strength of glass fibre reinforced self-leveling concrete and mixture proportions. According to the investigation's conclusions, the weighting factor is properly calibrated. They had a strong connection with the results of earlier experiments. Sun et al., (2021) investigated mechanical, chemical, and mechanical–chemical (combined) methods were used to activate glass powder in a mortar. The combined technique was shown to be the best for enhancing the expansion of flexural strength and decreasing alkali–silica reaction when using a mortar containing 30% waste glass powder. To investigate the activation impact on glass powder and mortar, a microstructure analysis was used. Furthermore, beetle antennae search in RF model hyperparameters was presented to accurately forecast flexural strength and alkali–silica reaction growth. Tavakoli et al., (2014) determine the experimental results which are utilized to create a computational model such as an ANN with a multi-layer perceptron. The neural model is then used to predict the influence of a variety of variables on the desired output, including tensile strength and flexural strength, structural rigidity, and compressive strength.

Several researches (Chopra et al., 2018, Singh et al., 2019, Tavakoli et al., 2014) on various engineering issues have been conducted throughout the years utilizing RF, bagging, stochastic, and M5P. Many studies have discovered that these approaches are as excellent as or better than the neural network method in terms of achieving satisfactory outcomes. The goal of this research is to look at various approaches in order to discover the optimal algorithm for predicting the flexural strength of a concrete mix with glass fiber, as well as the impact of input factors on that prediction.

4.2 MACHINE LEARNING MODELS

4.2.1 Random Forest (RF)

RF is a probabilistic classifier that employs a collection of different classifiers, each of which can be built using a bootstrapping sample of data. At each split, the candidate set of variables for tree construction is chosen at random. Another approach is to employ bagging, which allows for the successful combination of unsteady learners (Sun et al., 2019).

The construction of a trained dataset, computation of the mean value of a single decision tree outcome, and validation of predicted outcomes using a validating dataset are all examples of trained regression problems using a training dataset +e bootstrap data is used to create a new trained dataset from the original trained set in the RF approach (Khan et al., 2021). As a result, the RF regression is made up of a number of (*k*) trees (Singh et al., 2019). Moreover, each particular tree is extended to decrease categorization error, despite the fact that random selection emphasizes the outcome (Upadhya et al., 2021). As a result, every tree assigns a categorization, and the model selects the forest with the highest number of votes out of all the trees (E votes may be either 0 or 1.) The possibility of a prediction is computed as a percentage of the number of 1s that have been received (Chopra et al., 2018). As a result, to get the optimal RF model, RF models classify variables according to their significance (Nhu et al., 2020).

4.2.2 Bagging

The bagging technique was proposed by to (Breiman, 2001) enhance the prediction performance of weak learning machines. The proportion of the original database that will be substituted by recombination is determined by the bagging factor parameter (Moretti et al., 2015). The bootstrap resampled set of the observed dataset is used to create each model. Bagging algorithms may be broken down into stages: the bootstrap approach uses replacements to produce a fresh training set, which is then utilized the outcome of the fundamental classifiers is then associated using distributed voting not only does this method improve generalization capacity, but it also lowers classification variance (Pham et al., 2020). In the case of bagging, the training set in each iteration contains around 67% of the data from the initial training set while leaving out approximately one-third of the original training samples. Out-of-bag samples are the samples that were left out of the bag (Sepahvand et al., 2019). To increase the forecast accuracy of the models, a variety of models have been developed (Thakur et al., 2020). To take advantage of this method, the base classifier must be imbalanced, as little changes in the training set can result in significant changes in the classifier output otherwise, no classification will emerge. The k-nearest neighbor is a stable classifier, while the Multilayer Perceptron Classifier (MLPC) is an uneven classifier (Sahana et al., 2020). The RF model is employed using a bagged method in this analysis.

4.2.3 Stochastic

The stochastic meta-assembly approach was used to increase the accuracy of the standard techniques-based model. It's extensively utilized to solve difficult and nonlinear issues in engineering. In this study, the stochastic method is used to increase the accuracy of RF models (Thakur et al., 2020). The designing process is fraught with uncertainties, such as the natural randomness of physical quantities used in the design, model assumptions owing to inadequacies in the computation model in comparison to real structural data, and statistical assumptions during the identification of a quantity due to a lack of data (Venclovský et al., 2021). A stochastic process is a set of arbitrary variables whose values are determined by the state vector which is represented by the dataset Q. Another set Z, or a set of indexes, is used to index the gathering. Non-negative real numbers and natural numbers $Z=1, 2\ldots$ and $Z=[0,]$ are the two most frequent index sets for discrete and continuous times, respectively. As a result, the first index set produces a series of arbitrary variables $X0, X1, X2\ldots$, and the second index set produces a collection of arbitrary variables $X(v)$, $v0$, one for each v. For the construction of numerical models, stochastic processes are commonly used (Thakur et al., 2020).

4.2.4 M5P Tree

Model trees are based on a decision tree, which has linear functions at their leaves. They are comparable to linear models in small bit form. An M5P tree-based model is a binary decision tree that can estimate continuous mathematical characteristics and has linear equations at the terminal (leaf) nodes (Thakur et al., 2020). The first stage entails dividing criteria in order to create a decision tree. This method's splitting criteria are depending on the group's standard error value. When a node is split, the child node has a lower standard deviation than the parent node, making it even purer. M5P model tree chooses the split that maximizes error reductions out of all feasible splits (Singh et al., 2019). If the values of all examples reaching a node differ slightly or there are just a few instances left, the splitting ends and the changeability is determined by the predicted decrease in error as a consequence of checking each variable at that node is defined as the standard deviation of the values that reach that node from the roots to the branches (Upadhya et al., 2020).

4.3 METHODOLOGY AND DATASET

This study approach includes four key steps: (1) data gathering, (2) model building, (3) model validation and comparison, and (4) assessment of the best model utilizing soft computing techniques. The observations utilized in the investigation are listed in detail in Table 4.1. A total of 102 observations were taken from the literature for this investigation in which 76 data were chosen at random for training and 33 for testing the models. To assess the accuracy of the best model, three techniques, including M5P, RF, and to enhance RF hybrid technique, such as bagging and stochastic, were utilized in this study, which was implemented using Weka 3.9 software for the input parameters such as cement, fine aggregate, coarse aggregate, water, and superplasticizer, fiber length, glass fiber, and curing days are the parameters in the dataset used as input parameters for calculating flexural strength which is taken as an output. Table 4.2 lists the characteristics of the total, training, and testing datasets. The correlation coefficient (CC), mean absolute error (MAE), root mean squared error (RMSE), relative absolute error (RAE), and root relative squared error (RRSE) were used to assess the performance of each model. These variables were beneficial in determining which model was the best. Better results are predicted by a higher CC and a lower value of error.

TABLE 4.1
Detail of Dataset[a]

S. No.	Cement (kg/m³)	Fine Aggregate (kg/m³)	Coarse Aggregate (kg/m³)	Water/ w/c (kg/m³)	SP/ HR/ WR	Fiber Length%	Glass Fiber %	Curing Days	Flexural Strength (kg/m³)	Author
1	450	407	1253.8	221.4	4.45	0	0	28	11.51	Kelestemur et al., (2014)
2	450	325.4	1253.8	221.4	4.45	0	0	28	12.38	
3	450	243.6	1253.8	222.8	4.45	0	0	28	11.71	
4	450	203.6	1253.8	222.8	4.45	0	0	28	12.1	
5	450	407	1253.8	221.4	4.45	6	0.25	28	11.84	
6	450	325.4	1253.8	221.4	4.45	6	0.25	28	12.69	
7	450	243.6	1253.8	222.8	4.45	6	0.25	28	12.27	
8	450	203.6	1253.8	222.8	4.45	6	0.25	28	12.46	
9	450	407	1253.8	221.4	4.45	6	0.5	28	11.43	
10	450	325.4	1253.8	221.4	4.45	6	0.5	28	11.96	
11	450	243.6	1253.8	222.8	4.45	6	0.5	28	13.56	
12	450	203.6	1253.8	222.8	4.45	6	0.5	28	13.76	
13	450	407	1253.8	221.4	4.45	6	0.75	28	11.98	
14	450	325.4	1253.8	221.4	4.45	6	0.75	28	12.52	
15	450	243.6	1253.8	222.8	4.45	6	0.75	28	13.73	
16	450	203.6	1253.8	222.8	4.45	6	0.75	28	13.98	
17	400	980	561	140	0.4	0	0.0	7	1.88	Akbari and Abed (2020)
18	400	980	561	140	1.2	12	0.3	7	1.67	
19	400	980	561	140	1.6	12	0.6	7	2.24	
20	400	980	561	140	2.2	12	0.9	7	1.88	
21	350	975	569	157.5	0.35	0	0	7	2.76	
22	350	975	569	157.5	0.7	12	0.3	7	2.13	
23	350	975	569	157.5	0.88	12	0.6	7	2.81	
24	350	975	569	157.5	1.23	12	0.9	7	2.65	
25	400	980	561	140	0.4	0	0	28	3.03	
26	400	980	561	140	1.2	12	0.3	28	2.49	
27	400	980	561	140	1.6	12	0.6	28	3.18	
28	400	980	561	140	2.2	12	0.9	28	2.55	
29	350	975	569	157.5	0.35	0	0	28	2.75	
30	350	975	569	157.5	0.7	12	0.3	28	2.89	
31	350	975	569	157.5	0.88	12	0.6	28	4.13	
32	350	975	569	157.5	1.23	12	0.9	28	3.31	
33	350	1100	740	3.5	0.5	24	0.5	28	5.99	Arslan (2016)
34	350	1100	740	3.5	0.5	24	1	28	6.85	
35	350	1100	740	3.5	0.5	24	2	28	5.88	
36	350	1100	740	3.5	0.5	24	3	28	5.11	
37	300	825	1100	126	0.6	0	0	3	1.79	Liu et al., (2019)
38	300	825	1100	126	0.6	12	0.5	3	1.91	
39	300	825	1100	126	0.6	12	1	3	1.75	
40	300	825	1100	126	0.6	12	1.5	3	1.65	

(Continued)

TABLE 4.1 (*Continued*)
Detail of Dataset[a]

S. No.	Cement (kg/m³)	Fine Aggregate (kg/m³)	Coarse Aggregate (kg/m³)	Water/ w/c (kg/m³)	SP/ HR/ WR	Fiber Length%	Glass Fiber %	Curing Days	Flexural Strength (kg/m³)	Author
41	300	825	1100	126	0.6	0	0	7	5.12	
42	300	825	1100	126	0.6	12	0.5	7	5.02	
43	300	825	1100	126	0.6	12	1	7	5.15	
44	300	825	1100	126	0.6	12	1.5	7	5.44	
45	300	825	1100	126	0.6	0	0	14	5.70	
46	300	825	1100	126	0.6	12	0.5	14	5.68	
47	300	825	1100	126	0.6	12	1	14	5.76	
48	300	825	1100	126	0.6	12	1.5	14	5.87	
49	300	825	1100	126	0.6	0	0	28	6.09	
50	300	825	1100	126	0.6	12	0.5	28	6.26	
51	300	825	1100	126	0.6	12	1	28	6.57	
52	300	825	1100	126	0.6	12	1.5	28	7.10	
53	280	655	1165	129	6.45	0	0	7	5.46	Yuan and Jia (2021)
54	280	655	1165	129	6.45	12	0.45	28	4.75	
55	280	655	1165	129	6.45	12	0.9	7	6.26	
56	240	677	1165	129	5.55	12	1.35	28	7.0	
57	240	677	1204	129	5.55	0	0	7	4.31	
58	240	677	1204	129	5.55	12	0.45	28	5.30	
59	280	655	1204	129	6.45	12	0.9	7	6.41	
60	280	655	1204	129	6.45	12	1.35	28	5.35	
61	600	484	1068	0.37	2	0	0	7	4.84	Hilles and Ziara (2020)
62	600	484	1068	0.37	2	8	0.3	7	5.28	
63	600	484	1068	0.37	2	8	0.6	7	6.26	
64	600	484	1068	0.37	2	8	0.9	7	6.68	
65	600	484	1068	0.37	2	8	1.2	7	7.27	
66	600	484	1068	0.37	2	0	0	28	6.35	
67	600	484	1068	0.37	2	8	0.3	28	7.53	
68	600	484	1068	0.37	2	8	0.6	28	8.28	
69	600	484	1068	0.37	2	8	0.9	28	8.79	
70	600	484	1068	0.37	2	8	1.2	28	9.68	
71	318	732	1118	178	0	0	0	28	3.52	Chandra mouli et al., (2010)
72	318	732	1118	178	0	0	0	56	3.96	
73	318	732	1118	178	0	0	0	90	4.18	
74	318	732	1118	178	0	0	0	180	4.29	
75	318	732	1118	178	0	12	0.03	28	4.08	
76	318	732	1118	178	0	12	0.03	56	4.59	
77	318	732	1118	178	0	12	0.03	90	4.85	
78	318	732	1118	178	0	12	0.03	180	5.02	
79	350	686	1137	178	0	0	0	28	4.12	

(*Continued*)

TABLE 4.1 (*Continued*)
Detail of Dataset[a]

S. No.	Cement (kg/m³)	Fine Aggregate (kg/m³)	Coarse Aggregate (kg/m³)	Water/ w/c (kg/m³)	SP/ HR/ WR	Fiber Length%	Glass Fiber %	Curing Days	Flexural Strength (kg/m³)	Author
80	350	686	1137	178	0	0	0	56	4.57	
81	350	686	1137	178	0	0	0	90	4.96	
82	350	686	1137	178	0	0	0	180	4.98	
83	350	686	1137	178	0	12	0.03	28	4.78	
84	350	686	1137	178	0	12	0.03	56	5.39	
85	350	686	1137	178	0	12	0.03	90	5.62	
86	350	686	1137	178	0	12	0.03	180	5.93	
87	400	604	1170	164	0	0	0	28	4.72	
88	400	604	1170	164	0	0	0	56	5.28	
89	400	604	1170	164	0	0	0	90	5.42	
90	400	604	1170	164	0	0	0	180	5.91	
91	400	604	1170	164	0	12	0.03	28	5.52	
92	400	604	1170	164	0	12	0.03	56	6.18	
93	400	604	1170	164	0	12	0.03	90	6.40	
94	400	604	1170	164	0	12	0.03	180	6.95	
95	450	590	1142	163	0	0	0	28	5.42	
96	450	590	1142	163	0	0	0	56	5.80	
97	450	590	1142	163	0	0	0	90	6.43	
98	450	590	1142	163	0	0	0	180	6.57	
99	450	590	1142	163	0	12	0.03	28	6.23	
100	450	590	1142	163	0	12	0.03	56	6.79	
101	450	590	1142	163	0	12	0.03	90	7.52	
102	450	590	1142	163	0	12	0.03	180	7.56	

[a] Cement = C, Fine Aggregate = FA, Coarse Aggregate = CA, Water = W, Superplasticizer/ High-Range Water Reducer = SP/HR/WR, FibeR length = FL, Glass Fiber = GF, Curing Days = CD, Flexural Strength = FS.

TABLE 4.2
Statistics Features of Dataset

	Cement (kg/m^3)	Fine Aggregate (kg/m^3)	Coarse Aggregate (kg/m^3)	Water/w/c (kg/m^3)	SP/HR/WR	Fiber Length %	Glass Fiber %	Curing Days	FT (kg/m^3)
Training Dataset									
Mean	392.50704	672.6422	1050.3915	139.3797	1.645211	7.09859	0.340985	43.14084	6.253943
Standard deviation	96.46300	222.5418	221.64211	66.80227	2.029705	6.385597	0.446702	51.04600	3.030378
Kurtosis	0.116639	−0.584787	0.7773760	0.423090	−0.22148	−0.07195	1.695585	3.373556	0.305363
Skewness	0.792496	−0.028507	−1.4909285	−1.086304	1.10061	0.478443	1.388497	2.163770	0.967254
Minimum	240	243.6	561	0.37	0	0	0	3	1.67
Maximum	600	1100	1253.8	222.8	6.45	24	2	180	13.73
Confidence level (95.0%)	22.83241	52.67479	52.461833	15.81184	0.48042	1.511446	0.105732	12.08239	0.717278
Testing Dataset									
Mean	379.225	685.9612	1025	147.3335	1.648709	9.032258	0.62483	36.5161	6.10419
Standard deviation	84.6698	256.6986	246.7194	58.24493	1.933888	5.51049	0.672011	33.4054	3.55489
Kurtosis	1.16772	−0.398632	−0.109751	2.077983	0.016002	0.731852	3.801346	−0.8929	0.23366
Skewness	0.98355	−0.588082	−1.236421	−1.281411	1.135738	−0.151438	1.613574	0.88329	0.97777
Minimum	240	203.6	561	0.37	0	0	0	3	1.65
Maximum	600	1100	1253.8	222.8	6.45	24	3	90	13.98
Confidence level (95.0%)	31.0571	94.15783	90.4974	21.36441	0.709356	2.02126	0.246495	12.2532	1.30394

4.4 MODEL EVALUATION

To check the accuracy of the estimated value for both the training and testing stages, quantitative indices are calculated to assess the preciseness of the estimated model. The value of CC is always between −1 and 1, with 1 or −1 indicating full correlation (in this case, all points are on a straight line). A coefficient value/R value near 0 indicates that the variables are independent. The greater the CC number, the more accurate the predictions i.e., lower values of evaluation parameters such as RMSE, MAE, RAE, and RRSE. These are the statistics that compare the error among predicted and experimental values that describe the same behavior, i.e., if the computed error is low, it predicts better output results. This equation can be expressed as:

$$CC = = \frac{\sum_{i=1}^{m}\left(T_i - \overline{T}\right)\left(X_i - \overline{X}\right)}{\sqrt{\sum_{i=1}^{m}\left(T_i - \overline{T}\right)^2}\sqrt{\sum_{i=1}^{N}\left(X_i - \overline{X}\right)^2}} \tag{4.1}$$

$$\text{RMSE} = \sqrt{\frac{1}{m}\sum_{i=1}^{m}(X - T)^2} \tag{4.2}$$

$$\text{MAE} = \frac{1}{x}\left(\sum_{i=1}^{x}|X - T|\right) \tag{4.3}$$

$$\text{RAE} = \frac{\sum_{i=1}^{m}|T - X|}{\sum_{i=1}^{m}\left(\left|T - \overline{T}\right|\right)} \tag{4.4}$$

$$\text{RRSE} = \sqrt{\frac{\sum_{i=1}^{m}(T - X)^2}{\sum_{i=1}^{m}\left(\left|X - \overline{X}\right|\right)^2}} \tag{4.5}$$

T = Observed values
$\overline{T}$ = Average of the observed value
X = Predicted values
m = Number of observations

4.5 RESULT AND DISCUSSION

4.5.1 Random Forest (RF) Model Evaluation

The flexural strength of concrete is predicted using an RF-based model, which is based on a trial and error approach. In this technique, eight attributes are used. Cement, fine aggregate, coarse aggregate, water, superplasticizer/HRWR, glass fiber, fiber length, and curing days were used as input factors, whereas flexural strength was used as an output parameter listed in Table 4.3. The five different statistical factors were utilized to build models for both the testing and training stages. The results show that the RF-based model performs better in predicting the flexural strength of concrete using glass fiber with CC (0.9964, 0.9877), MAE (0.2117, 0.5227), RMSE (0.2644, 0.6066), RAE (9.52, 19.72), and RRSE (8.79, 17.33) values at both training and testing stages. Figure 4.1 illustrates the training and testing stages, the agreement graph plotting actual and predicted values using RF-based models. The majority of the points in these graphs lie near the line of the perfect agreement representing the best possible match between actual and projected outcomes attributes which define more accuracy. It has also been found that the majority of the novel model's predicted values fall within the ±15% error zone.

4.5.2 Bagging Random Forest Model Evaluation (BRF)

In this study, bagging is used with RF to increase the accuracy of the dataset and to figure out which model is the most effective. It has been observed that the bagging gives less result than RF and stochastic random forest (SRF) with CC (0.9891, 0.9813), MAE (0.3133, 0.5867), RMSE (0.4544, 0.744), RAE (14.08, 22.13), and RRSE (15.10, 21.25) values at both training and testing stages. Table 4.3 shows that the glass fiber is more effective in predicting the flexural strength of concrete. Figure 4.2 illustrates the training and testing stages, the agreement graph plotting actual and predicted values using RF-based models s using bagging random forest (BRF)-based models. It has also been found that the majority of the novel model's predicted values fall within the ±15% error zone.

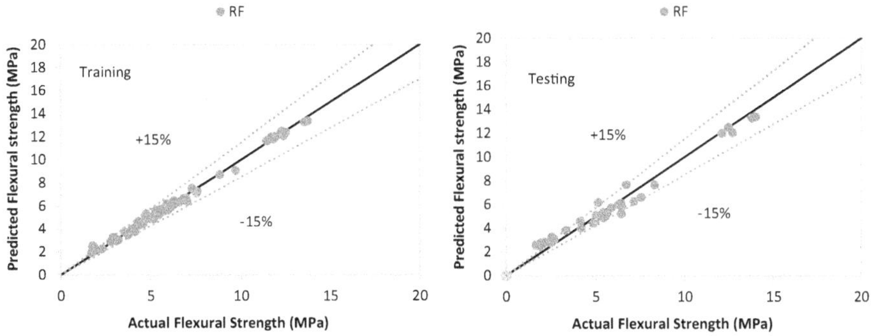

FIGURE 4.1 Agreement graph plots among actual and predicted values using RF-based models for training and testing stages.

TABLE 4.3
Assessment of Performance for Various Models

	Training					Testing				
Model No.	CC	MAE (MPa)	RMSE (MPa)	RAE (%)	RRSE (%)	CC	MAE (MPa)	RMSE (MPa)	RAE (%)	RRSE (%)
RF	0.9964	0.2117	0.2644	9.52	8.79	0.9877	0.5227	0.6066	19.72	17.33
BRF	0.9891	0.3133	0.4544	14.08	15.10	0.9813	0.5867	0.744	22.13	21.25
SRF	1	0.0018	0.0027	0.08	0.09	0.991	0.4059	0.503	15.31	14.37
M5P	0.9735	0.5884	0.7357	26.45	24.45	0.9777	0.6713	0.8858	25.32	25.31

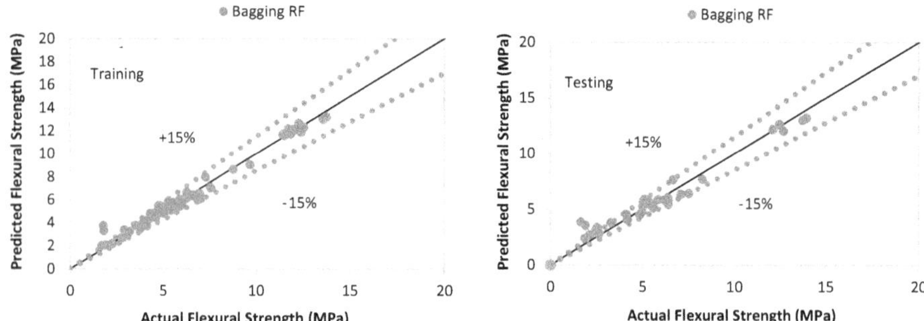

FIGURE 4.2 An agreement graph shows actual and predicted values based on BRF-based models for training and testing stages.

4.5.3 Stochastic Random Forest Model Evaluation (SRF)

It's a method for fitting an ensemble model of base classifiers using statistics. In this study, SRF is utilized in this work to improve predictive performance by bagging. As demonstrated in Table 4.3, the SRF model outperforms the RF and BRF models with greater CC (1,0.991), MAE (0.0018, 0.4059), RMSE (0.0027, 0.503), RAE (0.08, 15.31), and RRSE (0.09, 14.37) values at both the training and testing stages, as well as reduced error. Figure 4.3 illustrates the training and testing stages, the agreement graph plotting actual and predicted values using SRF-based models for training and testing stages. It also indicates that the majority of the predicted values for both the model and the perfect agreement are within a ±5% error zone.

4.5.4 M5P Tree Model Evaluation

The input variables were cement, fine aggregate, coarse aggregate, water, superplasticizer/HRWR, fiber length, glass fiber, and curing days, while the output parameter was flexural strength. The pruned model equations for input and output parameters are provided in equations (4.6)–(4.8), and (4.10), as shown below:

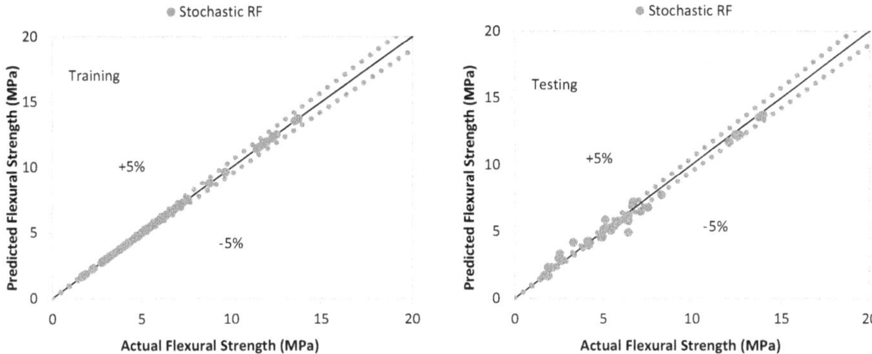

FIGURE 4.3 An agreement graph shows actual and predicted values using SRF-based models for training and testing stages.

Fine Aggregate (kg/m^3) ≤ 537: LM1 (19/13.763%)
Fine Aggregate (kg/m^3) > 537:
| Curing days ≤ 10.5:
| | Cement (kg/m^3) ≤ 325: LM2 (6/7.066%)
| | Cement (kg/m^3) > 325: LM3 (5/6.082%)
| Curing days > 10.5:
| | Water/w/c (kg/m^3) ≤ 127.5: LM4 (9/9.221%)
| | Water/w/c (kg/m^3) > 127.5: LM5 (32/17.759%)

LM1:

$$\begin{aligned} \text{F.T}\left(\text{kg/m}^3\right) = & -0.013 * \text{cement}\left(\text{kg/m}^3\right) - 0.0088 * \text{Fine Aggregate}\left(\text{kg/m}^3\right) \\ & + 0.0039 * \text{Water/w/c}\left(\text{kg/m}^3\right) + 0.0275 * \text{fiber length}\% \\ & + 1.8957 * \text{Glass fiber}\ \% + 0.054 * \text{curing days} - 17.088 \end{aligned} \tag{4.6}$$

LM2:

$$\begin{aligned} \text{F.T}\left(\text{kg/m}^3\right) = & -0.004 * \text{cement}\left(\text{kg/m}^3\right) - 0.0059 * \text{Fine Aggregate}\left(\text{kg/m}^3\right) \\ & - 0.0086 * \text{Water/w/c}\left(\text{kg/m}^3\right) + 0.0339 * \text{fiber length}\% \\ & + 0.4094 * \text{Glass fiber}\ \% + 0.234 * \text{curing days} - 9.6553 \end{aligned} \tag{4.7}$$

LM3:

$$\begin{aligned} \text{F.T}\left(\text{kg/m}^3\right) = & -0.007 * \text{cement}\left(\text{kg/m}^3\right) - 0.0059 * \text{Fine Aggregate}\left(\text{kg/m}^3\right) \\ & - 0.0086 * \text{Water/w/c}\left(\text{kg/m}^3\right) + 0.0339 * \text{fiber length}\% \\ & + 0.4094 * \text{Glass fiber}\ \% + 0.0119 * \text{curing days} - 10.9218 \end{aligned} \tag{4.8}$$

LM4:

$$\text{F.T}\left(\text{kg/m}^3\right) = -0.0064 * \text{Fine Aggregate}\left(\text{kg/m}^3\right) - 0.011 * \text{Water/w/c}\left(\text{kg/m}^3\right) + 0.0345 * \text{fiber length\%} + 0.4094 * \text{Glass fiber \%} + 0.0119 * \text{curing days} - 11.0647 \quad (4.9)$$

LM5:

$$\text{F.T}\left(\text{kg/m}^3\right) = 0.0016 * \text{cement}\left(\text{kg/m}^3\right) - 0.0077 * \text{Fine Aggregate}\left(\text{kg/m}^3\right) - 0.007 * \text{Water/w/c}\left(\text{kg/m}^3\right) \quad (4.10)$$

To generate the pruned model tree shown in the picture, smoothed logistic regressions were utilized (Figure 4.4).

The performance assessment of the M5P model is shown in Table 4.3. The results represent that the M5P model have better in predicting the flexural strength of concrete using glass fiber with CC (0.9735, 0.9777), MAE (0.5884, 0.6713), RMSE (0.7357, 0.8858), RAE (26.45, 25.32), and RRSE (24.45, 25.31) values at both training and testing stages with the less error. Figure 4.5 illustrates the training and testing

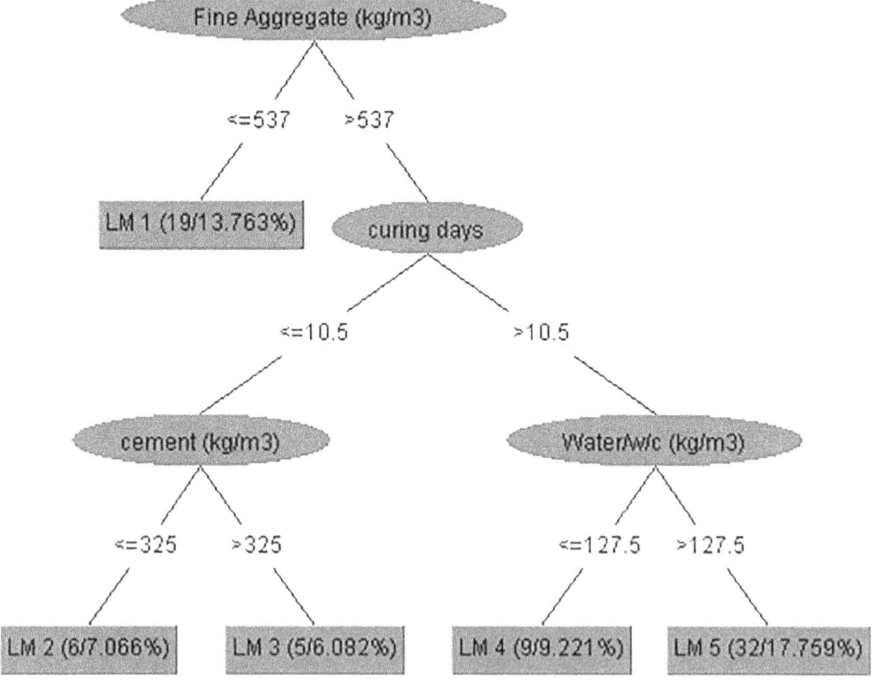

FIGURE 4.4 M5P model structure.

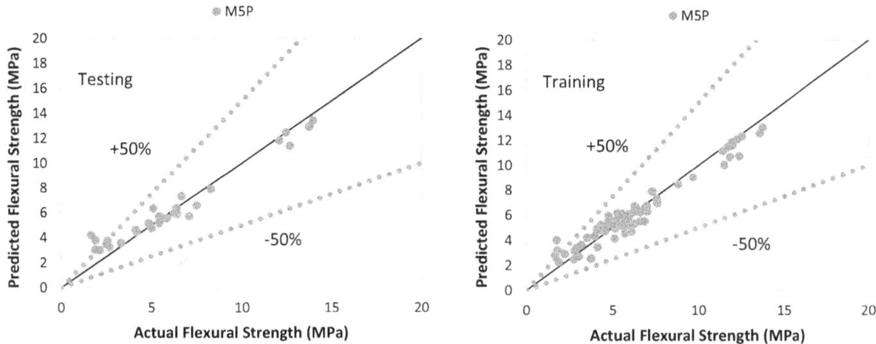

FIGURE 4.5 An agreement graph shows actual and predicted values using M5P tree-based models for training and testing stages.

stages, the agreement graph plotting actual and predicted values for the prediction of flexural strength of concrete mix using glass fiber. It also has been seen that most of the predicted values of the novel model lie under the range of ±50% error band.

4.6 RESULTS COMPARISON

In this research work, the flexural strength of concrete using glass fiber was predicted using a soft computing approach. As shown in Table 4.3, the performance of all five statistical parameters was assessed for both the training and testing stages. When all these models are examined, the SRF model appears to outperform the others on both the training and testing datasets. With the highest CC value for training and testing datasets i.e., (1, and 0.991) respectively also, for the testing dataset, the lowest error values were MAE (0.4059), RMSE (0.503), RAE (15.31%), and RRSE (14.37%). Figure 4.6 illustrates the differences between the anticipated and real datasets of machine learning algorithms that have been implemented. It also demonstrates that, when compared to other models, the SRF model anticipated readings are near to the actual data, resulting in a small error bandwidth and use glass fiber to evaluate the flexural strength of concrete with more precision. It also indicates that the majority of the predicted values for all the applied models and the perfect agreement are within a ±45% error zone.

4.7 SENSITIVITY ANALYSIS

For the prediction of flexural strength of concrete using glass sensitivity analysis is used to find the most relevant parameter among input factors. The highest performing model, SRF, was identified for this sensitivity study, and training data was organized and completed by removing a single input parameter. Statistical assessment metrics such as CC, MAE, RMSE, RAE, and RRSE were used to assess each model's performance. Table 4.4 demonstrates that the number of curing days is critical in predicting the flexural strength of a concrete mix.

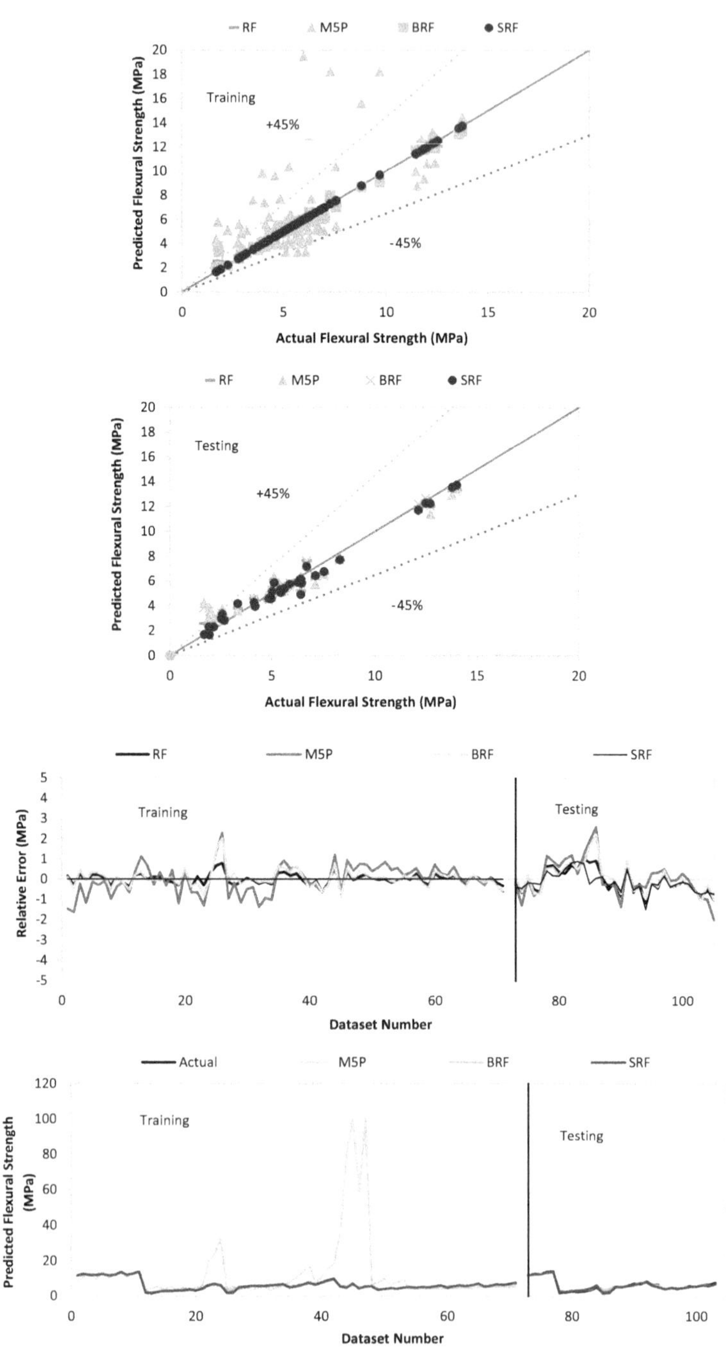

FIGURE 4.6 Comparison between predicted and actual flexural strength values by RF, BRF, SRF, M5P.

TABLE 4.4
Sensitivity Analysis Using SRF Model

Inputs Variables								Output	Stochastic Random Forest		
Cement (kg/m³)	Fine Aggregate (kg/m³)	Coarse Aggregate (kg/m³)	Water/w/c (kg/m³)	Super-plasticizer/High-Range Water Reducer SP/HRWR (kg/m³)	Fiber Length%	Glass Fiber %	Curing Days	Flexural Strength (MPa)	CC	MAE	RMSE
									0.9817	0.5289	0.7213
									0.9838	0.5076	0.6728
									0.977	0.5724	0.791
									0.9814	0.5265	0.7192
									0.9846	0.5266	0.6734
									0.9816	0.5046	0.7129
									0.9847	0.51	0.6328
									0.9251	0.9584	1.3575

4.8 CONCLUSION

In this study, the authors employed RF, BRF, SRF, and M5P tree-based models with glass fiber to compare machine learning approaches for predicting concrete strength. The input variables are used to predict the flexural strength of concrete using glass fiber, i.e., cement; fine aggregate; coarse aggregate; water/w/c; superplasticizer/HRWR; fiber length; glass fiber; curing days and flexural strength as an output variable. CC, MAE, RMSE, RAE, and RRSE were used to assess the performance of these models. The following are a summary of the findings of this study:

1. The SRF model's results were shown to be the best for predicting concrete flexural strength using glass fiber.
2. According to statistical analysis, SRF predicts superior results, with a higher CC (0.991), MAE (0.4059), RMSE (0.503), RAE (15.31%), and RRSE (14.37%).
3. The scatter graph plotted among actual and predicted values reveals that the SRF has a confined error band and is a good fit for predicting the output.
4. Results of the sensitivity analysis also showed that, when compared to other input factors for this dataset, curing days had a substantial impact on flexural strength prediction of concrete mixes utilizing glass fiber with an SRF-based model.

REFERENCES

Akbari, J. and Abed, A., 2020. Experimental evaluation of effects of steel and glass fibers on engineering properties of concrete: Engineering properties of concrete. *Fratturaed Integrità Strutturale*, *14*(54), pp. 116–127. Doi: 10.3221/IGF-ESIS.54.08.

Arslan, M.E., 2016. Effects of basalt and glass chopped fibers addition on fracture energy and mechanical properties of ordinary concrete: CMOD measurement. *Construction and Building. Materials*, *114*, pp. 383–391. Doi: 10.1016/j.conbuildmat.2016.03.176.

Atewi, Y.R., Hasan, M.F. and Güneyisi, E., 2019. Fracture and permeability properties of glass fiber reinforced self-compacting concrete with and without nanosilica. *Construction and Building Materials*, *226*, pp. 993–1005. Doi: 10.1016/j.conbuildmat.2019.08.029.

Breiman, L. 2001. Random forests. *Machine Learning*. 45(0): 5–32.

Chopra, P., Sharma, R.K., Kumar, M. and Chopra, T., 2018. Comparison of machine learning techniques for the prediction of compressive strength of concrete. *Advances in Civil Engineering*, *2018*. Doi: 10.1155/2018/5481705.

Emiroğlu, M., Beycioğlu, A. and Yildiz, S., 2012. ANFIS and statistical based approach to prediction the peak pressure load of concrete pipes including glass fiber. *Expert Systems with Applications*, *39*(3), pp. 2877–2883. Doi: 10.1016/j.eswa.2011.08.149.

Ghugal, Y.M., and Deshmukh, S.B., 2006. Performance of alkali-resistant glass fiber reinforced concrete. *Journal of Reinforced Plastics and Composites*, *25*(6), pp. 617–630. Doi:10.1177/0731684405058273.

Hilles, M.M. and Ziara, M.M., 2019. Mechanical behaviour of high strength concrete reinforced with glass fiber engineering science and technology, *An International Journal*, *22*, pp. 920–928. Doi: 10.1016/J.Jestch.2019.01.003.

Kelestemur, O., Yildiz, S., Gokcer, B. and Arici, E., 2014. Statistical analysis for freeze–thaw resistance of cement mortars containing marble dust and glass fiber. *Materials and Design*. *60* 548–555. Doi: 10.1016/J.Matdes.2014.04.013.

Khan, M.A., Memon, S.A., Farooq, F., Javed, M.F., Aslam, F. and Alyousef, R., 2021. Compressive strength of fly-ash-based geopolymer concrete by gene expression programming and random forest. *Advances in Civil Engineering, 2021.* Doi: 10.1155/2021/6618407.

Kizilkanat, A.B., Kabay, N., Akyüncü, V., Chowdhury, S. and Akça, A.H., 2015. Mechanical properties and fracture behavior of basalt and glass fiber reinforced concrete: An experimental study. *Construction and Building Materials, 100*, pp. 218–224. Doi: 10.1016/j.conbuildmat.2015.10.006.

Kumar, R.S., Vijayan, D.S., Manzoor, P.M., Subinjith, N. and Santhosh, S., 2020, September. Effect of silica fume on strength of glass fiber incorporated concrete. In *AIP Conference Proceedings, 2271*(1), p. 030020. Doi: 10.1063/5.0024775.

Liu, J., Jia, Y., and Wang, J., 2019. Experimental study on mechanical and durability properties of glass and polypropylene fiber reinforced concrete. *Fibers And Polymers, 20*(9), pp. 1900–1908. Doi: 10.1007/S12221-019-1028-9.

Małek, M., Jackowski, M., Łasica, W., Kadela, M. and Wachowski, M., 2021. Mechanical and material properties of mortar reinforced with glass fiber: An experimental study. *Materials, 14*(3), p. 698. Doi: 10.3390/ma14030698.

Moretti, F., Pizzuti, S., Panzieri, S. and Annunziato, M., 2015. Urban traffic flow forecasting through statistical and neural network bagging ensemble hybrid modeling. *Neurocomputing, 167*, pp. 3–7. Doi: 10.1016/j.neucom.2014.08.100.

Nataraja, M.C., Nagaraj, T.S. and Basavaraja, S.B., 2005. Reproportioning of steel fibre reinforced concrete mixes and their impact resistance. *Cement and Concrete Research, 35*(12), pp. 2350–2359. Doi: 10.1016/j.cemconres.2005.06.011.

Nematollahi, B., Sanjayan, J., Chai, J.X.H. and Lu, T.M., 2014. Properties of fresh and hardened glass fiber reinforced fly ash based geopolymer concrete. In *Key Engineering Materials* (Vol. 594, pp. 629–633). Trans Tech Publications Ltd. Doi: 10.4028/www.scientific.net/KEM.594-595.629.

Nhu, V.H., Shahabi, H., Nohani, E., Shirzadi, A., Al-Ansari, N., Bahrami, S., Miraki, S., Geertsema, M. and Nguyen, H., 2020. Daily water level prediction of Zrebar lake (Iran): A Comparison between M5P, random forest, random tree and reduced error pruning trees algorithms. *ISPRS International Journal of Geo-Information, 9*(8), p. 479. Doi: 10.3390/ijgi9080479.

Pham, B.T., Van Phong, T., Nguyen-Thoi, T., Trinh, P.T., Tran, Q.C., Ho, L.S., Singh, S.K., Duyen, T.T.T., Nguyen, L.T., Le, H.Q. and Van Le, H., 2020. GIS-based ensemble soft computing models for landslide susceptibility mapping. *Advances in Space Research, 66*(6), pp. 1303–1320. Doi: 10.1016/j.asr.2020.05.016.

Sahana, M., Pham, B.T., Shukla, M., Costache, R., Thu, D.X., Chakrabortty, R., Satyam, N., Nguyen, H.D., Phong, T.V., Le, H.V. and Pal, S.C., 2020. Rainfall induced landslide susceptibility mapping using novel hybrid soft computing methods based on multi-layer perceptron neural network classifier. *Geocarto International*, pp. 1–25. Doi: 10.1080/10106049.2020.1837262.

Sepahvand, A., Singh, B., Sihag, P., Nazari Samani, A., Ahmadi, H. and Fiz Nia, S., 2019. Assessment of the various soft computing techniques to predict sodium absorption ratio (SAR). *ISH Journal of Hydraulic Engineering*, pp. 1–12. Doi: 10.1080/09715010.2019.1595185.

Singh, B., Sihag, P., Tomar, A. and Sehgal, A., 2019. Estimation of compressive strength of high-strength concrete by random forest and M5P model tree approaches. *Journal of Materials and Engineering Structures, 6*(4), pp. 583–592.

Sivakumar, V.R., Kavitha, O.R., Arulraj, G.P. and Srisanthi, V.G., 2017. An experimental study on combined effects of glass fiber and Metakaolin on the rheological, mechanical, and durability properties of self-compacting concrete. *Applied Clay Science, 147*, pp. 123–127. Doi: 10.1016/j.clay.2017.07.015.

Srinivasa Rao, P., SeshadriSekhar, T. and Sravana, P., 2010, April. Strength properties of glass fibre concrete. *ARPN Journal of Engineering and Applied Sciences*, *5*(4), pp. 1–6.

Sun, J., Wang, Y., Yao, X., Ren, Z., Zhang, G., Zhang, C., Chen, X., Ma, W. and Wang, X., 2021. Machine-learning-aided prediction of flexural strength and ASR expansion for waste glass cementitious composite. *Applied Sciences*, *11*(15), p. 6686. Doi: 10.3390/app11156686.

Sun, Y., Li, G., Zhang, J. and Qian, D., 2019. Prediction of the strength of rubberized concrete by an evolved random forest model. *Advances in Civil Engineering, 2019*. Doi: 10.1155/2019/5198583.

Tassew, S.T. and Lubell, A.S., 2014. Mechanical properties of glass fiber reinforced ceramic concrete. *Construction and Building Materials*, *51*, pp. 215–224. Doi: 10.1016/j.conbuildmat.2013.10.046.

Tavakoli, H.R., Omran, O.L., Shiade, M.F. and Kutanaei, S.S., 2014. Prediction of combined effects of fibers and nanosilica on the mechanical properties of self-compacting concrete using artificial neural network. *Latin American Journal of Solids and Structures*, *11*, pp. 1906–1923.

Thakur, M.S., Pandhiani, S.M., Kashyap, V., Upadhya, A. and Sihag, P., 2021. Predicting bond strength of FRP bars in concrete using soft computing techniques. *Arabian Journal for Science and Engineering*, *46*(5), pp. 4951–4969. Doi: 10.1007/s13369-020-05314-8.

Upadhya, A., Thakur, M.S., Pandhian, S.M. and Tayal, S., 2020. Estimation of Marshall stability of asphalt concrete mix using neural network and M5P tree. In *Computational Technologies in Materials Science* (pp. 223–236). CRC Press. Doi: 10.1201/9781003121954-11.

Upadhya, A., Thakur, M.S., Sharma, N. et al., 2021. Assessment of soft computing-based techniques for the prediction of Marshall stability of asphalt concrete reinforced with glass fiber. *International Journal of Pavement Research and Technology*. Doi: 10.1007/s42947-021-00094-2.

Venclovský, J., Štěpánek, P. and Laníková, I., 2021. Stochastic optimization of a reinforced concrete element with regard to ultimate and serviceability limit states. In *Recent Advances in Soft Computing and Cybernetics* (pp. 157–169). Springer, Cham. Doi: 10.1007/978-3-030-61659-5_14.

Wagh, M., Bhandari, P. and Sengupta, A., 2020. The strength assessment of polymer fiber-induced concrete. *Open Journal of Science and Technology*, *3*(4), pp. 345–354.

Xiaochun, Q., Xiaoming, L. and Xiaopei, C., 2017. The applicability of alkaline-resistant glass fiber in cement mortar of road pavement: Corrosion mechanism and performance analysis. *International Journal of Pavement Research and Technology*, *10*(6), pp. 536–544. Doi: 10.1016/j.ijprt.2017.06.003.

Yıldızel, S.A. and Öztürk, A.U., 2016. A study on the estimation of prefabricated glass fiber reinforced concrete panel strength values with an artificial neural network model. *CMC: Computers, Materials & Continua*, *52*(1), pp. 41–52.

Yuan, Z. and Jia, Y., 2021. Mechanical properties and microstructure of glass fiber and polypropylene fiber reinforced concrete: An experimental study. *Construction and Buildingmaterials*, *266*, p. 121048. Doi: 10.1016/J.Conbuildmat.2020.121048.

5 A Comparative Study Using ANFIS and ANN for Determining the Compressive Strength of Concrete

Veena Kashyap, Arunava Poddar, and Navsal Kumar
Shoolini University

Rabee Rustum
Heriot-Watt University

CONTENTS

5.1 INTRODUCTION

Concrete is highly durable and commonly used for many applications for hundreds of years in dams, bridges, buildings, highways, irrigation channels, and other infrastructures (Sekar et al., 2018). Developing countries are also consuming concrete in a wide range. Concrete is commonly used as building material and is generally strong in compression, while tension is weak. Concrete acquires very minimal tensile strength,

DOI: 10.1201/9781003184331-5

offers low resistance to cracking, and has less ductility. The various mechanical and physical properties of concrete play a significant role in determining the deformation when subjected to working load to determine the compressive strength (CS) (Ahmadi-Nedushan, 2012; Poddar et al., 2017). Concrete consumes a large number of natural resources during its manufacturing. Hence, it is necessary to implement engineering practices that focus on the preservation of non-renewable resources as well as energy to ensure the readiness of resources for future generations.

Internal microcracks are naturally found in concrete, and their propagation enables the concrete to lose tensile strength, inevitably leading to a brittle failure (Golewski, 2021). Microcracks develop after loading, and stress accumulation causes additional cracks in areas where there are small defects in the concrete. Inelastic deformations in concrete are primarily caused by the creation of such microcracks. The addition of thin, widely spaced, and evenly scattered fibers to concrete has long been known as a crack arrester.

Nowadays, researchers have been developing innovations for the building industry that are more sustainable. As a result, massive amounts of commercial, agricultural, and domestic wastes are recycled to replace cement or aggregate in concrete (Batayneh et al., 2007). One of the solutions is to use alternate materials like by-products and solid waste materials as construction materials due to the increase of environmental issues and demand for the development of eco-friendly and energy-efficient materials. For the last two decades, waste materials are used as construction materials such as rice husk ash, fly ash, and other fibers. Among these waste materials, coconut fiber is also used as waste material in various constructions. The use of coconut fiber as waste material in concrete mixes offers numerous gains, i.e., improving the various engineering properties, controlling the cracks that occur due to plastic and drying shrinkage, to improve the overall strength of the concrete.

Several investigations were performed to assess the impact of coconut fiber on the concrete properties. Khan et al., (2018) studied the impact of superplasticizer on the variation of concrete properties mixed with coconut fiber, and improvement in CS, splitting-tensile strength, and flexural strength is reported with the addition of 2% coconut fiber and 1% plasticizer. Hwang et al., (2016) evaluated the effect of short coconut fiber (0%, 1%, 2.5%, and 4%) on several cementitious properties, mechanical properties, plastic cracking, and impact resistance of concrete. Furthermore, ground blast furnace slag (GBFS) and fly ash (FA) were also added to these composites. The results showed that the higher volume of coconut fiber tends to reduce the density and increased the flexural strength of concrete at 28 days from 5.2 to 7.4 MPa. Khan and Ali (2019) studied the improvement in the concrete behavior by adding silica fume (15%), FA (0%, 5%, 10%, 15%) and coconut fiber (length 50 mm and content 2%). The mechanical properties of FA-SPC (fly ash silica-fume plain concrete) and FA-SCFRC (fly ash silica-fume coconut fiber reinforced concrete) were studied and found that FA-SCFRC generally enhanced the properties and 10% FA content provides the best mechanical properties than FA-SPC. Kumar et al., (2019) investigated the structural properties like compressive, flexural, and tensile strength by the addition of coconut fiber (5%) and coconut fiber ash (15%) in M20 mix concrete. In addition, seawater replaced the conventional ones in the mix, and results found that there was an optimistic increase in the basic properties of the concrete mix by mixing coconut fibers and ash components. Syed et al., (2020) studied the effect of concrete strength parameters like bending, compression, and tensile by the addition of different coconut coir fiber content with a proportion of (0.6% and 1.2%) of the total weight of a volume of concrete and reported the improvement in the overall compression strength with 0.40 value as ideal water–cement ratio and with the addition of fiber content the tensile strength increases at a maximum of 5% and with further addition a decrement of tensile stress was observed.

Coconut fiber is one of the most popular agricultural solid wastes in many tropical countries and is one such alternative (Asim et al., 2015; Poddar et al., 2021b). It is introduced as a viable material for concrete production and its aggregate characteristics in structural and non-structural elements were transferred. It is also an intriguing cellulosic fiber that has the potential to be applied as reinforcement in cement-fiber composites growth (Ardanuy et al., 2015). Furthermore, it can be used as a low-cost, readily accessible natural fiber that did not harm the environment (Agrawal et al., 2014). Hence, the effects of adding coconut fiber to the concrete are investigated in this article. After the addition of coconut fiber, mechanical properties like CS are studied, and the results are recorded.

5.2 SOFT COMPUTING TECHNIQUES

For the last two decades soft computing techniques (SCTs) are implemented in various construction-related activities (Kumar et al., 2021a, b). Recently various techniques like artificial neural network (ANN), Gaussian process regression (GPR), multiple regression analysis (MRA), M5P, grey wolf optimizer (GWO), support vector regression (SVR), radial basis function neural network (RBFNN), artificial intelligence (AI) techniques, biological neural networks (BNN), back-propagation neural networks (BPNN), fuzzy inference system (FIS), particle swarm optimization (PSO), support vector machine (SVM), machine learning (ML), fuzzy logic (FL), adaptive neuro-fuzzy inference system (ANFIS) are widely used (Kumar et al., 2019a, 2020a, b; Singh et al., 2021; Poddar et al., 2020a, 2021b; Kumari et al., 2021). Several researches emphasized the application of various SCTs for approximating the concrete properties. Apostolopoulou et al., (2020) investigated the performance of the ANN and ANFIS model for the determination of CS of cement mortar with or without metakaolin and it was found that ANFIS was able to determine the CS of mortars with higher accuracy than the ANN model. As a result, the constructed ANFIS model was proposed as the most effective projecting approach for resolving the problem of mortar CS. Akkurt et al., (2004) predicted the CS of cement mortar by using the FL model. The input and output variables were fuzzified by ANN Mamdani rules, for the fuzzy subsets and triangle membership functions were used. The results showed that the FL technique is more user-friendly. Armaghani et al., (2019) investigated the estimation of shear strength of reinforced concrete beams by using SCT based on findings of experimental tests found in the relevant literature. The findings showed that ANNs can accurately forecast the shear strength of reinforced concrete beams. Hoang et al., (2016) carried out a comparative study for estimating the CS of High-Performance Concrete (HPC) with GPR. A data set encompassing 239 HPC experimental tests was acquired for this study to train and test the prediction model. The GPR model's prediction results were superior to those of the Least Squares ANN and the SVM based on experimental data. For assessing HPC strength, the GPR model was highly suggested. Chithra et al., (2016) constructed ANN models to predict the CS of HPC containing copper slag and nano-silica. They obtained the CS experimentally for specimens containing nano-silica (0%–3%) and of copper slag (0%–50%) as a replacement (partially) of cement and aggregate (fine) respectively at curing days of 1, 3, 7, 28, 56 and 90 days, with 264 observations. Both techniques were used for modeling the three sets of data. The results showed that for CS prediction, ANN proposes the best fit model rather than MRA. Behnood et al., (2017) studied the M5P-modeled tree algorithm to predict the CS of normal cement (NC) and HPC. The model was developed on 1912 distinct data gathered from the literature. The findings showed that using the M5P model tree to forecast the CS of NC and HPC is a better alternative. Golafshani et al., (2019) studied the CS of NC and high strength concrete by using ANN and ANFIS hybridized with GWO, a complete data set of 2817 different data records were gathered to develop six ANN and three ANFIS models. Some models were constructed by hybridization with GWO in both ANN and ANFIS. The results showed that the hybridization of GWO models enhances the generalization and training capabilities of both models. Sultana et al., (2020) investigated the three different SCTs – RSM (Response Surface Methodology), ANN, and SVR – for determining the compressive and tensile strength of jute fiber reinforced concrete composites. Based on performance measuring parameters, i.e., CC, RMSE, mean absolute error, and fractional bias, the findings showed that the SVR model outperforms the ANN and RSM models for the projection of concrete's mechanical properties. Thakur et al., (2021) studied the bond strength of FRP bars using various soft computing techniques such as random forests, random tree, M5P, stochastic-M5P, bagged-M5P tree, multilinear regression, and Gaussian process and found that the bagged M5P model is outperforming better than other developed models. Khademi et al., (2016) considered three different models of the multiple linear regression (MLR) model, ANN, and ANFIS for the determination of CS of concrete with 173 different mix designs using MATLAB®. These three models are compared, suggesting that the ANN and ANFIS models may be used to evaluate concrete's CS with various mix compositions. However, the MLR model is not appropriately feasible for degerming concrete's CS. Dao et al., (2019) studied two artificial intelligence approaches namely ANN and ANFIS to predict the CS of geopolymer concrete. Some content of waste steel slag, fly ash, sodium hydroxide was added to the mixture. The total number of 210 samples were prepared and the model's performance was assessed based on CC, RMSE, and MAE. The study revealed that both ANN and ANFIS models have a huge amount of potential to forecast Geopolymer Concrete (GPC) CS but ANFIS performed better than ANN. After performing an extensive literature review, very few investigations were found on modeling the CS of concrete by mixing coconut fiber using ANFIS and ANN. Hence, this chapter determines the CS of concrete with coconut fiber using ANFIS and ANN.

5.3 ANN

The ANN is known as a machine learning method for analyzing and processing information that is widely employed for a numerical forecast. It is a computing system that is motivated through the nervous system and brain architecture (Kumar et al., 2019b). An ANN has one layer of input and output with an additional hidden layer. Every layer is made up of a certain number of nodes, and the weighted link between them reflects the relationship among them. The input layer has the same number of nodes as the number of input parameters and allocates the data to the network. This layer is followed by one or more hidden layers that facilitate data processing. The final processing unit is the output layer. The values are multiplied by the associated weight when an input layer is exposed to an input value that moves between the interconnections nodes and added together to get the output (P_j) to the unit using the following equation:

$$P_j = \sum_i s_{ij} \times b_i \tag{5.1}$$

where s_{ij} = the weight of interconnection among units i to j, b_i = the input value at the input layer, P_j = output acquired by activation function to generate an output for unit j.

5.4 ANFIS

ANFIS is a technique in which the number of hybrid combinations of FIS and ANN is used to find out the various aspects and prepare the model forecast for future responses. This inference system can combine the benefits of FL with neural network principles in a single framework to solve problems. ANFIS is a hybrid machine learning technique that transforms a given input into a target output by using a fuzzy interference system model. It also enables the analysis easier. The ANFIS model is built by using training and test data set and evaluating the built model's prediction ability during the model building process, respectively. The parameters decide the performances of a fuzzy system which describes the membership function (MFs) and the rule-based use in the system specifies how often a fuzzy system performs. The various parameters can be modified and can be combined with a fuzzy system and ANNs to build a neuro-fuzzy system (Ahmadi-Nedushan, 2012; Poddar et al., 2018, 2020b; Kumar et al., 2020c). The benefits of both approaches are combined in neuro-fuzzy systems, which merge the natural language explanation of fuzzy systems with the features of ANNs. ANFIS is a composite in which the parameters of the fuzzy system are fixed using an adaptive backpropagation learning algorithm. By using the fuzzy sets, ANFIS combines the human-like reasoning structure of fuzzy systems. ANFIS is a feed-forward network with multiple layers where every node performs a specific role on incoming signals (node function). Two inputs, x and y, and one output, z, can be considered in this process. During the ANFIS operation mainly five processing stages take place. The aim of ANFIS is to combine the best aspects of neural networks and fuzzy systems. Figure 5.1 illustrates the structural layout of the first-order Sugeno fuzzy model of ANFIS giving 2 inputs and 1 output x, y, and z, respectively.

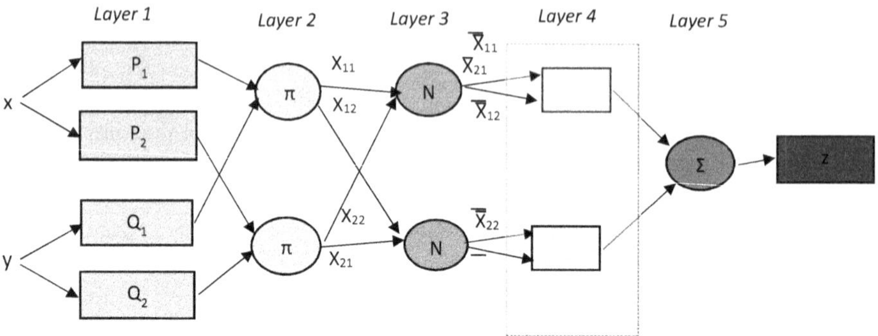

FIGURE 5.1 ANFIS structure: Sugeno model with two inputs.

Layer 1: Every node is a self-adapting node and the input and output grades of membership are given by this layer are:

$$O^1P_i = \mu_{Pi}(x), i = 1,2,3 \tag{5.2}$$

$$O^1Q_j = \mu_{Qj}(y), j = 1,2,3 \tag{5.3}$$

where x; y = optimum inputs, X_i; Y_j = fuzzy sets of appropriate MFs such as low, medium, and high and can be of any shape such as triangular, trapezoidal, generalized bell, and Gaussian.

Layer 2: In this layer, all nodes are fixed nodes characterized as π worked as a multiplier, and the output is represented as:

$$O_{ij}^2 = X_{ij} = \mu_{Pi}(x)\mu_{Qi}(y), i, j = 1,2,3 \tag{5.4}$$

Layer 3: In this layer, all nodes are fixed nodes described as N, which serves as a normalization role for the entire network, and output is represented as:

$$O_{ij}^3 = \bar{X}_{ij} = \frac{Xij}{X_{11} + X_{12} + X_{12} + X_{12}} \quad i, j = 1,2,3,s \tag{5.5}$$

Layer 4: This layer is used to figure out how input and output values are related. It is also called the adaptive layer:

$$O_{ij}^4 = \bar{X}_{ij}\left(p_{ij}x + q_{ij}y + r_{ij}\right)i, j = 1,2,3 \tag{5.6}$$

where $\bar{X}_{ij}$ = output resulted from layer 3, and p_{ij}, q_{ij}, r_{ij} = parameter set.

Layer 5: The only node in this layer is a fixed node, and it's termed the defuzzification layer and labeled as $\sum$ that combined all input signals to get the overall output as the sum of all input signals:

$$O_{ij}^5 = \sum_{ij} \overline{X_{ij}} f_{ij} = \frac{\sum iX_{ij} f_{ij}}{\sum iX_{ij}} \tag{5.7}$$

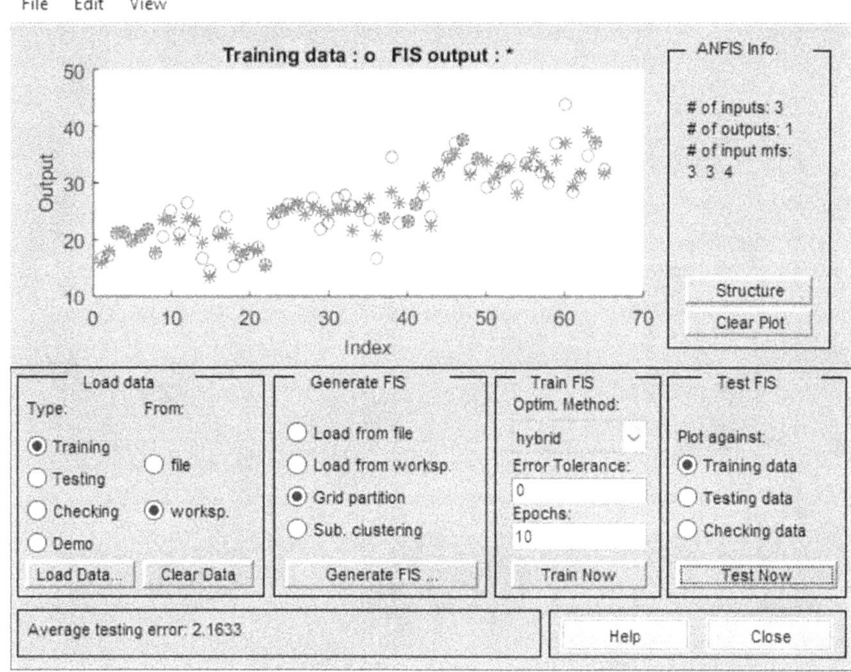

FIGURE 5.2 Actual and predicted values of compressive strength concrete for the training data set.

Figure 5.2 illustrates the actual and predicted values observed from the literature and ANFIS technique, respectively.

ANFIS approach mainly uses the membership function parameters which can be adjusted using either alone or in combination with various types of methods. One of the common methods for computing the output is the Sugeno type fuzzy model in which the fuzzy preposition worked upon the 'If' rule condition. Sugeno type model is one of the simplest methods to detect because it has fewer rules and its parameters can be estimated from numerical data using optimization approaches as shown in Figure 5.3.

5.5 PERFORMANCE ASSESSMENT INDICES

The accuracy of the ANFIS and ANN-based models for determining the CS of concrete were evaluated based on three performance assessment indices, i.e., CC, RMSE, and MAE. CC has a range of −1–1, whereas the value obtained close to −1 shows a perfect negative correlation and value obtained close to 1 shows a perfect positive correlation while MAE and RMSE have a range of zero to infinity and the lower values obtained are the best for both cases.

To analyze the CS of concrete by using ANFIS and ANN, the CC, RMSE, and MAE values are calculated using the testing and training data sets.

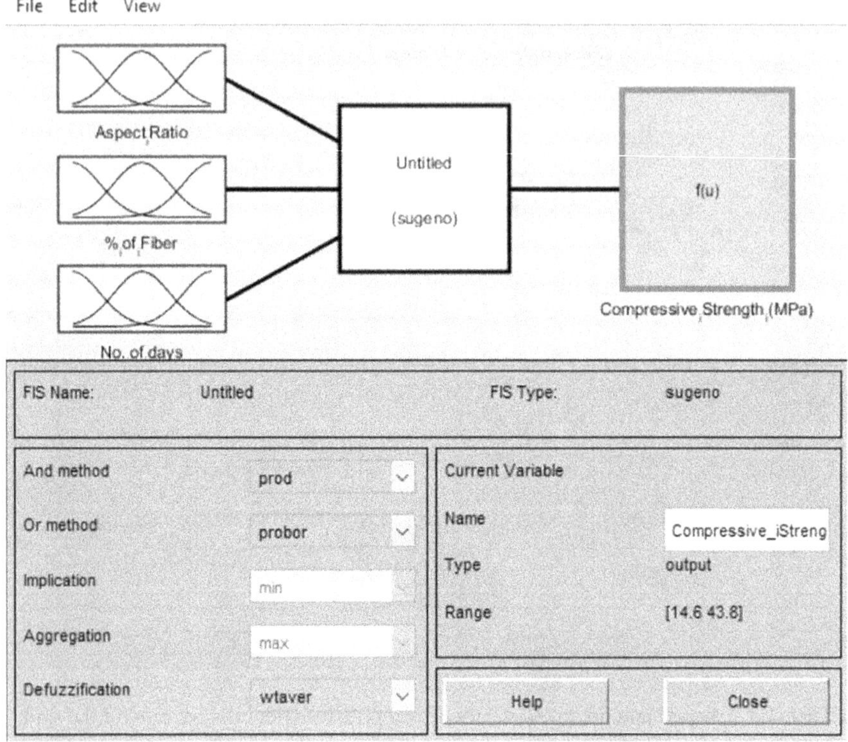

FIGURE 5.3 Sugeno-type approach of ANFIS.

$$\mathrm{CC} = \frac{n\sum PQ - \left(\sum P\right)\left(\sum Q\right)}{\sqrt{\left[n\left(\sum P^2\right) - \left(\sum P\right)^2\right]\left[n\left(\sum Q^2\right) - \left(\sum Q\right)^2\right]}} \tag{5.8}$$

$$\mathrm{RMSE} = \sqrt{\frac{1}{n}\sum_{i=1}^{n}(P-Q)^2} \tag{5.9}$$

$$\mathrm{MAE} = \frac{1}{n}\sum_{i=1}^{n}\left|P-Q\right|^2 \tag{5.10}$$

where P = Actual values, Q = Predicted values, n = number of observations.

Data Set: The model was built and validated using a total of 93 observations of concrete CS. The entire data set was split into two categories. The technique for splitting was random. Sixty-five samples for training were chosen at random from a total of 93 samples, while the remaining 28 samples were used for testing. The input

TABLE 5.1
Data Characteristics Used to Validate the Model

	Input Parameters			Output Parameter	
Type of Data Set	Aspect Ratio	Percentage of Fiber	No. of Days	Compressive Strength (MPa)	Statistics
Complete	16.67	1	3	14.6	Minimum
Training	16.67	1	3	14.6	
Testing	16.67	1	3	14.9	
Complete	56.9894	3.0968	12.6667	25.8994	Mean
Training	58.2052	3.0000	12.7538	26.0526	
Testing	54.1668	3.3214	12.4643	25.5436	
Complete	100	6	28	43.8	Maximum
Training	100	6	28	43.8	
Testing	100	6	28	39.5	
Complete	29.1087	1.4968	11.0240	6.6510	Standard deviation
Training	29.4802	1.4790	11.1117	6.6549	
Testing	28.5531	1.5409	11.0168	6.7498	
Complete	5.9949	0.3083	2.2704	1.3697	Confidence level (95.0%)
Training	7.3048	0.3665	2.7533	1.6490	
Testing	11.0717	0.5975	4.2719	2.6173	

data set consists of the aspect ratio of coconut shell concrete, percentage of Coconut fiber, and the number of curing days whereas CS (MPa) was used as output for the model's formulation and validation objective. Table 5.1 shows the parameters of the data utilized for model growth and validation.

5.6 RESULTS AND DISCUSSION

In this chapter, the building of an ANFIS and ANN for the prediction of CS of concrete has been performed. The relationship between inputs and outputs was modeled using ANFIS and ANN. The model was constructed in MATLAB® (2015) using the fuzzy logic toolkit, Sugeno type approach. The membership functions trimf, trapmf, gbellmf and gaussmf for input are used to obtain the appropriate fit in the fuzzy model as shown in Figure 5.4. Each input and output parameter had the number of membership functions added to it one by one, and the ANFIS model was then trained and evaluated. For each input, the 'trimf' membership functions were found to be the best. The ANFIS model was trained using a hybrid learning approach.

WEKA 3.9 was used to evaluate the model's performance through ANN modeling. The training data set was used to build the model, while the validation data set is used to test it. For the varied shapes of MF-based ANN models, the findings show from

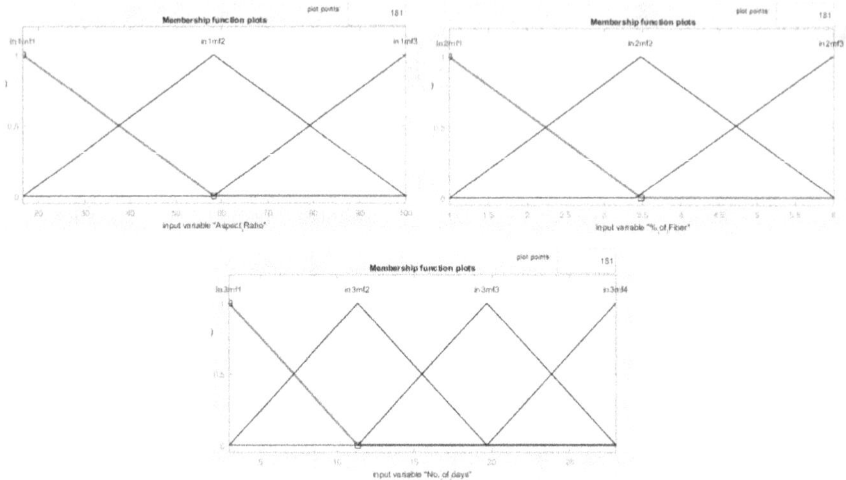

FIGURE 5.4 Final Gaussian membership functions for three inputs derived from the training process.

Table 5.2 and Figure 5.5 and 5.6 , values of CC, RMSE and MAE are 0.922, 3.693, and 2.972 for the training stage and 0.825, 4.881, and 4.087 for the testing stage.

5.6.1 Results of ANFIS Training-Based Model

The actual and predicted values of CS of concrete (MPa) for the training data set of ANFIS Triangular shape and ANFIS Trapezoidal shape are shown in Figures 5.7 and 5.8, respectively.

The actual and predicted values of CS of concrete (MPa) for the training data set of ANFIS Gbell shape are shown in Figure 5.9, and for ANFIS Gauss shape is shown in Figure 5.10.

TABLE 5.2
Performance Indices for ANFIS and ANN-Based Models

Approaches	ANFIS Tri	ANFIS Trap	ANFIS Gbell	ANFIS Gauss	ANN
		Training Data Set			
CC	0.975	0.978	0.977	0.977	0.922
RMSE	1.555	1.437	1.457	1.467	3.693
MAE	1.000	0.798	0.874	0.918	2.972
		Testing Data Set			
CC	0.841	0.803	0.825	0.833	0.825
RMSE	3.879	4.306	4.053	3.985	4.881
MAE	2.701	2.872	2.806	2.755	4.087

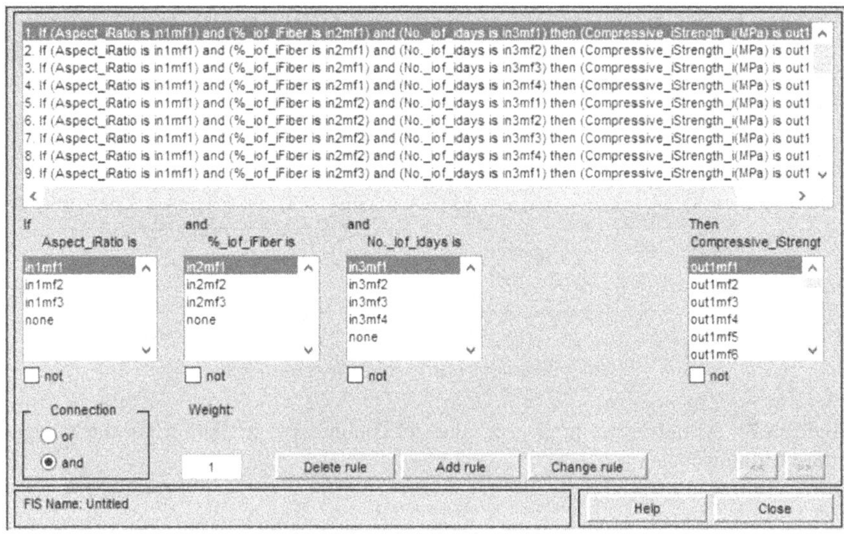

FIGURE 5.5 Rule viewer for compressive strength by using triangular MFs.

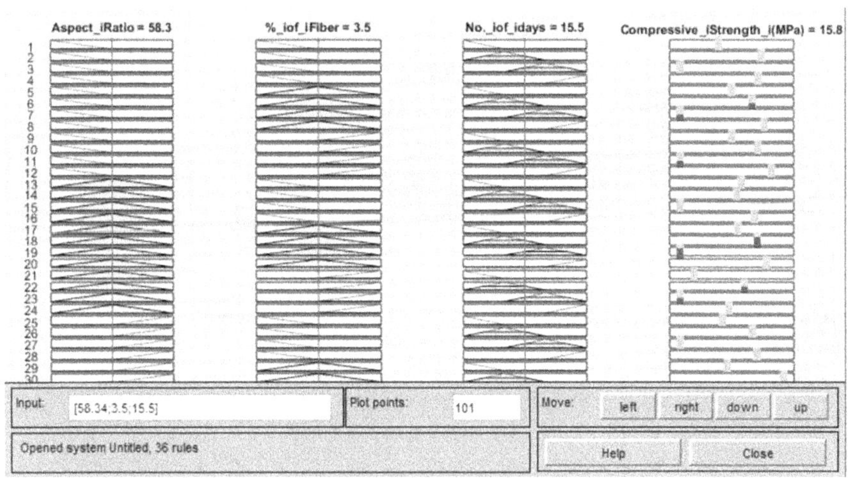

FIGURE 5.6 Rules for compressive strength using triangular MFs.

5.6.2 Results of ANFIS Testing-Based Model

The actual and predicted values of CS of concrete (MPa) for the testing data set of ANFIS triangular shape is shown in Figure 5.11, and for ANFIS trapezoidal shape is shown in Figure 5.12.

The actual and predicted values of CS of concrete (MPa) for the testing data set of ANFIS

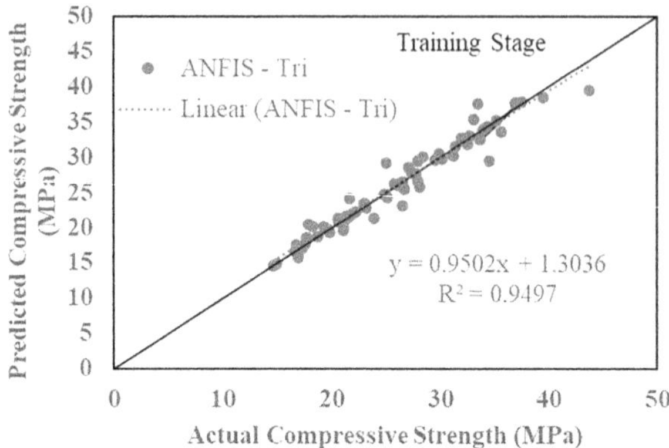

FIGURE 5.7 Actual versus predicted values of compressive strength using the triangular shape of training data.

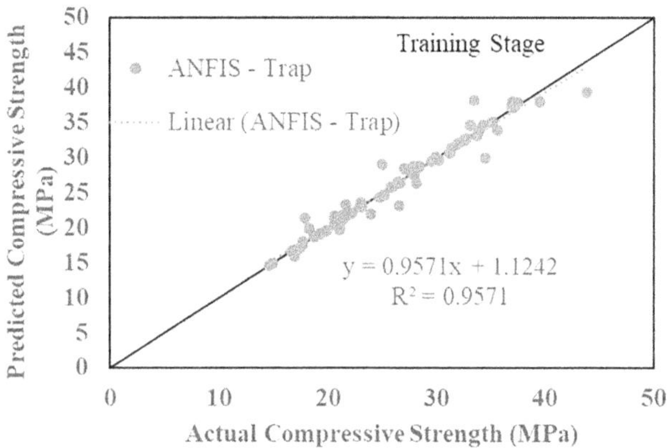

FIGURE 5.8 Actual versus predicted values of compressive strength using the trapezoidal shape of training data.

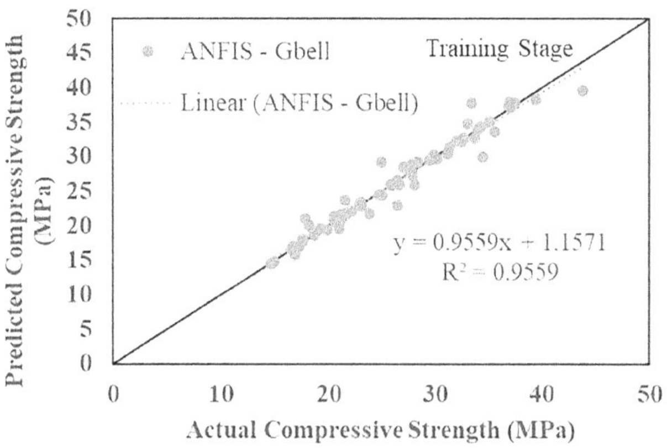

FIGURE 5.9 Actual versus predicted values of compressive strength using Gauss-bell shape of training data.

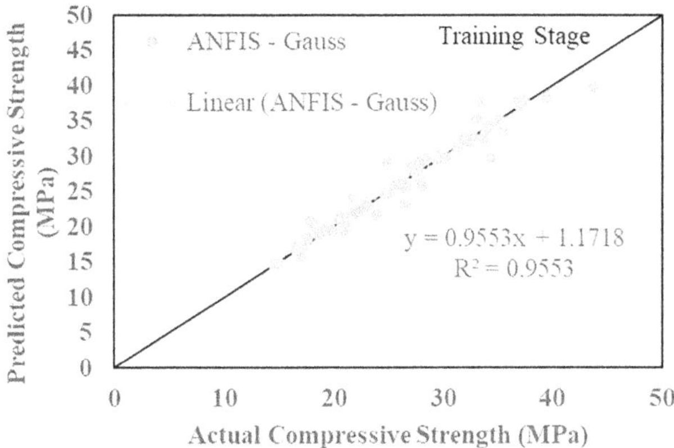

FIGURE 5.10 Actual versus predicted values of compressive strength using Gaussian shape of training data.

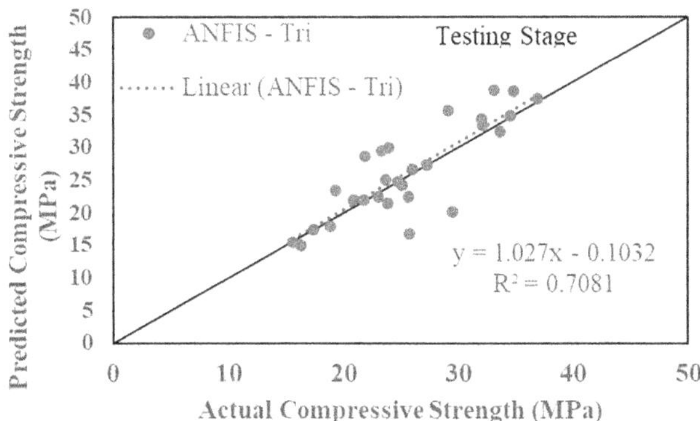

FIGURE 5.11 Actual versus predicted values of compressive strength using the triangular shape of testing data.

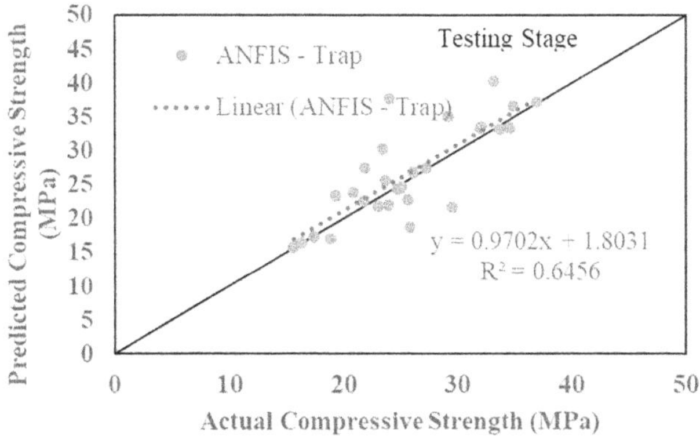

FIGURE 5.12 Actual versus predicted values of compressive strength using the trapezoidal shape of testing data.

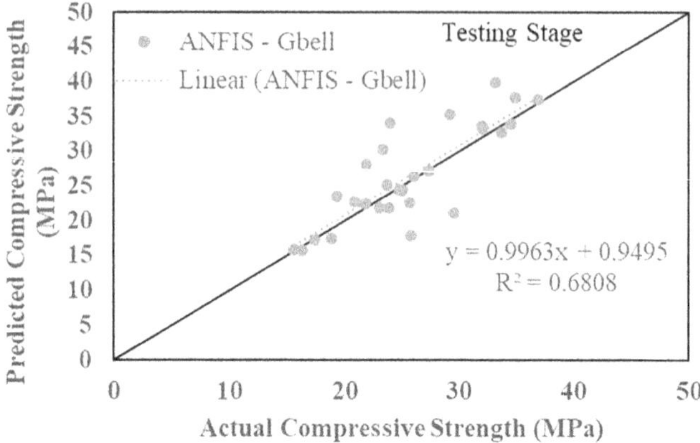

FIGURE 5.13 Actual versus predicted values of compressive strength using Gauss-bell shape of testing data.

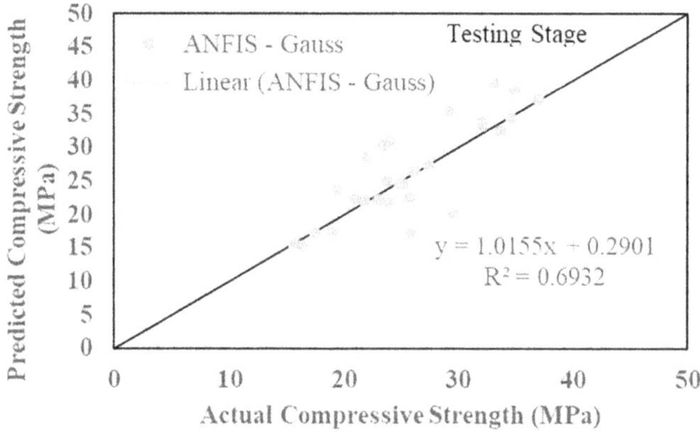

FIGURE 5.14 Actual versus predicted values of compressive strength using Gaussian shape of testing data.

Gbell shape is shown in Figure 5.13, and for ANFIS Gauss shape is shown in Figure 5.14.

From the above figures, it is found that the triangular shape performed well as compared to all other shapes in both training and testing data sets.

5.6.3 Results of ANN-Based Model

It involves a lot of trial and error to progress an ANN-based model. The model consists of one hidden layer. The performance of the actual and predicted CS of concrete is shown in Figures 5.15 and 5.16 for the training and testing data sets, respectively. Overall, examining the following figures reveals that the ANN model's performance is appropriate for predicting concrete CS.

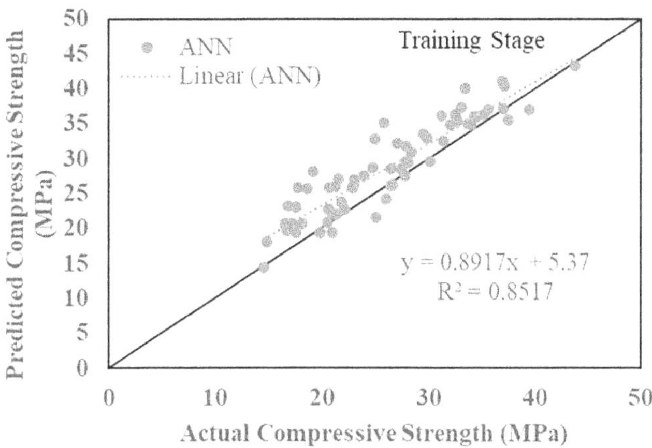

FIGURE 5.15 Actual versus predicted values of compressive strength for training data.

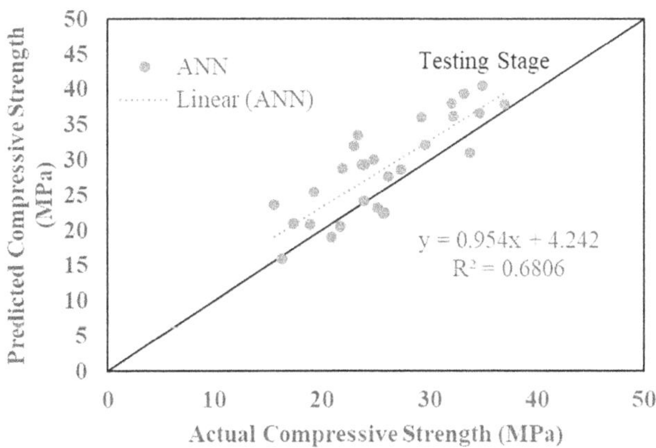

FIGURE 5.16 Actual versus predicted values of compressive strength for testing data.

5.7 COMPARISON OF MODELS

When comparing soft computing models (Table 5.2), it is clear that the ANFIS model outperforms the ANN-based model. Table 5.2 also indicates that triangular MFs based on the ANFIS model work well than other MFs such as trapezoidal, Gbell, and Gauss. Keeping in view the improved performance of MFs of ANFIS models an agreement is plotted between training and testing data sets as shown in Figure 5.17. Among all the membership functions ANFIS triangular performs well for the expectation of CS of concrete with CC, RMSE, and MAE values of 0.97, 1.56, and 1.01 for the training data set and 0.84, 3.88, and 2.70 for the testing stage respectively.

Total Data Set: The variation of input parameters with the actual CS (MPa) is shown in Figure 5.18. The relationship is drawn between the different aspect ratios of coconut fiber with the CS of concrete and with the percentage of fiber.

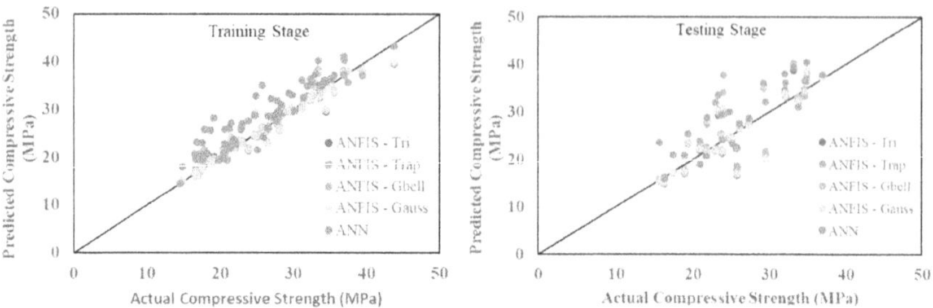

FIGURE 5.17 Comparison of actual versus predicted compressive strength of concrete using MFs of ANFIS-based models for training and testing stages.

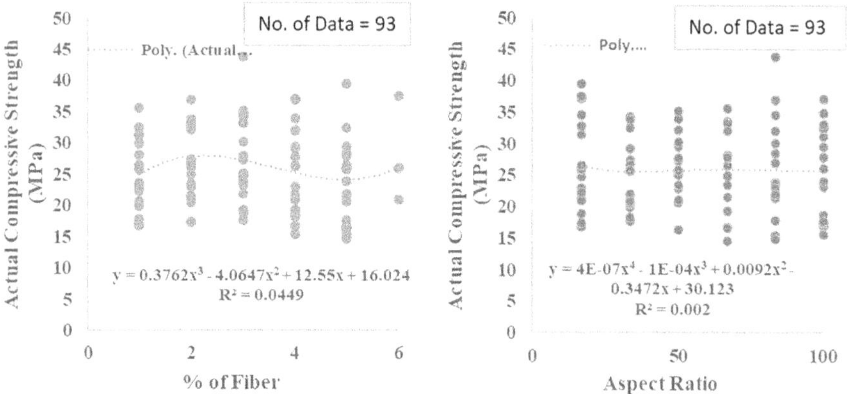

FIGURE 5.18 Comparison of actual compressive strength of concrete versus input parameters.

The variation of actual CS of concrete with the number of curing days is shown in Figure 5.19. It shows in Figure 5.20, that with the increasing curing period the CS of concrete increases.

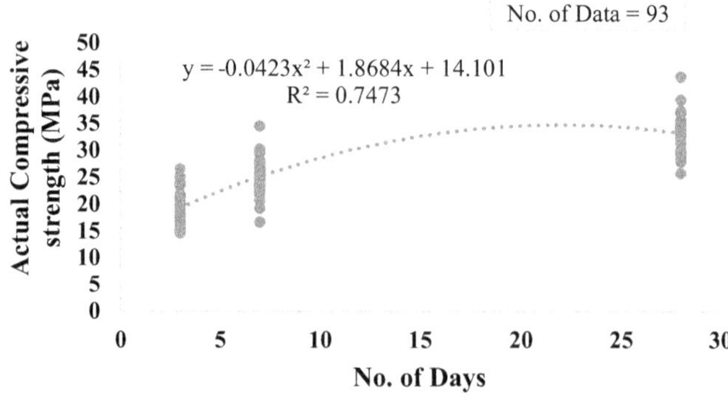

FIGURE 5.19 Comparison of actual compressive strength of concrete versus the number of curing days.

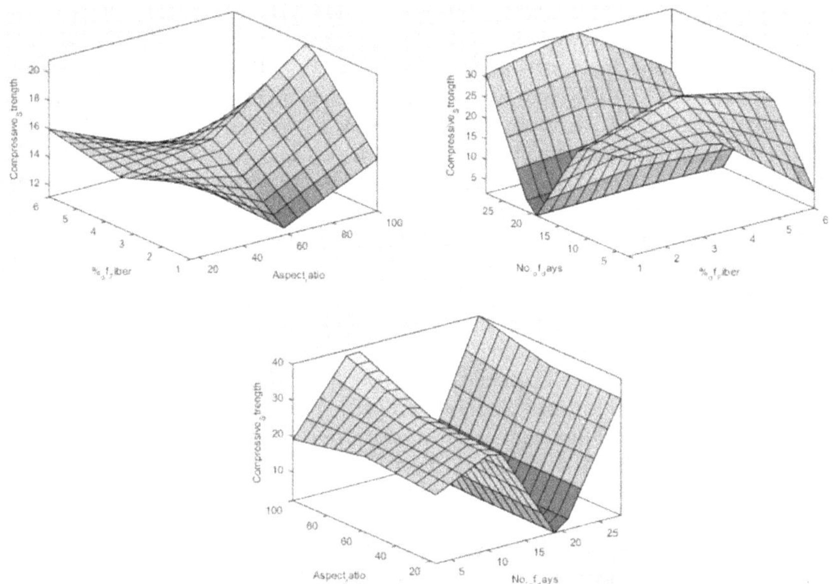

FIGURE 5.20 Surface diagram of the relationship among different input variables and output (compressive strength).

5.8 CONCLUSION

In this chapter, an attempt has been made to determine the CS of concrete by using coconut fiber in different aspect ratios using two SCTs, i.e., ANFIS and ANN approaches. According to the findings, it can be concluded that both approaches perform well with this data set. From the comparison of performance evaluation parameters, it has been observed that the ANFIS model works well with CC values of 0.9745 and 0.841 for the training and testing stage as compared to ANN. Another conclusion from this chapter can be drawn that the performance of the triangular membership function-based ANFIS is better than the other three membership functions. The surface diagrams indicate that the input parameters of the concrete CS have a nonlinear relationship.

REFERENCES

Agrawal, S., & Rathore, P. (2014). Nanotechnology pros and cons to agriculture: A review. *Int J Curr Microbiol App Sci*, 3(3), 43–55.

Ahmadi-Nedushan, B. (2012). Prediction of elastic modulus of normal and high strength concrete using ANFIS and optimal nonlinear regression models. *Const Build Mater*, 36, 665–673.

Akkurt, S., Tayfur, G., & Can, S. (2004). Fuzzy logic model for the prediction of cement compressive strength. *Cement Concrete Res*, 34(8), 1429–1433.

Apostolopoulou, M., Asteris, P. G., Armaghani, D. J., Douvika, M. G., Lourenço, P. B., Cavaleri, L., … Moropoulou, A. (2020). Mapping and holistic design of natural hydraulic lime mortars. *Cement Concrete Res*, 136, 106167.

Ardanuy, M., Claramunt, J., & Toledo Filho, R. D. (2015). Cellulosic fiber reinforced cement-based composites: A review of recent research. *Construct Build Mater*, 79, 115–128.
Armaghani, D. J., Hatzigeorgiou, G. D., Karamani, C., Skentou, A., Zoumpoulaki, I., & Asteris, P. G. (2019). Soft computing-based techniques for concrete beams shear strength. *Procedia Struct Integrity*, 17, 924–933.
Asim, N., Emdadi, Z., Mohammad, M., Yarmo, M. A., & Sopian, K. (2015). Agricultural solid wastes for green desiccant applications: An overview of research achievements, opportunities and perspectives. *J Clean Prod*, 91, 26–35.
Batayneh, M., Marie, I., & Asi, I. (2007). Use of selected waste materials in concrete mixes. *Waste Manage*, 27(12), 1870–1876.
Behnood, A., Behnood, V., Gharehveran, M. M., & Alyamac, K. E. (2017). Prediction of the compressive strength of normal and high-performance concretes using M5P model tree algorithm. *Construct Build Mater*, 142, 199–207.
Chithra, S., Kumar, S. S., & Chinnaraju, K. (2016). The effect of colloidal nano-silica on workability, mechanical and durability properties of high performance concrete with copper slag as partial fine aggregate. *Construct Build Mater*, 113, 794–804.
Dao, D. V., Ly, H. B., Trinh, S. H., Le, T. T., & Pham, B. T. (2019). Artificial intelligence approaches for prediction of compressive strength of geopolymer concrete. *Materials*, 12(6), 983.
Golafshani, E. M., Behnood, A., & Arashpour, M. (2020). Predicting the compressive strength of normal and high-performance concretes using ANN and ANFIS hybridized with grey wolf optimizer. *Const Build Mater*, 232, 117266.
Golewski, G. L. (2021). The beneficial effect of the addition of fly ash on reduction of the size of microcracks in the ITZ of concrete composites under dynamic loading. *Energies*, 14(3), 668.
Hoang, N. D., Pham, A. D., Nguyen, Q. L., & Pham, Q. N. (2016). Estimating compressive strength of high performance concrete with Gaussian process regression model. *Adv Civil Eng*, 2016. Article ID 2861380, 8 pages.
Hwang, C. L., Tran, V. A., Hong, J. W., & Hsieh, Y. C. (2016). Effects of short coconut fiber on the mechanical properties, plastic cracking behavior, and impact resistance of cementitious composites. *Const Build Mater*, 127, 984–992.
Khademi, F., Akbari, M., Jamal, S. M., & Nikoo, M. (2017). Multiple linear regression, artificial neural network, and fuzzy logic prediction of 28 days compressive strength of concrete. *Front Struct Civil Eng*, 11(1), 90–99.
Khan, M., & Ali, M. (2019). Improvement in concrete behavior with fly ash, silica-fume and coconut fibers. *Construct Build Mater*, 203, 174–187.
Khan, M., Cao, M., & Ali, M. (2018). Effect of basalt fibers on mechanical properties of calcium carbonate whisker-steel fiber reinforced concrete. *Construct Build Mater*, 192, 742–753.
Kumar, N., Poddar, A., & Shankar, V. (2019a, August). Optimizing irrigation through environmental canopy sensing–A proposed automated approach. In *AIP Conference Proceedings* (Vol. 2134, No. 1, p. 060003). AIP Publishing LLC.
Kumar, N., Poddar, A., Dobhal, A., & Shankar, V. (2019b). Performance assessment of PSO and GA in estimating soil hydraulic properties using near-surface soil moisture observations. *Compusoft*, 8(8), 3294–3301.
Kumar, N., Poddar, A., & Shankar, V. (2020a). Nonlinear regression for identifying the optimal soil hydraulic model parameters. *Num Optimiz Eng Sci*, Springer, Singapore, 25–34.
Kumar, N., Poddar, A., Shankar, V., Ojha, C. S. P., & Adeloye, A. J. (2020b). Crop water stress index for scheduling irrigation of Indian mustard (Brassica juncea) based on water use efficiency considerations. *J Agron Crop Sci*, 206(1), 148–159.

Kumar, N., Rustum, R., Shankar, V., & Adeloye, A. J. (2021a). Self-organizing map estimator for the crop water stress index. *Comput Electron Agri*, 187, 106232.

Kumar, N., Shankar, V., & Poddar, A. (2020c). Agro-hydrologic modelling for simulating soil moisture dynamics in the root zone of potato based on crop coefficient approach under limited climatic data. *ISH J Hydra Eng*, 28(1) 1–17.

Kumar, N., Shankar, V., Rustum, R., & Adeloye, A. J. (2021b). Evaluating the performance of self-organizing maps to estimate well-watered canopy temperature for calculating crop water stress index in Indian mustard (Brassica juncea). *J Irrigation Drain Eng*, 147(2), 04020040.

Kumari, S., Poddar, A., Kumar, N., & Shankar, V. (2021). Delineation of groundwater recharge potential zones using the modeling based on remote sensing, GIS and MIF techniques: A study of Hamirpur District, Himachal Pradesh, India. *Model Earth Syst and Environ*, 2021, 1–12.

Poddar, A., Kumar, N., Kumar, R., & Shankar, V. (2021b). Application of regression modeling for the prediction of field crop coefficients in a humid sub-tropical agro-climate: A study in Hamirpur district of Himachal Pradesh (India). *Model Earth Syst Environ*, 2021, 1–13.

Poddar, A., Kumar, N., & Shankar, V. (2018). Effect of capillary rise on irrigation requirements for wheat. In *Proceedings of International Conference on Sustainable Technologies for Intelligent Water Management (STIWM-2018)*, IIT Roorkee, India.

Poddar, A., Kumar, N., & Shankar, V. (2021a). Evaluation of two irrigation scheduling methodologies for potato (*Solanum tuberosum* L.) in north-western mid-hills of India. *ISH J Hydra Eng*, 27(1), 90–99.

Poddar, A., Kumar, N., Kumar, R., Shankar, V., & Jat, M. K. (2020a). Evaluation of non-linear root water uptake model under different agro-climates. *Curr Sci*, 119(3), 485.

Poddar, A., Preeti, N. K., & Shankar, V. (2020b). Artificial Ground Water Recharge Planning Using Geospatial Techniques in Hamirpur Himachal Pradesh, India. *Roorkee Water Conclave*. Organized by Indian Institute of Technology Roorkee and National Institute of Hydrology, Roorkee during February 26-28, 2020.

Poddar, A., Sharma, A., & Shankar, V. (2017). Irrigation Scheduling for Potato (solanum tuberosum l.) Based on daily crop coefficient approach in a sub-humid sub-tropical region. In *Proceedings of hydro-2017 International*, LD College of Engineering Ahmedabad, India.

Sekar, A., & Kandasamy, G. (2018). Optimization of coconut fiber in coconut shell concrete and its mechanical and bond properties. *Materials*, 11(9), 1726.

Singh, B., Sihag, P., Singh, V. P., Sepahvand, A., & Singh, K. (2021). Soft computing techniques-based prediction of water quality index. *Water Supply*, 21(8), 4015–4029. https://doi.org/10.2166/ws.2021.157

Sultana, N., Hossain, S. Z., Alam, M. S., Islam, M. S., & Al Abtah, M. A. (2020). Soft computing approaches for comparative prediction of the mechanical properties of jute fiber reinforced concrete. *Adv Eng Soft*, 149, 102887.

Syed, H., Nerella, R., & Madduru, S. R. C. (2020). Role of coconut coir fiber in concrete. *Mater Today: Proc*, 27, 1104–1110.

Thakur, M. S., Pandhiani, S. M., Kashyap, V., Upadhya, A., & Sihag, P. (2021). Predicting bond strength of FRP bars in concrete using soft computing techniques. *Arab J Sci Eng*, 46(5), 4951–4969.

6 Prediction of Concrete Mix Compressive Strength Using Waste Marble Powder

A Comparison of ANN, RF, RT, and LR Models

Nitisha Sharma, Mohindra Singh Thakur, and Ankita Upadhya
Shoolini University

Parveen Sihag
Chandigarh University

CONTENTS

DOI: 10.1201/9781003184331-6

6.1 INTRODUCTION

Compressive strength is one of the main factors to analyze the performance of the concrete. Concrete has primary constituents such as cement, fine aggregate, coarse aggregate, and water, and these constituents have been used since early ages to give the desired performance (Zongjin, 2011). As the input parameter for desired compressive strength is never certain (Chopra et al., 2018). Also due to growth in industrialization and urbanization concrete becomes the most widely used material in the construction industry. The recent scenario shows that due to the negative impact of industrial waste on the environment researchers use different industrial waste to improve the strength properties of concrete mix (Sharma et al., 2020, Missan et al., 2019, Gao et al., 2020). Marble waste is one such industrial waste with high calcium oxide content which also is a cementing property (Ashish et al., 2016). Literature shows that the prediction of compressive strength of concrete mix has been generated using previously known data or laboratory tests. Compressive strength of concrete mix was predicted for 1, 3, 7, 28, 56, 91, and 180 days by linear mathematical model on parameters like cement, lime, water, silica-fume, metakaolin, sand, aggregate, and superplasticizer to reach the conclusion that predicted model and experimental values for concrete compressive strength had a high correlation and provided mathematical models were capable of predicting required output (Ahmed, 2012). Topcu and Saridemir (2008) designed a model with the help of neural network (NN) and fuzzy logic (FL) by adding fly-ash to the input parameters, namely, cement, water, sand, aggregate, recycled aggregate, superplasticizer, and silica fume and collected data from the previous study and found both NN and FL predict the outcome of concrete compressive strength with high accuracy but NN provides better results compared to FL. GMDH-type neural networks and ANFIS models are used to predict compressive strength (Madandoust et al., 2012). These approaches have resulted in a model that can be used to estimate concrete compressive strength. Deepa et al., (2010) performs different algorithms, multilayer perceptron, M5P tree, and LR, to the dataset to find out the best model performance. It was concluded in the result that tree-based models' performance was way better than Multilayer Perceptron and Linear regression because of the high correlation coefficient and lower RMSE values.

Based on the literature study, the goal of this article is to create models for forecasting concrete compressive strength at a 28-day age. To achieve this aim, machine learning techniques including artificial neural networks (ANN), random forest (RF), random tree (RT), and linear regression (LR) were used. To get the output of compressive strength for concrete, input factors included cement, fine aggregate, coarse aggregate, water/cement ratio, and marble powder.

6.2 CONVENTIONAL MODELS

Researchers conducted experiments and presented an empirical equation for predicting the compressive strength of concrete mixes. Equations (1.1) and (1.2) are such examples of the empirical formulae carried out by Singh et al., (2019), and Chopra et al., (2016), respectively.

$$\text{CS} = \frac{195.434}{79.43^{\beta}} \tag{6.1}$$

$$\text{where, } \beta = \frac{w}{c + \text{MP}} + 0.085\frac{\text{MP}}{c}$$

where w = water content, c = cement content, MP = marble powder, and CS = compressive strength.

$$\text{CS} = \sqrt{\text{CA}}\ W + \left(\text{FA} * \sqrt{\left(\left(\left((\text{FA} + \text{CA}) + \text{FA}\right) + \text{FA}\right) * \text{FA}\right)}\right)$$
$$+ \left(\text{FA} + \left(\sqrt{\text{CA}} + ((W + \text{CA}) + W)\right)\right) + \text{CA} + \left(\left(\sqrt{\text{FA}} + \text{FA}\right) + (W + \text{FA})^{-1}\right) \tag{6.2}$$

where CS = compressive strength, CA = coarse aggregate, FA = fine aggregate, and W = water content.

6.3 SOFT COMPUTING TECHNIQUES

Soft computing techniques were helpful for finding a solution for complex engineering problems. In the construction industry, concrete's strength is important to reach the target strength. A significant influence is played by cement, fine aggregates, coarse aggregates, the water/cement ratio, and waste marble powder. Various approaches for predicting the mechanical characteristics of concrete have been developed (Chopra et al., 2018, Sepahvand et al., 2019, Han et al., 2019, Goldberg, 2017, Nhu et al., 2020). These methods provide good information and insights about inputs parameters. Soft computing techniques have been effectively applied to solve a variety of engineering-related challenges, such as Topcu and Saridemir (2007)'s use of ANN and FL to forecast concrete mechanical characteristics. Using tree-based models, Deepa et al. (2010) aid in the prediction of concrete compressive strength. In this paper, some of the soft computing techniques such as LR, RF, RT, and ANN were used for understanding the relation of operating parameters to the target strength of the concrete mix.

6.3.1 Artificial Neural Network (ANN)

The input is processed using an ANN. In an ANN, inputs are represented by neurons that are linked to determine the optimal output estimation. There are several layers in ANNs, including an input layer, a hidden layer, and an output layer. The hidden layer comprises a vast number of neurons that create a complex structure when connected to the input layer. This study's data is divided into two sub-datasets: training and testing. The training dataset is chosen for model creation, while the testing dataset is utilized to validate the model. By optimizing the user-defined parameters, the best model was created. The best user-defined parameters were discovered through a trial-and-error method in this investigation. The best user-defined parameter values for best models are learning rate = 0.2, momentum = 0.1, no. of iteration = 5000 and hidden layer = 1 with no. of neurons = 20.

6.3.2 Random Forest (RF)

It was first proposed by Bremen, 2001 for classification and regression for the solution of complex problems. RF model was prepared by ensemble process; where the number of trees grew together and form a forest where data was assigned to each tree randomly. The data frame is designed by randomly selecting 2/3rd of the training dataset and assigning value to each tree. This process is known as bagging. For the validation of the data, the testing dataset was used (Chopra et al., 2018). For the regression process, input parameters were used at each node for the growth of a tree, and for the best split on each node only selected variables were found (Sepahvand et al., 2019). By randomly replacing the individual variable, the importance of input parameters was determined. Model quality can be evaluated by altering the error for the testing dataset. Using Weka 3.8.4 software for the random forest, the model was prepared with the number of iterations = 2000.

6.3.3 Random Tree (RT)

Tree-based models are able to handle the missing data variables and are good at nonlinear relationships findings. It will also be helpful for the interpretation of the outcomes. It was presented by Leo Breiman and Adele Cutler. It is a learning algorithm which deals with complex problems related to classification and regression. A stochastic process is used which results in the development of random trees. It is the combination of two algorithms, i.e., a single tree model with the idea of random forest. According to Nhu et al., (2020), a single tree model has low prediction accuracy and becomes unstable so random tree modeling is needed. Without pruning Random Tree, the algorithm constructs a tree with the number of random features (k) at each node. A random tree is developed by selecting a set of possible trees randomly.

6.3.4 Linear Regression (LR)

One of the most popular types of regression analysis is Linear Regression. It has the linear approach towards the development of the relationship between scalar and explanatory variables (Sobhani et al., 2010). The least-square techniques are used for the model development using LSTAT software. Linear mathematical equation is formed using dependent variable (t) and independent variable ($r_1, r_2, r_3, r_4, r_5, r_6 \ldots r_n$) (Eq. (6.3)). The performance of the developed equation can be checked by the performance evaluation parameter such as coefficient of correlation, RMSE, MAE, RAE, and RRSE. The developed equation is as follows:

$$t = a_0 + a_1 r_1 + a_2 r_2 + a_3 r_3 + a_4 r_4 + a_5 r_5 + a_6 r_6 + \ldots + a_n r_n \tag{6.3}$$

6.4 METHODOLOGY AND DATASET

6.4.1 Dataset

The selection of variables for the model development is very important. Dataset was collected from past published publications. The detail of the data collection is listed in Table 6.1. A total of 49 observations were collected. Out of 49, 34 were selected for training, and the rest 15 were used for model validation. The features of training and testing datasets are listed in Table 6.2. For model development, different input combination-based models were prepared. The input parameters were cement,

TABLE 6.1
Detail of Dataset

S. No.	1	2	3	4	5	6	7	8	9	10	Total
Author's name	Soliman	Ergun	Dhiman et al.,	Topcu et al.,	Talah et al.,	Sounthararajan et al.,	Chavhan and Bhole	Vaidevi	Sakalkale et al.,	Alyamac and Ince	
Year	2013	2011	2018	2009	2015	2013	2014	2013	2014	2009	
Total dataset	6	7	3	1	1	4	11	3	3	10	49

TABLE 6.2
Features of Dataset

Statistics/ Parameters[a]	Cement (kg/m³)	Fine Aggregate (kg/m³)	Coarse Aggregate (kg/m³)	w/c	Marble Powder (kg/m³)	Compressive Strength (MPa)
			Training Dataset			
Minimum	255.00	0.00	578.00	0.30	0.00	21.32
Maximum	500.00	1157.00	1721.40	0.70	860.00	59.00
Mean	375.12	723.09	1025.39	0.48	98.05	34.57
Standard deviation	57.34	310.71	359.08	0.08	158.17	9.59
Kurtosis	−0.15	−0.79	−0.81	2.41	16.48	0.41
Skewness	−0.03	−0.15	0.27	1.00	3.62	0.96
			Training Dataset			
Minimum	240.00	312.30	578.00	0.30	0.00	20.30
Maximum	500.00	1157.00	1721.40	0.63	316.47	64.50
Mean	360.90	596.18	1227.99	0.49	75.67	36.77
Standard deviation	77.79	288.00	391.57	0.09	100.05	9.89
Kurtosis	−0.98	−0.80	−0.78	0.05	1.45	4.08
Skewness	−0.16	0.67	−0.23	−0.15	1.58	1.37

[a] Cement = C, Fine Aggregate = FA, Coarse Aggregate = CA, Water/cement = w/c, Marble Powder = MP, Compressive Strength = CS

TABLE 6.3
Combination of Various Variable Subset

Sr. No.	Input Combinations	Output	Model No.
1	C + FA + CA + w/c + MP	CS	M1
2	C + FA + CA + w/c	CS	M2
3	C + FA + CA + MP	CS	M3
4	C + FA + w/c + MP	CS	M4
5	C + CA + w/c + MP	CS	M5
6	FA + CA + w/c + MP	CS	M6
7	C + w/c + MP	CS	M7
8	C + FA + w/c	CS	M8
9	C + FA + MP	CS	M9
10	C + CA + w/c	CS	M10

fine aggregate, coarse aggregate, water/cement ratio, and waste marble powder, and the goal was compressive strength. Table 6.3 lists the details of these various input combination-based models.

All these combinations are applied to construct a model for the prediction of compressive strength as an output.

6.4.2 Model Evaluation

After preparing models, evaluation has to be done by evaluating the accuracy parameters such as coefficient of correlation (CC), mean absolute error (MAE), root mean square error (RMSE), relative absolute error (RAE), and root relative squared error (RRSE) are calculated.

$$\mathrm{CC} = \frac{a\left(\sum_{i=1}^{a} LK\right) - \left(\sum_{i=1}^{a} L\right)\left(\sum_{i=1}^{a} K\right)}{\sqrt{\left[a\sum_{i=1}^{a} L^2 - \left(\sum_{i=1}^{a} L\right)^2\right]}\sqrt{\left[a\sum_{i=1}^{a} K^2 - \left(\sum_{i=1}^{a} K\right)^2\right]}} \tag{6.4}$$

$$\mathrm{RMSE} = \sqrt{\frac{1}{a}\left(\sum_{i=1}^{a}(K-L)^2\right)} \tag{6.5}$$

$$\mathrm{RAE} = \frac{\sum_{i=1}^{a} |L-K|}{\sum_{i=1}^{a} \left(\left|L-\bar{L}\right|\right)} \tag{6.6}$$

$$\text{RRSE} = \sqrt{\frac{\sum_{i=1}^{a}(L-K)^2}{\sum_{i=1}^{a}\left(\left|L-\bar{L}\right|\right)^2}} \tag{6.7}$$

$$\text{MAE} = \frac{1}{a}\left(\sum_{i=1}^{a}\left|K-L\right|\right) \tag{6.8}$$

where L = observed values, $\bar{L}$ = average of observed value, K = predicted values, and a = number of observations.

The strength of the model link between two variables is described by the coefficient of correlation. It has a range of −1.0–1.0. If value increases or decreases it implies to an error in the relation. Higher the value of the coefficient of correlation implies a better prediction of the model. RMSE and MAE are helpful for measuring the errors present in the prediction of the output. Minimum the value of errors more chances for the prediction of a better model (Nhu et al., 2020). RAE and RRSE are the error in percentage. If the relative error is large the measured value is very small, i.e., the prediction is not good for the desired output and vice versa.

6.5 RESULT ANALYSIS

6.5.1 Assessment of Empirical Formula

To assess the performance of the empirical formulae, performance evaluation parameters were calculated on the total dataset. Performance of parameters was listed in Table 6.4, whereas Figure 6.1 shows the agreement plot between actual and predicted compressive strength. According to Table 6.4 results, formulae based on Singh et al., (2019) and Chopra et al., (2016) were not suitable for the prediction of compressive strength as correlation coefficients are very low (0.24 and −0.15, respectively).

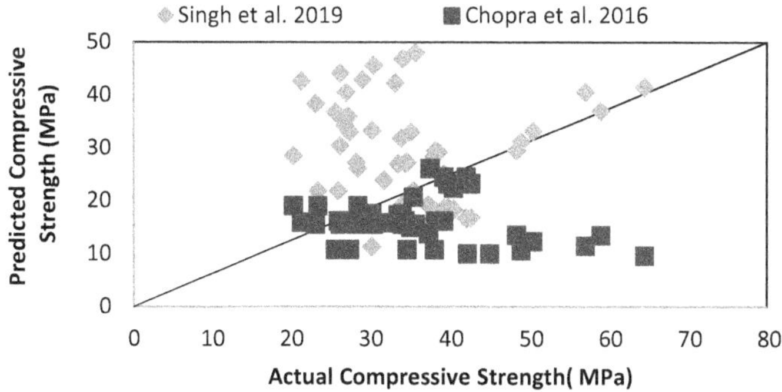

FIGURE 6.1 Performance of conventional formulae for predicted compressive strength.

TABLE 6.4
Performance by Empirical Formula

Sr. No.	Author	CC	MAE (MPa)	RMSE (MPa)	RAE (%)	RRSE (%)
1.	Singh et al. (2019)	0.04	12.42	14.16	1.7	1.48
2.	Chopra et al. (2016)	−0.15	19.42	22.27	2.66	2.33

6.5.2 Assessment of ANN Based Model

To assess the performance of the ANN, different input combination based models, five statistic performance evaluation measures, including CC, MAE, RMSE, RAE, and RRSE, were calculated for training and testing dataset. Performance evaluation measures values were listed in Table 6.5 for both stages. With CC of 0.9575, MAE of 2.6096, RMSE of 3.0388, RAE of 39.4920%, and RRSE of 30.9990%, M6 outperforms other input combination-based models. For the comparison of real and predicted values using the M6 ANN model, an agreement plot is generated (Figure 6.2). This figure suggests that the M6 ANN model may be used to estimate the compressive strength of concrete that has been combined with waste marble powder.

6.5.3 Assessment of RF Based Model

To assess the performance of the RF, different input combination based models, five statistic performance evaluation measures, including CC, MAE, RMSE, RAE, and RRSE, were calculated for training and testing dataset. Performance evaluation measures values are listed in Table 6.6 for both stages. With CC of 0.8633, MAE of 3.0309, RMSE of 5.3504, RAE of 45.8674%, and RRSE of 54.5797%, this table

TABLE 6.5
Statistic Quantitative Results of ANN

Model No.	Training					Testing				
	CC	MAE (MPa)	RMSE (MPa)	RAE (%)	RRSE (%)	CC	MAE (MPa)	RMSE (MPa)	RAE (%)	RRSE (%)
M1	0.9935	0.9761	1.2756	13.0359	13.5018	0.9369	3.4530	4.5501	52.2555	46.4154
M2	0.9600	1.7429	2.6774	23.2767	28.3388	0.7617	4.6367	6.2937	70.1684	64.2024
M3	0.9819	1.6103	2.1034	21.5056	22.2629	0.6662	5.0869	7.1305	76.9806	72.7384
M4	0.9908	1.1593	1.5316	15.4824	16.2111	0.7984	4.6266	5.9167	70.0158	60.3565
M5	0.9938	0.7723	1.0661	10.3139	11.2836	0.8255	4.0955	5.5935	61.9781	57.0596
M6	0.9618	2.2314	2.7029	29.8009	28.6083	0.9575	2.6096	3.0388	39.4920	30.9990
M7	0.9355	2.6353	3.4198	35.1950	36.1972	0.9234	4.1819	5.5020	63.2852	56.1264
M8	0.9794	1.8483	2.2673	24.6852	23.9984	0.8340	4.6438	5.9975	70.2755	61.1805
M9	0.9639	1.9517	2.6678	26.0656	28.2370	0.6728	5.4995	7.4864	83.2246	76.3694
M10	0.9103	3.3809	4.1952	45.1529	44.4035	0.5117	6.1052	8.6802	92.3911	88.5466

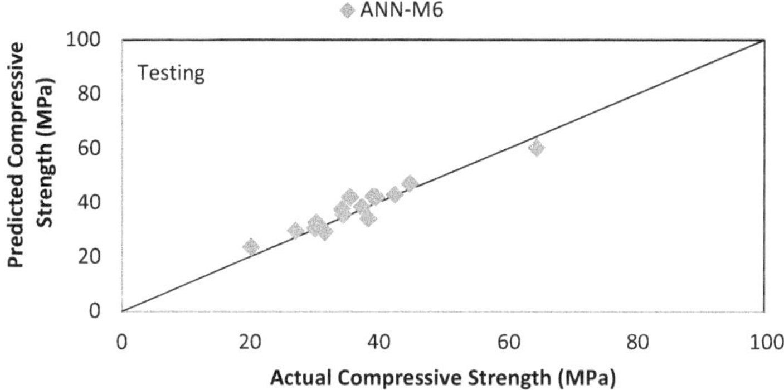

FIGURE 6.2 Scatter graphs showing the best model's observed and predicted compressive strength by ANN technique.

TABLE 6.6
Statistic Quantitative Results of RF

	Training					Testing				
Model No.	CC	MAE (MPa)	RMSE (MPa)	RAE (%)	RRSE (%)	CC	MAE (MPa)	RMSE (MPa)	RAE (%)	RRSE (%)
M1	0.9792	1.7820	2.6231	23.7989	27.7637	0.8691	3.3867	5.4542	51.2509	55.6389
M2	0.9727	1.7629	2.7028	23.5445	28.6078	0.8381	3.7660	5.4976	56.9913	56.0812
M3	0.9745	1.8162	2.6963	24.2560	28.5391	0.8255	3.9499	5.9813	59.7741	61.0151
M4	0.9738	2.1190	3.1455	28.3006	33.2935	0.8626	3.5525	5.6890	53.7607	58.0332
M5	0.9781	1.8754	2.7747	25.0472	29.3687	0.8323	3.8835	5.7328	58.7700	58.4810
M6	0.9703	2.1346	2.9759	28.5078	31.4986	0.8633	3.0309	5.3504	45.8674	54.5797
M7	0.9747	2.2723	2.9540	30.3468	31.2667	0.8500	3.6295	5.5604	54.9259	56.7222
M8	0.9750	1.8851	2.7249	25.1757	28.8411	0.8418	3.9112	5.5097	59.1894	56.2042
M9	0.9787	1.8966	2.6400	25.3294	27.9431	0.8018	4.1143	6.3876	62.2621	65.1600
M10	0.9585	2.2822	3.2927	30.4792	34.8518	0.7983	4.3752	5.9087	66.2113	60.2750

suggests that M6 outperforms other input combination based models. An agreement plot was drawn for the comparison among actual and predicted values using the M6 RF model (Figure 6.3). This figure suggests that the M6 RF model may be used to estimate the compressive strength of concrete that has been combined with waste marble powder.

6.5.4 Assessment of RT Based Model

To assess the performance of the RT, different input combination based models, five statistic performance evaluation measures, including CC, MAE, RMSE, RAE, and RRSE, were calculated for training and testing dataset. Performance evaluation measures values were listed in Table 6.7 for both stages. The results suggest that M5 is

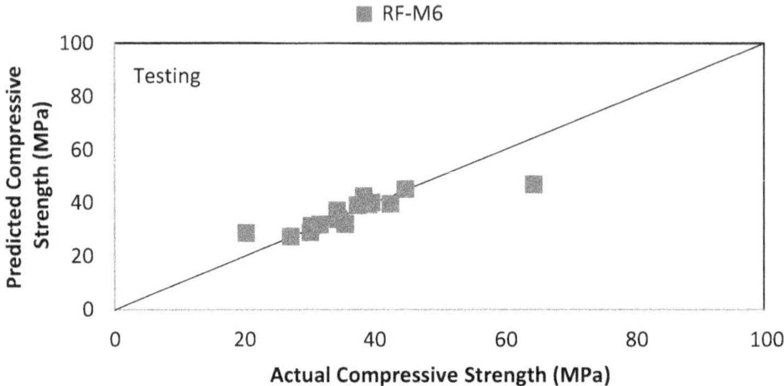

FIGURE 6.3 Scatter graphs showing the best model's observed and predicted compressive strength by RF technique.

TABLE 6.7
Statistic Quantitative Results of RT

	Training					Testing				
Model No.	**CC**	**MAE (MPa)**	**RMSE (MPa)**	**RAE (%)**	**RRSE (%)**	**CC**	**MAE (MPa)**	**RMSE (MPa)**	**RAE (%)**	**RRSE (%)**
M1	1.0000	0.0059	0.0243	0.0786	0.2567	0.8263	4.3247	5.5389	65.4459	56.5026
M2	1.0000	0.0059	0.0243	0.0786	0.2567	0.8239	4.3647	5.5619	66.0513	56.7369
M3	1.0000	0.0229	0.0744	0.3064	0.7875	0.7934	4.8727	5.9203	73.7389	60.3928
M4	1.0000	0.0206	0.0653	0.2750	0.6912	0.8358	4.4353	5.3630	67.1207	54.7086
M5	1.0000	0.0059	0.0243	0.0786	0.2567	0.8895	3.9760	4.4210	60.1695	45.0983
M6	1.0000	0.0059	0.0243	0.0786	0.2567	0.8222	4.2800	5.5363	64.7700	56.4757
M7	0.9951	0.2324	0.9341	3.1032	9.8867	0.8195	4.4713	5.5906	67.6655	57.0296
M8	1.0000	0.0206	0.0653	0.2750	0.6912	0.8344	4.3727	5.5468	66.1723	56.5833
M9	1.0000	0.0206	0.0653	0.2750	0.6912	0.8189	4.4453	5.6056	67.2720	57.1835
M10	0.9770	0.8532	2.0168	11.3952	21.3466	0.8110	4.7177	5.9971	71.3933	61.1770

performing better than other input combination-based models with CC being 0.8895, MAE being 3.9760, RMSE being 4.4210, RAE being 60.1695%, and RRSE being 45.0983%. An agreement plot was drawn for the comparison among actual and predicted values using the M5 RT model (Figure 6.4). This figure suggests that the M5 RT model may be used to estimate the compressive strength of concrete that has been mixed with waste marble powder.

6.5.5 Assessment of LR Based Model

To assess the performance of the LR, different input combination based models, five statistic performance evaluation measures, including CC, MAE, RMSE, RAE, and RRSE, were calculated for training and testing dataset. The equation generated

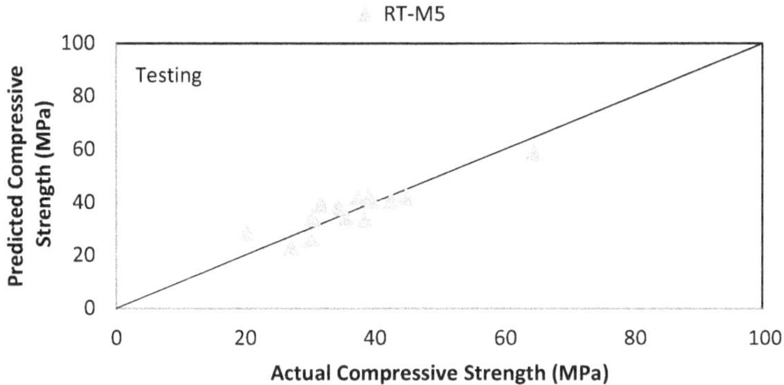

FIGURE 6.4 Scatter graphs showing the best model's observed and predicted compressive strength in the RT testing.

TABLE 6.8
Mathematical Formulas for the Predictions by LR

Model No.	Equation
M1	$CS = 84.33136 + 0.01613 * C - 0.03521 * FA - 0.02929 * CA + 9.08984 * \frac{w}{c} - 0.04817 * MP$
M2	$CS = 6.79652 + 0.04530 * C + 0.00706 * FA + 0.00320 * CA + 4.93782 * \frac{w}{c}$
M3	$CS = 90.81566 + 0.00906 * C - 0.03470 * FA - 0.02915 * CA - 0.04757 * MP$
M4	$CS = 15.49109 + 0.04295 * C + 0.00068 * FA + 8.12181 * \frac{w}{c} - 0.01492 * MP$
M5	$CS = 20.45755 + 0.04015 * C - 0.00242 * CA + 6.35988 * \frac{w}{c} - 0.01578 * MP$
M6	$CS = 99.53076 - 0.03898 * FA - 0.03244 * CA + 3.21376 * \frac{w}{c} - 0.05168 * MP$
M7	$CS = 15.70858 + 0.04312 * C + 8.67275 * \frac{w}{c} - 0.01548 * MP$
M8	$CS = 13.93699 + 0.04219 * C + 0.00393 * FA + 4.05257 * w / c$
M9	$CS = 21.57498 + 0.03652 * C + 0.00099 * FA - 0.01452 * MP$
M10	$CS = 19.00904 + 0.04072 * C - 0.00202 * CA + 4.87372 * \frac{w}{c}$

by the LR technique is listed in Table 6.8. Performance evaluation measures values are listed in Table 6.9 for both stages. With CC of 0.4740, MAE of 6.8569, RMSE of 8.8431, RAE of 103.8300%, and RRSE of 92.5600%, Table 6.9 suggests that M6 outperforms other input combination based models. An agreement plot was drawn for the comparison among actual and predicted values using the M6 LR model (Figure 6.5). The M6 LR model appears to be suitable for predicting the compressive strength of concrete combined with waste marble powder, as shown in the diagram (Figure 6.5).

TABLE 6.9
Statistic Quantitative Results of LR

	Training					Testing				
Model No.	CC	MAE (MPa)	RMSE (MPa)	RAE (%)	RRSE (%)	CC	MAE (MPa)	RMSE (MPa)	RAE (%)	RRSE (%)
M1	0.4396	6.4897	8.4858	86.6723	89.8100	0.4466	7.1940	8.9905	108.9000	94.1000
M2	0.2690	7.4645	9.0996	99.6900	96.3100	0.1462	8.1769	10.1014	123.8300	105.7300
M3	0.4359	6.5380	8.5031	87.3100	90.0000	0.4474	7.2181	8.9820	109.3000	94.0100
M4	0.3435	7.0297	8.8728	93.8800	93.1900	0.2351	7.8663	9.6295	119.1200	100.7900
M5	0.3543	7.0511	8.8348	94.1600	93.5100	0.2631	8.0187	9.6719	121.1400	101.2300
M6	0.4341	6.5037	8.5114	86.8500	90.0800	0.4740	6.8569	8.8431	103.8300	92.5600
M7	0.3430	7.0001	8.8748	93.4800	93.9300	0.2339	7.8040	9.6024	118.1800	100.5100
M8	0.2618	7.4889	9.1184	100.0100	96.5100	0.1688	8.1731	10.1427	123.7700	106.1600
M9	0.3397	7.0553	8.8861	94.2200	94.0500	0.2447	7.8856	9.6135	119.4100	100.6200
M10	0.2413	7.5337	9.1686	100.6100	97.0400	0.1658	8.0222	10.0793	121.4800	105.5000

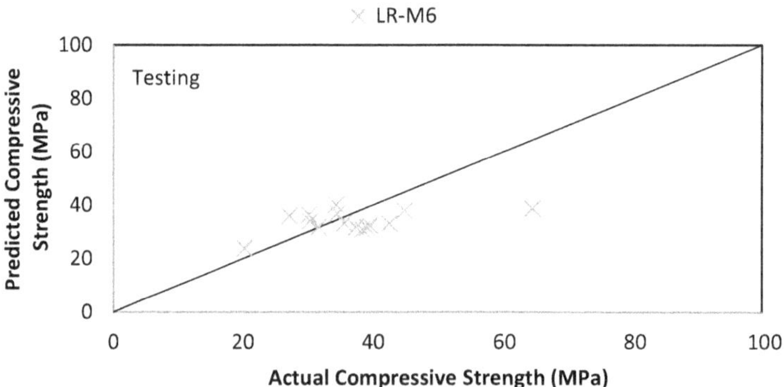

FIGURE 6.5 Scatter graphs showing the best model's observed and predicted compressive strength by LR technique.

6.5.6 Comparison among Best Developed Models

In this paper, the compressive strength of concrete mix with waste marble powder was predicted by applying soft computing techniques. Performance was checked for both stages by five performance statistics evaluation criteria listed in Table 6.10. For both phases, a scatter plot was created to compare the actual and predicted compressive strength of the best models from ANN, RF, RT, and LR (Figure 6.6). Also, this figure shows that relative error was minimum in the ANN-M6 model for both training and testing datasets. According to the statistical analysis, ANN-M6 predicts better results among all the models as listed in Table 6.10. The coefficient of correlation indicated that all techniques are adequate for predicting the compressive strength of concrete mix, with the exception of linear regression, which has a lower CC value

TABLE 6.10
Comparison of Model Prediction

Models	CC	MAE (MPa)	RMSE (MPa)	RAE (%)	RRSE (%)	Order
ANN-M6	0.9575	2.6096	3.0388	39.4920	30.9990	1
RT-M5	0.8895	3.9760	4.4210	60.1695	45.0983	2
RF-M6	0.8633	3.0309	5.3504	45.8674	54.5797	3
LR-M6	0.4740	6.8569	8.8431	103.8300	92.5600	4

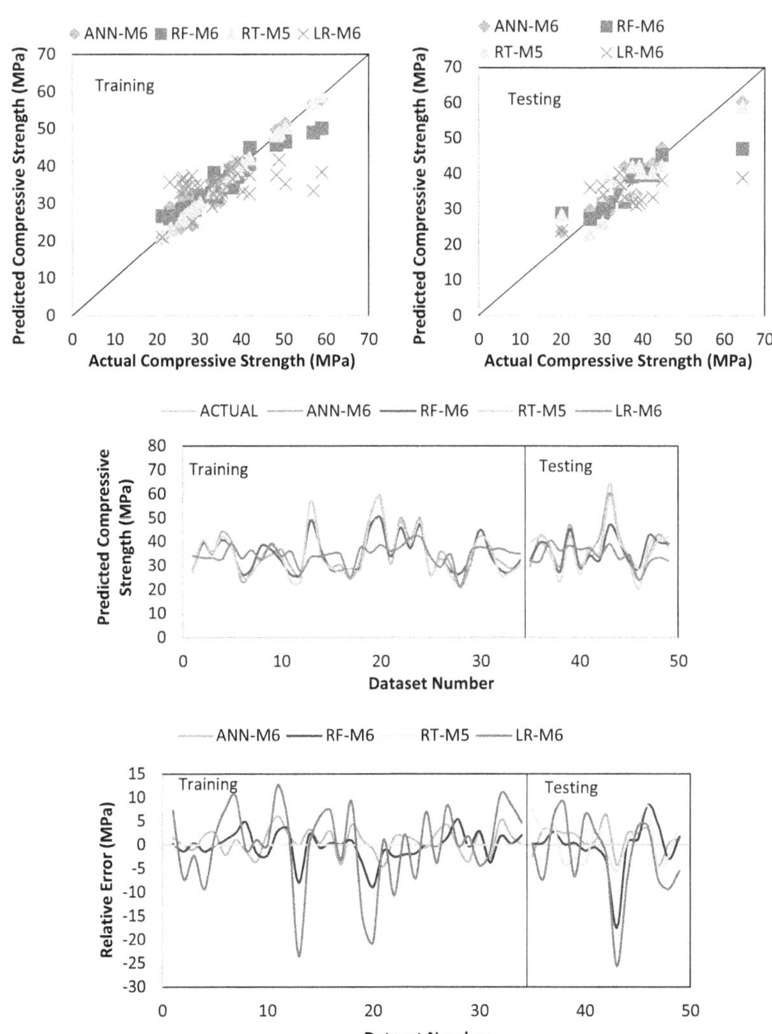

FIGURE 6.6 Comparison between ANN-M6, RF-M6, RT-M5, and LR-M6 predicted and actual values.

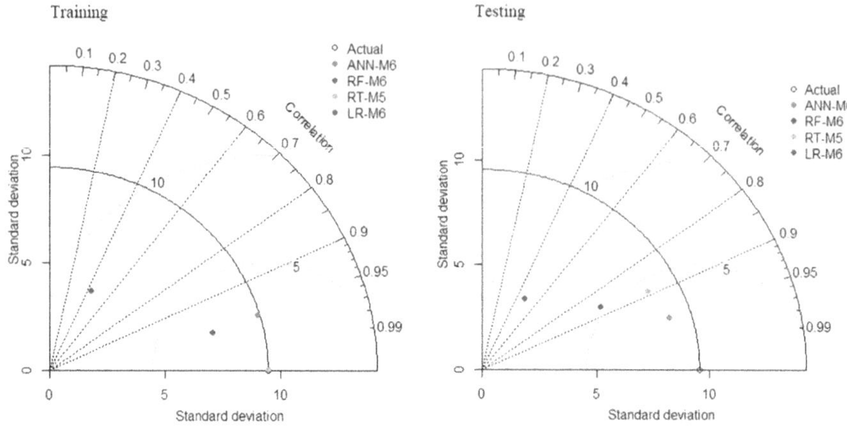

FIGURE 6.7 Taylor diagram for best result.

(less than 0.5) and higher value of errors. Out of four techniques, ANN-M6 presents better results with the highest value of CC (0.9575) followed by RT-M5 with CC (0.8895), RF-M6 with (0.8633), and LR-M6 with (0.4740) also RMSE results shows ANN-M6 has the best values for accurate data with the minimum error value of 3.0388 followed by RT-M5 (4.4210), RF-M6 (5.3504), and LR-M6 (8.8431). Similarly, RRSE has a minimum error value for ANN-M6 (30.9990%) followed by RT-M5 (45.0983), RF-M6 (54.5797), and LR-M6 (92.5600).

Also for the graphical comparison for the performance of best-suited predictions, the Taylor diagram was plotted for both training and testing datasets (Figure 6.7). The Taylor diagram for best models shows that all models performed well in the training dataset, as all values were near the hollow black circle, whereas in the testing dataset, it indicates that ANN-M6 predicts the best outcome with high accuracy compared to other models for predicting the compressive strength of concrete mix mixed with waste marble powder.

6.6 CONCLUSION

Accurate prediction of 28-day compressive strength of concrete mix is helpful for sustainable development of strength property of concrete mix. To forecast the proper compressive strength of the marble powder concrete mix, a number of models were constructed by altering the input parameters and utilizing soft computing techniques such as ANN, RF, RT, and LR. For all ten models, CC, RMSE, MAE, RAE, and RRSE were used to assess the results produced by these approaches. Results for this study are summarized below:

1. The best results of the ANN model were found in model six, which used fine aggregate, coarse aggregate, water/cement ratio, and marble powder as input parameters to forecast the 28-day compressive strength of the concrete mix.

2. Results showed that the best performance was given by the ANN-M6 model with the highest CC value (0.9575) and lower MAE (2.6096) and RMSE (3.0388) values followed by the RT-M5 model where CC = 0.8895, MAE = 3.9760, and RMSE = 4.4210.
3. The scatter diagram shows that the ANN-M6 model is a good fit for the observed data and thus gives a promising application in the prediction of the strength property of concrete mix.

REFERENCES

Ahmed, M.S. 2012. Statistical modelling and prediction of compressive strength of concrete. *Concrete Research Letters*. 3(2). Doi: 10.6084/M9.FIGSHARE.105905.

Alyamac, K.E., and Ince, R. 2009. A preliminary concrete mix design for SCC with marble powders. *Construction and Building Materials*. 23(2009): 1201–10. Doi: 10.1016/j.conbuildmat.2008.08.012.

Ashish, D.K., Verma, S.K., Kumar, R., and Sharma, N. 2016. Properties of concrete incorporating sand and cement with waste marble powder. *Advances in Concrete Construction*. 4(2): 145–60. Doi: 10.12989/acc.2016.4.2.145

Chavhan, P.J., and Bhole, S.D. 2014. To study the behaviour of marble powder as supplementry cementitious material in concrete. *International Journal of Engineering Research and Applications*. 4(4): 377–81.

Chopra P., Sharma R.K., and Kumar M. 2016. Prediction of compressive strength of concrete using artificial neural network and genetic programming. *Advances in Materials Science and Engineering*. 2016, Article ID 7648467,10 pages. Doi: 10.1155/2016/7648467.

Chopra, P., Sharma, R.K., Kumar, M., and Chopra, T. 2018. Comparison of machine learning techniques for the prediction of compressive strength of concrete. *Advances in Civil Engineering*. Doi: 10.1155/2018/5481705.2.

Deepa, C., SathiyaKumari, K., and Pream Sudha, V. 2010. Prediction of the compressive strength of high performance concrete mix using tree based modeling. *International Journal of Computer Application*. 6(5): 18–24.

Dhiman, H., and Bhardwaj, S. 2018. Partial replacement of cement with marble dust powder and addition of polypropylene fibres. *International Research Journal of Engineering and Technology (IRJET)*. 5(11).

Ergun, A. 2011. Effects of the usage of diatomite and waste marble powder as partial replacement of cement on the mechanical properties of concrete. *Construction and Building Materials*. 25(2): 806–12.

Gao, S., Chen, Y., Chen, L., Cheng, X., and Chen, G. 2020. Experimental and field study on treatment waste mud by in situ solidification. *Proceedings of the Institution of Civil Engineers- Municipal Engineer*. 173(4): 237–45. Doi: 10.1680/jmuen.18.00021.

Goldberg, Y. 2017. Neural network methods for natural language processing. *Synthesis Lectures on Human Language Technologies*. 10(1): 1–309. Doi: 10.2200/S00762ED1V01Y201703HLT037.

Han, Q., Gui, C., Xu, J., and Lacidogna, G. 2019. A generalized method to predict the compressive strength of high-performance concrete by improved random forest algorithm. *Construction and Building Materials*. 226(0): 734–42. Doi: 10.1016/j.conbuildmat.2019.07.315.

Madandoust, R., Bungey, J.H., and Ghavidel, R. 2012. Prediction of the concrete compressive strength by means of core testing using GMDH-type neural network and ANFIS models. *Computational Materials Science*. 51(2012): 261–71. Doi: 10.1016/j.commatsci.2011.07.053.

Missan, N., Sharma, N., Sharma, H., and Mor, N. 2019. Stabilization of clayey soil using bagasse ash as a stabilizing material. *Journal of Emerging Technologies and Innovative Research*. 6(6).

Nhu, V.H., Shahabi, H., Nohani, E., Shirzadi, A., Ansari, N.A., Bahrami, S., Miraki, S., Geertsema, M., and Nguyen, H. 2020. Daily water level prediction of zrebar lake(Iran): A comparison between M5P, random forest, random tree and reduced error pruning trees algorithms. *International Journal of Geo-Information*. 9: 479. Doi: 10.3390/ijgi9080479.

Sakalkalel, A.D., Dhawale, G.D., and Kedar, R.S. 2014. Experimental study on use of waste marble dust in concrete. *Journal of Engineering Research and Applications*. 4(10): 44–50.

Sepahvand, A., Singh, B., Sihag, P., Samani, A.N., Ahmadi, H., and Nia, S.F. 2019. Assessment of the various soft computing techniques to predict sodium absorption ratio (SAR). *ISH Journal of Hydraulic Engineering*. Doi: 10.1080/09715010.2019.1595185.

Sharma, N., Thakur, M. S., Goel, P.L., and Sihag, P. 2020. A review: Sustainable compressive strength properties of concrete mix with replacement by marble powder. *Journal of Achievements in Materials and Manufacturing Engineering*. 98(1): 11–23. Doi: 10.5604/01.3001.0014.0813.

Singh, B., Sihag, P., Tomar, A., and Sehgad, A. 2019. Estimation of compressive strength of high-strength concrete by random forest and M5P model tree approaches. *Journal of Materials and Engineering Structures*. 6(2019): 583–92.

Singh, M., Srivastava, A., and Bhunia, D. 2019. Analytical and experimental investigations on using waste marble powder in concrete. *Journal of Materials in Civil Engineering*. 31(4): 04019011. Doi: 10.1061/ (ASCE)MT.1943-5533.0002631.

Sobhani, J., Najimi, M., Pourkhorshidi, A.R., and Parhizkar, T. 2010. Prediction of the compressive strength of no-slump concrete: A comparative study of regression, neural network and ANFIS models. *Construction and Building Materials*. 24(2010): 709–18. Doi: 10.1016/j.conbuildmat.2009.10.037.

Soliman, N. 2013. Effect of using marble powder in concrete mixes on the behavior and strength. *International Journal of Current Engineering and Technology*. 3(5): 1863–70.

Sounthararajan, V.M., and Sivakumar, A., 2013. Effect of the lime content in marble powder for producing high strength concrete. *ARPN Journal of Engineering and Applied Sciences*. 8(4): 260–64 ISSN: 1819-6608.

Talah, A., Kharchi, F., and Chaid, R. 2015. Influence of marble powder on high performance concrete behavior. *Procedia Engineering*. 114(0): 685–90.

Topcu, I.B., Bilir, T., and Uygunoglu, T. 2009. Effect of waste marble dust content as filler on properties of self-compacting concrete. *Construction and Building Materials journal*. 23(2009): 1947–53. Doi: 10.1016/j.conbuildmat.2008.09.007.

Topcu, I.K., and Saridemir, M. 2008. Prediction of mechanical properties of recycled aggregate concretes containing silica fume using artificial neural networks and fuzzy logic. *Computational Materials Science*. 42(2008): 74–82. Doi: 10.1016/j.commatsci.2007.06.011.

Vaidevi, C. 2013. Engineering study on marble dust as partial replacement of cement in concrete. *Indian Journal of Engineering*. 4(7): 14–16.

Zongjin, L. 2011 *Advanced Concrete Technology* John Wiley and Sons, Hoboken, New Jersey.

7 Using GA to Predict the Compressive Strength of Concrete Containing Nano-Silica

Sakshi Gupta
ASET, Amity University

Deepika Garg
G.D Goenka University

CONTENTS

7.1 INTRODUCTION

Concrete is said to be one of the oldest and the most useful materials in the construction industry which is a mixture of paste and aggregates. Production of concrete is a complex process that involves the effect of several processing criteria on the quality control of concrete pertaining to workability, strength, durability, etc. These parameters are all effective in producing a single strength quantity of compressive strength. Nanotechnology creates new possibilities to improve material properties for civil construction. Attracting civil engineers to adopt nanotechnology could empower them to provide a pioneering solution to the tangled problems of construction today. It is well known that materials such as concrete; are the core elements of the construction industry and these materials could be developed by using nanotechnology. Nano-silica is a highly productive pozzolanic material and normally, it consists of very fine vitreous particles approximately 1000 times smaller than the average cement particles. It has proven to be an eminent admixture for cement to improve

DOI: 10.1201/9781003184331-7

strength and durability and decrease permeability [1]. The widespread of concrete materials in structural engineering in recent decades has led to many different optimization problems improving the design and overall performance of concrete. This could enhance the application of concrete in various practices thereby reducing the amount of hazardous material and improving its use in construction.

The present study was envisaged to develop a relation between various input parameters and an output parameter, i.e., 28-day compressive strength, using the genetic algorithm (GA) technique. The objective was to study the application of GA for predicting the 28-day compressive strength of concrete containing nano-silica which is a partial replacement of cement, with data obtained from various literatures. Over the last two decades, different data mining methods such as fuzzy logic, GA, and artificial neural network have become popular and have been used by many researchers for a variety of engineering applications. Over the years, the ability to reason has been developed on the basis of the evidence available to achieve the required goals. To deal with the problem of uncertainty, the theory of probability had been established and successfully applied to many areas of engineering and technology [2–5].

The purpose of this article is to provide a methodology for predicting the strength of concrete where the partial replacement of cement is done with nano-silica which uses the features of GA and is presented as an improved approach.

7.2 GENETIC ALGORITHM (GA)

Data mining has as goal to extract knowledge from large databases. GA is an adaptive heuristic search algorithm premised on the evolutionary ideas of natural selection and genetic. The basic concept of GA is designed to stimulate processes in the natural system necessary for evolution, especially those that follow the principles first laid down by Charles Darwin of survival of fittest. As a result, they constitute a problem-solving intelligent exploration of a random explore inside a defined search space. GA is a method of "breeding" computer programs and solutions to optimization or search problems by means of stimulated evolution, processes based on natural selection, crossover, and mutation are repeatedly applied to a population of binary strings which represent potential solutions. Over time, the numbers of above-average individuals are created, until a good solution to the problem at hand is found. However, GA has its own shortcomings such as lower local convergence speed inkling to premature convergence, etc. GA belongs to the larger class of evolutionary algorithms, which generate solutions to optimization problems using techniques inspired by natural evolution, such as inheritance, mutation, selection, and crossover. GA finds application in bioinformatics, computational science, engineering, economics, chemistry, manufacturing, mathematics, physics, and other fields.

7.3 DATABASE

The database was collected from available literature on concrete containing nano-silica, as summarized in Table 7.1. The success of the models depends upon the comprehensiveness of the data. Thus, large varieties of data were collected, and a total of 32 datasets have been used with the following input and output variables.

TABLE 7.1
Details of Data Used in Modeling

S.No.	Cement (kg/m³) (x1)	FA (kg/m³) (x2)	CA (kg/m³) (x3)	W/B Ratio (x4)	SP (kg/m³) (x5)	NS (kg/m³) (x6)	D (nm) (x7)	28-day CS (MPa) (y)	Researcher (Year)
1	396.6	826	722	0.37	7	16.5	15	75.2	Beigi et al. [6]
2	380	826	722	0.35	7	33	15	86.1	
3	363.5	826	722	0.33	7	49.6	15	85.4	
4	318.4	840	1040	0.5	2.71	1.6	15	36.8	Heidari [12]
5	316.8	840	1040	0.5	4.75	3.2	15	40.2	
6	390	783	1175	0.4	1.78	23.4	35	70	Said et al. [9]
7	390	774	1162	0.4	3.56	46.8	35	76	
8	390	769	1154	0.4	1.27	23.4	35	60	
9	390	762	1143	0.4	2.54	46.8	35	66	
10	356.4	650	1260	0.42	5.4	3.6	10	66.36	Zhang et al. [10]
11	349.2	650	1260	0.42	7.2	10.8	10	61.16	
12	447.75	492	1148	0.4	0	2.25	80	39.2	Givi et al. [13]
13	445.5	492	1148	0.4	0	4.5	80	40.3	
14	443.25	492	1148	0.4	0	6.75	80	41.2	
15	441	492	1148	0.4	0	9	80	38.1	
16	447.75	492	1148	0.4	0	2.25	15	42.7	
17	445.5	492	1148	0.4	0	4.5	15	43.6	
18	443.25	492	1148	0.4	0	6.75	15	42.9	
19	441	492	1148	0.4	0	9	15	39.7	
20	394	811	915	0.45	1.68	12	15	53.8	
21	388	811	915	0.45	2.32	24	15	56.5	
22	382	811	915	0.45	3	36	15	60	Nili et al. [14]
23	247.5	625	0	0.5	4.5	7.5	40	54.3	Jo et al. [7]
24	240.6	626	0	0.5	5.8	14.4	40	61.9	
25	241.8	627	0	0.5	7	23.2	40	68.2	
26	227.7	628	0	0.5	7.5	27.3	40	68.8	
27	370	647	1088	0.49	13.5	13.9	15	44	Ji [11]
28	568.36	1757.8	0	0.5	8.85	17.5	15	32.9	Li et al. [15]
29	556.64	1757.8	0	0.5	14.58	29.3	15	33.8	
30	527.34	1757.8	0	0.5	29.3	58.59	15	36.4	
31	556.64	1757.8	0	0.5	10.28	11.71	15	35.4	
32	480	647	1140	0.28	10	20	10	75.8	Li [8]

Note: All types of SP have been considered to be the same.

The basic parameters considered in this study were cement content, fine aggregate content, coarse aggregate content, nano-silica content, diameter of nano-silica, water-to-binder ratio, and superplasticizer dosage. The criteria for data identification were determined by the omission of one or more tangible properties in certain

TABLE 7.2
Input and Output Variables

Variables	Parameter	Abbreviation	Database Range	
			Minimum	Maximum
Input	Cement (kg/m^3)	Cement	227.70	568.36
	Fine aggregate (kg/m^3)	FA	492	1757.80
	Coarse aggregate (kg/m^3)	CA	0	1260
	Water-to-binder ratio	W/B ratio	0.28	0.50
	Superplasticizer (kg/m^3)	SP	0	29.30
	Nano-silica (kg/m^3)	NS	1.60	58.59
	Diameter of nano-silica (nm)	D	10	80
Output	28-day compressive strength (MPa)	28-day CS	32.90	86.10

research and the ambiguity of mixture proportions and testing techniques in others. The successful model to predict the 28-day compressive strength depends upon the magnitude of the training data. The predicted results were compared with the values obtained experimentally [1–16].

The ranges of various input and the output parameters used in data mining techniques are given in Table 7.2.

7.4 FUNCTION APPROXIMATION

With the help of the multiple regression, the value of an output variable based on the value of seven input variables is predicted. When y is a function of more than one predictor variable, the matrix equations that express the relationships among the variables must be expanded to accommodate the additional data. This is called multiple regression problems. Consider the task of determining the output variable y for independent variables x1, x2, x3, x4, x5, x6, and x7, i.e., forming a model of this data that is of the form "a0 + a1 * x1 + a2 * x2 + a3 * x3 + a4 * x4 + a5 * x5 + a6 * x6 + a7 *x7. The multiple regression solves for unknown coefficients $a0$, $a1$, $a2$, $a3$, $a4$, $a5$, $a6$, and $a7$ by minimizing the sum of the squares of the deviations of the data from the model (least-squares fit).

Construct and solve the set of simultaneous equations by forming a design matrix, X

$$X = \left[\text{ones }\left(\text{size}(x1)\right)\ x1\ x2\ x4\ x5\ x6\ x7\right];$$

Solving for the parameters by using the backslash operator:

$$a = X/y$$

To validate the model, find the maximum of the absolute value of the deviation of the data from the model:

$$Y = X * a;$$

$$\text{Validation Max Err} = \max\left(\text{abs}\left(Y - y\right)\right)$$

Solution:

$$a0 = 221.364265226088,\ a1 = -0.175546705711193,$$

$$a2 = 0.0218338476779627,\ a3 = -0.00411624974491635,$$

$$a4 = -260.894997850725,\ a5 = -0.207507335152316,$$

$$a6 = 0.232271784522169,\ a7 = -0.0731846372091189$$

and

$$y = 221.364265226088 + -0.175546705711193x1 + 0.0218338476779627x2 +$$
$$-0.00411624974491635x3 + -260.894997850725x4 + 0.207507335152316x5 +$$
$$0.232271784522169x6 + -0.0731846372091189x7$$

$$\text{Max Err} = 10.9937 * 10^{-1}$$

This value is much smaller than any of the data values, indicating that this model accurately follows the data.

7.5 OPTIMIZATION USING GENETIC ALGORITHM TECHNIQUE

Flow chart representation of the action of GA technique for parameter optimization in the present problem is as follows (see Figure 7.1):

$$\text{Maximize} \quad f(x), \quad \text{for} \quad x_i^{\min} \Leftarrow x_i \Leftarrow x_i^{\max}$$

where $i = 1,2,3 \ldots N$.

7.6 AVAILABILITY OPTIMIZATION USING GENETIC ALGORITHM

- Problem definition: Find the maximum value of 28-day compressive strength.

$$y = 221.364265226088 + -0.175546705711193x1 + 0.0218338476779627x2$$
$$+ 0.00411624974491635x3 + -260.894997850725x4 + -0.207507335152316x5$$
$$+ 0.232271784522169x6 + -0.0731846372091189x7$$

TABLE 7.3
Bound of Input Variables

Variable	Minimum	Maximum
Cement (*x*1)	227.70	568.36
FA (*x*2)	492	1757.80
CA (*x*3)	0	1260
W/B ratio (*x*4)	0.28	0.50
SP (*x*5)	0	29.30
NS (*x*6)	1.60	58.59
D (*x*7)	10	80

- Bounds of input variables are shown in Table 7.3.
- Element of a input vectors are *x*1, *x*2, *x*3, *x*4, *x*5, *x*6, and *x*7.

The best value of compressive strength related to a certain value of system parameters was estimated using a MATLAB® software for compressive strength-based optimization utilizing GA. The effect of the number of crossover, population size, and generation on the compressive strength of concrete is shown in Figures 7.2–7.4, respectively.

The optimum value of compressive strength of concrete is 151.523 kN/m^2. The best possible combination values are shown in Table 7.4.

7.7 CONCLUSIONS

In the present study, the GA model for 28-day compressive strength has been developed. The model was trained with input and output experimental data. As in this study, there are seven inputs and one output, as a result, compressive strength values of concrete can be predicted in GA models without attempting any experiments in a quite short period of time with tiny error rates. The GA results were compared with the random optimization where prediction of function was done using an artificial neural network for 28-day compressive strength [16]. It is found that the optimized value for 28-day compressive strength for concrete containing nano-silica using GA-based optimization is 151.523 MPa as compared to artificial neural network prediction which was 95.909 MPa. The nano-SiO_2 particles fill the voids and act as a nucleus to tightly bond with C–S–H gel particles, making the binding paste matrix denser and resulting in an increase in the long-term strength and durability of concrete. Thus, the use of nano-particle materials in concrete can add many benefits that are directly related to the durability of various cementitious materials, besides the fact that it is possible to reduce the quantities of cement in the composite thereby gaining the optimized cost benefits and material saving.

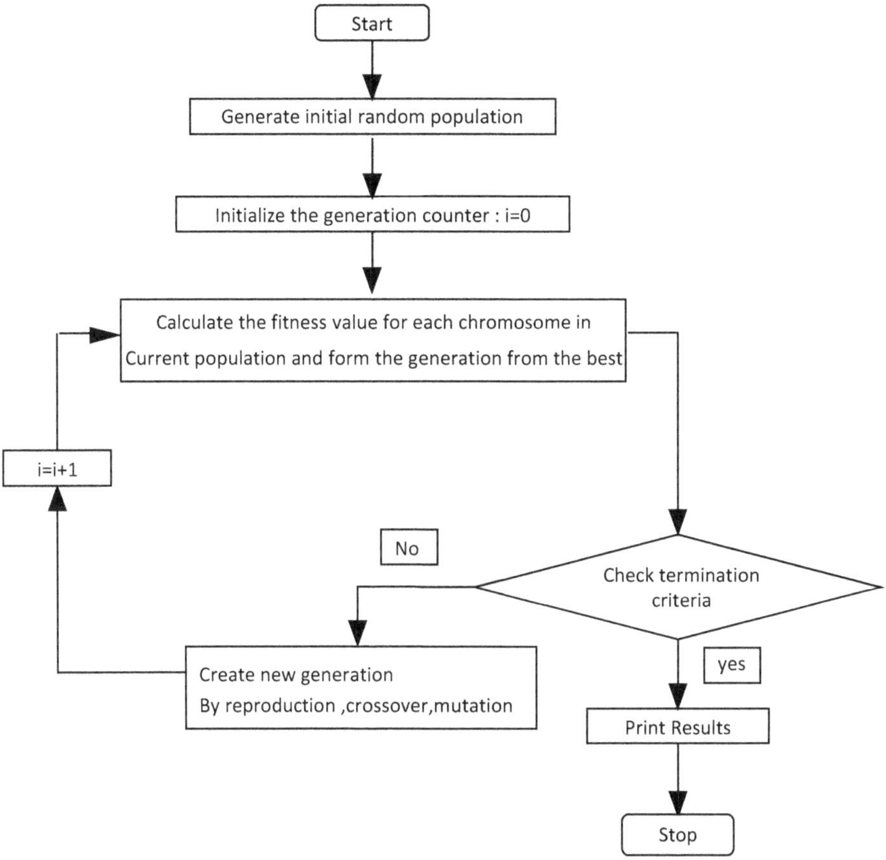

Start
Generate initial random population
Initialize the generation counter : i=0
Calculate the fitness value for each chromosome in
Current population and form the generation from the best
i=i+1
No
Check termination
criteria
Create new generation
By reproduction ,crossover,mutation
yes
Print Results
Stop

FIGURE 7.1 Flowchart depicting the action of GA technique.

Variation of 28-days Compressive strength w.r.t.
Crossover Fraction
Fitness Value
160
140
120
100
80
60
40
20
0
0 0.1 0.2 0.3 0.4 0.5 0.6 0.7 0.8 0.9 1
fitness value 120 147 140 135 145 140 151 142 139 134 141
Cross over rate

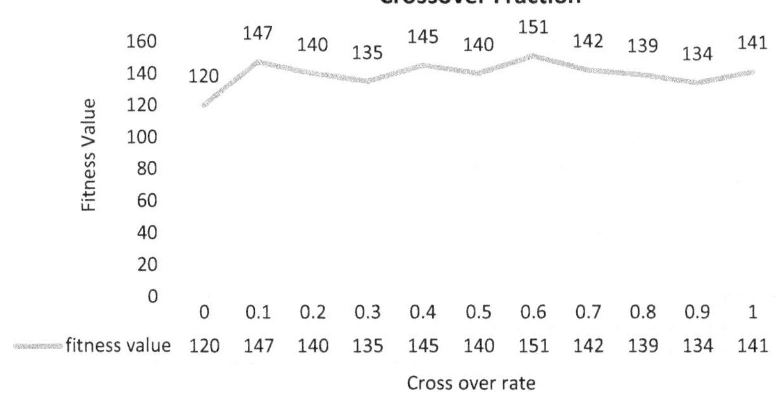

FIGURE 7.2 Variation of 28-day compressive strength w.r.t. crossover fraction.

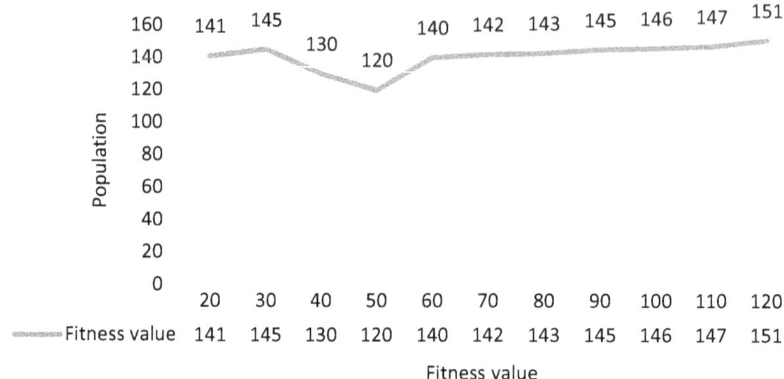

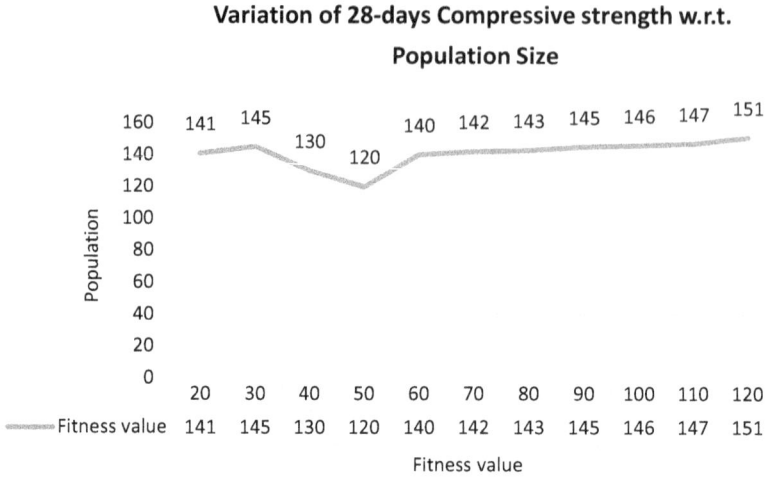

FIGURE 7.3 Variation of 28-day compressive strength w.r.t. population size.

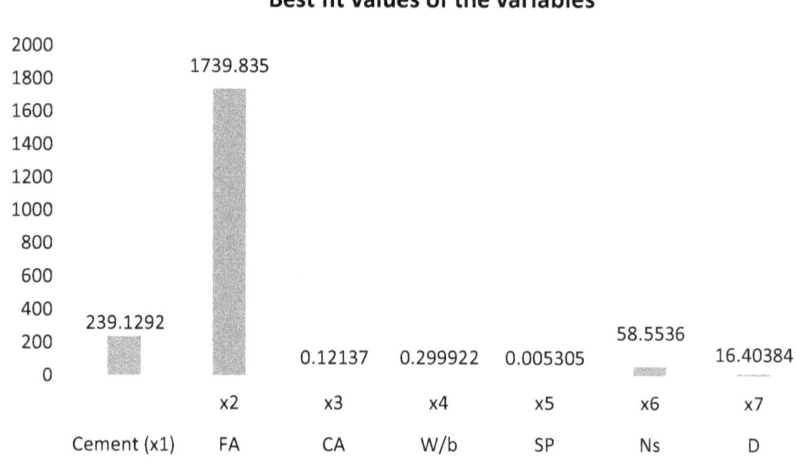

FIGURE 7.4 Variation of 28-day compressive strength w.r.t. generations.

TABLE 7.4
Best Fit Values of the Variables

	FA	CA	W/B	SP	NS	D
Cement ($x1$)	$x2$	$x3$	$x4$	$x5$	$x6$	$x7$
239.1292	1739.835	0.12137	0.299922	0.005305	58.5536	16.40384

REFERENCES

1. P. C. Aitcin, P.A. Hershey, and Pinsonneault. Effect of the addition of condensed silica fume on the compressive strength of mortars and concrete, *American Ceramic Society*, 1981, 22, pp. 286–290.
2. S. Akkurt, G. Tayfur, and S. Can, Fuzzy logic model for prediction of cement compressive strength, *Cement and Concrete Research*, 2004, 34(8), pp. 1429–1433.
3. O. Unal, F. Demir, and T. Uygunoglu, Fuzzy logic approach to predict stress–strain curves of steel fiber-reinforced concretes in compression, *Building and Environment*, 2007, 42(10), pp. 3589–3595.
4. F. Demir, A new way of prediction elastic modulus of normal and high strength concrete-fuzzy logic, *Cement and Concrete Research*, 2005, 35(8), pp. 1531–1538.
5. I.B. Topcu, and M. Sarıdemir, Prediction of mechanical properties of recycled aggregate concretes containing silica fume using artificial neural networks and fuzzy logic, *Computational Materials Science*, 2008, 42(1), pp. 74–82.
6. H.B. Morteza, B. Javad, O.O. Lotfi, N.A. Sadeghi, and I.M. Nikbin, An experimental survey on combined effects of fibers and nanosilica on the mechanical, rheological, and durability properties of self-compacting concrete. *Materials & Design*, 2013, 50, pp. 1019–1029.
7. B.W. Jo, C.H. Kim, G. Tae, J.B. Park, Characteristics of cement mortar with nano- SiO_2 particles. *Construction and Building Materials*, 2007, 21(6), pp. 1351–1355.
8. G. Li, Properties of high-volume fly ash concrete incorporating nano-SiO_2. *Cement and Concrete Research*, 2004, 34(6), pp. 1043–1049.
9. A.M. Said, M.S. Zeidan, M.T. Bassuoni, and Y. Tian, Properties of concrete incorporating nano-silica. *Construction and Building Materials*, 2012, 36, pp. 834–844.
10. M.-h. Zhang and H. Li, Pore structure and chloride permeability of concrete containing nano-particles for pavement. *Construction and Building Materials*, 2011, 25, pp. 608–616.
11. T. Ji, Preliminary study on the water permeability and microstructure of concrete incorporating nano-SiO_2. *Cement and Concrete Research*, 2005, 35, pp. 943–947.
12. A. Heidari and D. Tavakoli, A study of the mechanical properties of ground ceramic powder concrete incorporating nano-SiO_2 particles. *Construction and Building Materials*, 2013, 38, pp. 255–264.
13. A.N. Givi, S.A. Rashid, F.N.A. Aziz, and M.A.M. Salleh, Experimental investigation of the size effects of SiO_2 nano-particles on the mechanical properties of binary blended concrete. *Composites: Part B*, 2010, 41, pp. 673–677.
14. M. Nili, A. Ehsani, and K. Shabani, Influence of Nano-SiO_2 and Micro-silica on Concrete Performance. *Second International Conference on Sustainable Construction Materials and Technologies*, 2010.
15. H. Li, H. Gang, X. Jie, J. Yuan, and J. Ou, Microstructure of cement mortar with nanoparticles. *Composites Part B: Engineering*, 2004, 35(2), pp. 185–189.
16. S. Gupta, Using artificial neural network to predict the compressive strength of concrete containing nano-silica. *Civil Engineering and Architecture*, 2013, 1(3), pp. 96–102, Doi: 10.13189/cea. 2013.010306.

8 Evaluation of Models by Soft Computing Techniques for the Prediction of Compressive Strength of Concrete Using Steel Fibre

Nitisha Sharma, Mohindra Singh Thakur, and Ankita Upadhya
Shoolini University

Parveen Sihag
Chandigarh University

CONTENTS

DOI: 10.1201/9781003184331-8

8.1 INTRODUCTION

Concrete is an important part of the construction industry, which increases its demand. It plays an important role in building bridges, dams, roads, or any type of structural member. For ages, cement, fine aggregate, coarse aggregate, and water are the basic components of concrete (Zongjin, 2011). Concrete performs well in compression but poorly in tension. So, to enhance the strength properties of concrete; admixtures were used. Different approaches were applied by the researchers to improve the mechanical properties of concrete. As concrete is weak in tension, different types of fibre reinforcement were used (Haddadou et al., 2014). Steel fibre is one such type which consists of various good properties, for example, it is good in flexural and tensile strength. When steel fibres are mixed with concrete constituents, it reduces the growth of the cracks which forms due to the shrinkage or the applied loading and helps in increasing the strength properties of the concrete mix (Lau and Anson, 2006). Zhang et al., (2017), in the investigation, mechanical characteristics of a high-strength polyvinyl alcohol-steel hybrid fibre reinforced engineered cementitious composite were analyzed. The results show that with the increase in the amount of steel fibre it enhances the strength properties as well as the capacity of the tensile strain of the control mix. Ganesan et al., (2015) show in the study that fibre reinforced concrete is more durable as compared to conventional concrete. Sounthararajan and Sivakumar (2013), in the study, experiments were conducted with the steel fibre reinforced concrete as a replacement of finer sand with different percentages (0%, 10%, 20%, and 30%). The result showed that finer sand can be used as a replacement. Also due to industrialization production of waste material pollutes the environment which needs a solution. To solve the problem, waste generated by these industries can be utilized by using these wastes in the concrete mix either partial or full replacement with sand or cement (Siddique et al., 2020, Sharma et al., 2020).

According to prior studies, researchers utilize many sorts of algorithms to anticipate the outcome. The primary elements of concrete, such as cement, fine and coarse aggregates, and water, determine the concrete's strength properties. So, cement, fine and coarse aggregates, and water were considered as input variables, and strength (like compressive strength) was considered as an output parameter. Prediction of the output depends upon the regression techniques either linear or non-linear. Over the last few decades, different machine learning algorithms were used for the estimation of the mechanical properties of concrete. Algorithms such as artificial neural network (ANN), linear regression (LR), and cross-validation were applied to the dataset to compute the desired output. Researchers use concrete constituents as input parameters to prepare the models, and soft-computing techniques were applied such as neural network technique and adaptive neuro-fuzzy inference system for the desired outcome (Thakur et al., 2021, Sobhani et al., 2010). Sobhani et al., (2010) show that, compared to traditional regression analysis, neural network technique and adaptive neuro-fuzzy inference system predicts better, 28-day compressive strength. In Chopra et al., (2018), the research has been done by applying artificial intelligence (AI) techniques to estimate the strength property of concrete. Because of its great accuracy, the neural network model was shown to be more dependable for predicting the compressive strength of concrete. Vakharia and Gujar (2019) apply a ten-fold cross-validation procedure on different soft computing techniques such as ANN, random forest, and random trees and compare the results. It was observed from the results that ten-fold cross-validation predicts better results with a high coefficient of correlation and minimum error.

8.1.1 Objectives of the Study

The study's objective is to see how the input factors affect the outcome, e.g., effect of water, superplasticizer, fine aggregate, coarse aggregate, cement, fly ash, steel fibre, and curing days on the compressive strength of the concrete mix. Variations in the input parameters have been done to reach the goal with the help of LR, ANN, and *k*-fold cross-validation on the ANN algorithm. Furthermore, a comparison of these techniques is also analyzed by applying statistic measures.

8.2 SOFT COMPUTING TECHNIQUES

The soft computing technique is the computational technique based upon artificial intelligence. Many researchers use these techniques in their research to solve the complex design. Furthermore, it helps in the simplification of the complex problem in the engineering design (Laghari et al., 2019). Soft-computing techniques such as ANN, decision trees, and regression analysis are implemented in engineering design problems because they have the capability to solve linear, non-linear, and multi-dimensional complex engineering problems (Singh et al., 2019, Chopra et al., 2018, Sepahvand et al., 2019, Han et al., 2019, Goldberg 2017, Nhu et al., 2020). In this study, ANN, LR, and cross-validation algorithms are applied to the various models to achieve the target.

8.2.1 Artificial Neural Network (ANN)

A soft computing approach is known as an ANN which behaves like a human brain, i.e., it simulates the problem and processes it. An ANN technique can find a way to reach the goal for the complex problem. In this process, the input parameters behave like neurons. These neurons are interconnected and process the data to determine the best-predicted outcome. There are three main layers constructed in ANN: (1) input layer where all the input parameters are defined, (2) hidden layer in which the number of neurons has to be described, and (3) output layer which gives the desired output (see Figure 8.1). The main parameters used in ANN techniques which are user defined are momentum, hidden layers, number of neurons, and learning rate. The hidden layer consists of the number of neurons which form a complex structure when connected with the input layer. The entire dataset is split into two subsets: training (70%) and testing (30%). The best model is selected by trial-and-error method with user defined parameters as an example, learning rate as 0.2, momentum as 0.1, the number of iterations is 5000, and one hidden layer with the number of neuron 11.

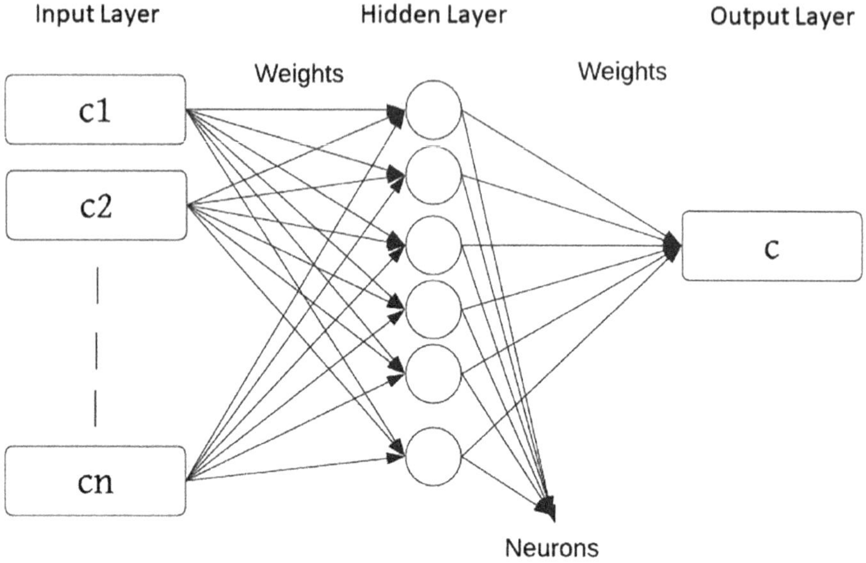

FIGURE 8.1 Artificial neural networks: a basic structure.

8.2.2 Artificial Neural Network–Cross-Validation

The soft computing technique has two main types, classification and regression, and both the types are used to predict the data from the supplied dataset. In regression, the output variable is continuous whereas in the classification it is discrete. In both classification and regression machine learning techniques, overfitting was observed. Generally, overfitting was observed where the dataset was randomly selected without any relationship with the output function. In the evaluation of data, cross-validation plays a crucial part in the machine learning algorithm. Data was well predicted using the algorithm for the independent dataset. Cross-validation helps to assess the capability of the independent dataset to predict the target function. Cross-validation is the assessing procedure which may be used for output prediction in both regression and classification. It is applied to the dataset but first, numbers of folds are required. There are k-folds in cross-validation, where the dataset is randomly divided into k number of equal subsets. For example, in tenfold cross-validation, 90% of the data is used as a training subset and 10% is used for validation, and the total dataset is divided into ten equal parts. In this study, weka 3.8.4 is used for cross-validation where ten folds are used, i.e., the total dataset is divided into ten equal subsets. In every fold, the learning algorithm is applied to every subset, and finally it is applied to the total dataset for the final prediction of the output. The benefit of this strategy is that every one of the qualities is considered for simultaneous testing and training, and every value is used for testing purposes (Vakharia and Gujar, 2019).

8.2.3 Linear Regression (LR)

Machine learning algorithms are used to predict the outcome with minimum error. LR is one of the supervised machine learning algorithms. LR analyzes the model in every possible way, it assumes that there is a linear relationship between input and output variables. Various techniques are applied to form a linear equation. Each input factor has been assigned a coefficient factor. More than two variables can form a linear equation which consists of dependent and independent variables (Sobhani et al., 2010). LR forms a complex equation with dependent, independent variables and a coefficient value assigned to each parameter to predict the value of the output. As shown in Eq. (8.1), there are two variables where "o" is considered as a dependent variable and "i" is considered as an independent variable, where "C" is the coefficient assigned to each parameter. The main objective to form LR model, is to draw a linear line which shows that the model behaves linearly, i.e., both dependent and independent variables are best suited to predict the output. Evaluation of the LR model can be checked through the coefficient of correlation; the more the value is nearer to one, the more the output is reliable (Deepa et al., 2010).

$$o = C_0 + C_1 i_1 + C_2 i_2 + C_3 i_3 + C_4 i_4 + C_5 i_5 + C_6 i_6 + \cdots \quad (8.1)$$

8.3 METHODOLOGY AND DATASET

8.3.1 Dataset

The collection of datasets plays a crucial role to predict the output. The dataset used in this study were collected from past publications. One hundred and seventeen observations were collected and analyzed by using weka 3.8.4. Details of the observations are recorded in Table 8.1. The total dataset of 117 observations was divided into two parts (70%–30%) as training and testing datasets. Table 8.2 lists the characteristics of the training and testing datasets. Models were prepared by adopting different input combinations to reach the desired goal (as listed in Table 8.3). In this study, input parameters such as water, superplasticizer, fine aggregate, coarse aggregate, cement, fly ash, steel fibre, and curing days were adopted to achieve the desired outcome, and compressive strength was taken as an output parameter. To understand the methodology, a flow chart was prepared (Figure 8.2).

All these 20 combinations were applied to construct a concrete strength prediction model as an output.

8.3.2 Model Evaluation

After a combination of input variables which results in the generation of 20 models, evaluating of parameters has to be applied on them for checking the accuracy. Evaluating parameters used in this study are coefficient of correlation (CC), mean absolute error (MAE), and root mean square error (RMSE).

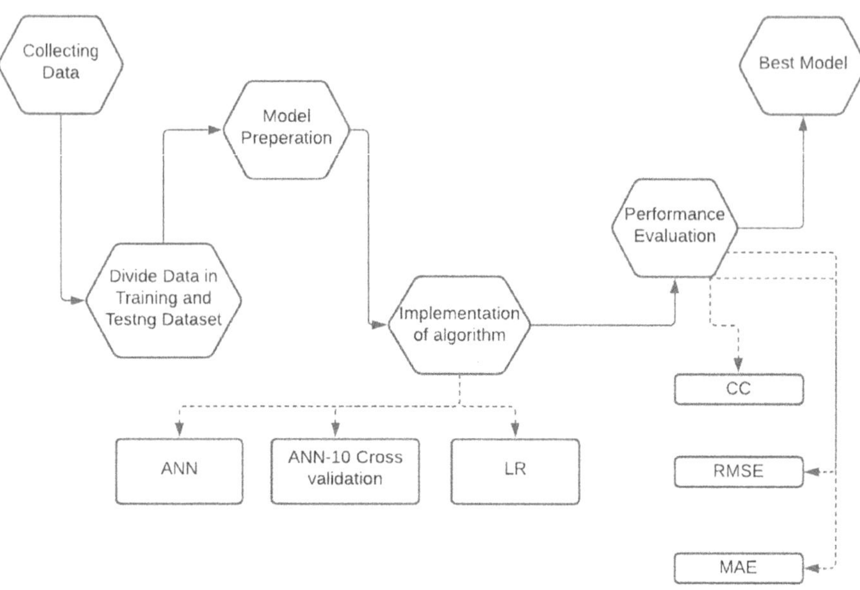

FIGURE 8.2 Flow chart of methodology.

$$\mathrm{CC} = \frac{l\left(\sum_{i=1}^{l} DC\right) - \left(\sum_{i=1}^{l} D\right)\left(\sum_{i=1}^{l} C\right)}{\sqrt{\left[l\sum_{i=1}^{l} D^2 - \left(\sum_{i=1}^{l} D\right)^2\right]}\sqrt{\left[l\sum_{i=1}^{l} C^2 - \left(\sum_{i=1}^{l} C\right)^2\right]}} \tag{8.2}$$

$$\mathrm{RMSE} = \sqrt{\frac{1}{l}\left(\sum_{i=1}^{l}(C-D)^2\right)} \tag{8.3}$$

$$\mathrm{MAE} = \frac{1}{l}\left(\sum_{i=1}^{l}|C-D|\right) \tag{8.4}$$

where D = observed values, $\bar{D}$ = average of observed value, C = predicted values, l = number of observations.

CC is one of the evaluating parameters which describe the performance of the individual model. It goes from −1 to +1. The higher worth of CC predicts better outcomes with the least mistake. For evaluating the errors, RMSE and MAE were calculated. The output shows better results if the relative error is minimum and vice versa, i.e., if a calculated error is low, it predicts better results for the desired output (Nhu et al., 2020).

TABLE 8.1
Detail of Dataset

Sr. No.	Cement (kg/m³)	Fine Aggregate (kg/m³)	Coarse Aggregate (kg/m³)	Water (kg/m³)	SP/HRWR	Fly Ash (kg/m³)	Steel %	Curing Days	CS (N/mm²)	Author Name & Year
1	315	919.00	649	256.5	9.5	135	0	7	18.34	Nalanth et al., (2014)
2	315	918.75	454.13	256.5	8	135	0	7	17.75	
3	315	918.75	389.25	256.5	8.5	135	0	7	14.71	
4	315	918.75	324.36	256.5	9.5	135	0	7	11.32	
5	315	918.75	454.13	256.5	8	135	0.5	7	16.89	
6	315	918.75	389.25	256.5	8.5	135	0.75	7	16.76	
7	315	918.75	324.36	256.5	9.5	135	1	7	13.25	
8	315	919.00	649	256.5	9.5	135	0	14	24.72	
9	315	918.75	454.13	256.5	8	135	0	14	21.12	
10	315	918.75	389.25	256.5	8.5	135	0	14	18.89	
11	315	918.75	324.36	256.5	9.5	135	0	14	15.56	
12	315	918.75	454.13	256.5	8	135	0.5	14	22.3	
13	315	918.75	389.25	256.5	8.5	135	0.75	14	20.24	
14	315	918.75	324.36	256.5	9.5	135	1	14	17.37	
15	315	919.00	649	256.5	9.5	135	0	28	32.45	
16	315	918.75	454.13	256.5	8	135	0	28	28.32	
17	315	918.75	389.25	256.5	8.5	135	0	28	23.14	
18	315	918.75	324.36	256.5	9.5	135	0	28	21.11	
19	315	918.75	454.13	256.5	8	135	0.5	28	29.5	
20	315	918.75	389.25	256.5	8.5	135	0.75	28	25.45	
21	315	918.75	324.36	256.5	9.5	135	1	28	23.22	

(Continued)

TABLE 8.1 (*Continued*)
Detail of Dataset

Sr. No.	Cement (kg/m³)	Fine Aggregate (kg/m³)	Coarse Aggregate (kg/m³)	Water (kg/m³)	SP/HRWR	Fly Ash (kg/m³)	Steel %	Curing Days	CS (N/mm²)	Author Name & Year
22	485	645	1052	170	4.85	0	1	28	56.52	Guo et al., (2014)
23	485	645	954	170	4.85	0	1	28	51.41	
24	485	625	954	170	4.85	0	1	28	49.06	
25	485	605	954	170	4.85	0	1	28	39.41	
26	485	585	954	170	4.85	0	1	28	37.61	
27	485	565	954	170	4.85	0	1	28	35.88	
28	473	672	1113	142	0	0	0	7	42.1	Sounthararajan and Sivakumar (2013)
29	425	605	1113	142	0	0	0.5	7	39.3	
30	425	605	1113	142	0	0	1	7	41.8	
31	425	605	1113	142	0	0	1.5	7	42.5	
32	378	538	1113	142	0	0	0.5	7	39.9	
33	378	538	1113	142	0	0	1	7	42.1	
34	378	538	1113	142	0	0	1.5	7	44.2	
35	331	471	1113	142	0	0	0.5	7	37.4	
36	331	471	1113	142	0	0	1	7	38.5	
37	331	471	1113	142	0	0	1.5	7	37.8	
38	473	672	1113	142	0	0	0	28	43.6	
39	425	605	1113	142	0	0	0.5	28	40.2	
40	425	605	1113	142	0	0	1	28	42.5	
41	425	605	1113	142	0	0	1.5	28	43.3	
42	378	538	1113	142	0	0	0.5	28	42.1	
43	378	538	1113	142	0	0	1	28	43.9	
44	378	538	1113	142	0	0	1.5	28	45.6	
45	331	471	1113	142	0	0	0.5	28	39.2	

(*Continued*)

TABLE 8.1 (*Continued*)
Detail of Dataset

Sr. No.	Cement (kg/m³)	Fine Aggregate (kg/m³)	Coarse Aggregate (kg/m³)	Water (kg/m³)	SP/HRWR	Fly Ash (kg/m³)	Steel %	Curing Days	CS (N/mm²)	Author Name & Year
46	331	471	1113	142	0	0	1	28	40.3	
47	331	471	1113	142	0	0	1.5	28	39.4	
48	473	672	1113	142	0	0	0	56	44.1	
49	425	605	1113	142	0	0	0.5	56	42.7	
50	425	605	1113	142	0	0	1	56	43.4	
51	425	605	1113	142	0	0	1.5	56	44.8	
52	378	538	1113	142	0	0	0.5	56	43.6	
53	378	538	1113	142	0	0	1	56	47.5	
54	378	538	1113	142	0	0	1.5	56	50.8	
55	331	471	1113	142	0	0	0.5	56	40.7	
56	331	471	1113	142	0	0	1	56	41.9	
57	331	471	1113	142	0	0	1.5	56	42.2	
58	605	483	0	338.6	3.6	726	0	28	65	Al-Gemeel et al., (2018)
59	544	435	0	304.8	3.3	653	0	28	55.5	
60	605	483	0	338.6	3.6	726	0.25	28	60.1	
61	544	435	0	304.8	3.3	653	0.25	28	58.8	
62	605	483	0	338.6	3.6	726	0.5	28	55.4	
63	544	435	0	304.8	3.3	653	0.5	28	52.1	
64	605	483	0	338.6	3.6	726	0.75	28	61.6	
65	544	435	0	304.8	3.3	653	0.75	28	54.3	
66	370	704	1120	144.3	3.8	93	0.5	28	78.04	Tadepalli et al., (2015)
67	370	704	1120	144.3	3.8	93	0.5	28	70.73	
68	370	704	1120	144.3	3.8	93	0.5	28	73.14	
69	370	704	1120	144.3	3.8	93	0.5	28	59.34	

(*Continued*)

TABLE 8.1 (*Continued*)
Detail of Dataset

Sr. No.	Cement (kg/m³)	Fine Aggregate (kg/m³)	Coarse Aggregate (kg/m³)	Water (kg/m³)	SP/HRWR	Fly Ash (kg/m³)	Steel %	Curing Days	CS (N/mm²)	Author Name & Year
70	370	704	1120	144.3	3.8	93	1.5	28	65.96	
71	370	704	1120	144.3	3.8	93	1.5	28	92.18	
72	370	704	1120	144.3	3.8	93	1.5	28	81.56	
73	370	704	1120	144.3	3.8	93	1.5	28	37.12	
74	370	704	1120	144.3	3.8	93	0	28	70.4	
75	340	894	952	204	0	0	0.5	28	40.78	
76	340	894	952	204	0	0	0.5	28	36.64	
77	340	894	952	204	0	0	1.5	28	43.13	
78	340	894	952	204	0	0	1.5	28	41.06	
79	365	977	695	223	0	0	0	7	18.02	Nataraja et al., (2005)
80	365	977	695	223	0	0	0.5	7	24.44	
81	365	977	695	223	0	0	1	7	17.59	
82	365	977	695	223	0	0	1.5	7	23.36	
83	365	977	695	223	0	0	0	28	33.15	
84	365	977	695	223	0	0	0.5	28	30.65	
85	365	977	695	223	0	0	1	28	27.98	
86	365	977	695	223	0	0	1.5	28	28.04	
87	557	822	695	223	0	0	0	7	37.41	
88	557	822	695	223	0	0	0.5	7	39.34	
89	557	822	695	223	0	0	1	7	32.99	
90	557	822	695	223	0	0	1.5	7	43.76	
91	557	822	695	223	0	0	0	28	50.62	
92	557	822	695	223	0	0	0.5	28	50.36	
93	557	822	695	223	0	0	1	28	47.25	

(*Continued*)

TABLE 8.1 (*Continued*)
Detail of Dataset

Sr. No.	Cement (kg/m³)	Fine Aggregate (kg/m³)	Coarse Aggregate (kg/m³)	Water (kg/m³)	SP/HRWR	Fly Ash (kg/m³)	Steel %	Curing Days	CS (N/mm²)	Author Name & Year
94	557	822	695	223	0	0	1.5	28	50.66	
95	250	876	956	168	0	70	0	28	38.3	Neves and Almeida (2005)
96	250	876	956	168	0	70	0.38	28	36.5	
97	250	876	956	168	0	70	0.38	28	40	
98	250	876	956	168	0	70	0.75	28	33.3	
99	250	876	956	168	0	70	0.75	28	40.1	
100	250	876	956	168	0	70	1.13	28	33.7	
101	250	876	956	168	0	70	1.13	28	44.4	
102	250	876	956	168	0	70	1.5	28	30.7	
103	450	682	1036	174	5.71	0	0	28	62.2	
104	450	682	1036	174	5.71	0	0.38	28	62.1	
105	450	682	1036	174	5.71	0	0.38	28	65.3	
106	450	682	1036	174	5.71	0	0.75	28	65	
107	450	682	1036	174	5.71	0	0.75	28	62.8	
108	450	682	1036	174	5.71	0	1.13	28	58.5	
109	450	682	1036	174	5.71	0	1.13	28	65.7	
110	450	682	1036	174	5.71	0	1.5	28	67.9	
111	0	600	1248	14.5	10.2	408	0	28	37	Ganesan et al., (2015)
112	0	600	1248	16	10.2	408	0.25	28	38.4	
113	0	600	1248	16	14.5	408	0.5	28	41.2	
114	0	600	1248	18	14.5	408	0.75	28	42.5	
115	0	600	1248	18	16	408	1	28	43.8	
116	360	598	1266	192	0	0	0	28	35	
117	360	598	1266	192	4	0	0.5	28	39.5	

TABLE 8.2
Features of Dataset

Statistics	Cement (kg/m3)	Fine Aggregate (kg/m³)	Coarse Aggregate (kg/m³)	Water (kg/m³)	SP/HRWR	Fly Ash (kg/m³)	Steel Fibre %	Curing Days	Compressive Strength (N/mm²)
Training Dataset									
Minimum	0.0000	435.0000	0.0000	14.5000	0.0000	0.0000	0.0000	7.0000	11.3200
Maximum	605.0000	977.0000	1266.0000	338.6000	16.0000	726.0000	1.5000	56.0000	92.1800
Mean	375.3659	720.4085	843.2252	186.8585	3.5300	111.8171	0.6589	24.7561	42.4351
Standard deviation	135.0748	169.9693	352.6851	70.3307	4.1278	195.7483	0.5005	12.4539	16.0913
Kurtosis	1.7724	−1.3505	0.1684	0.7704	0.3003	4.0041	−0.9899	1.0673	0.3488
Skewness	−0.8670	−0.0213	−1.0483	−0.1734	1.0012	2.2132	0.2360	0.5182	0.4898
Testing Dataset									
Minimum	250.0000	435.0000	0.0000	142.0000	0.0000	0.0000	0.0000	7.0000	14.7100
Maximum	557.0000	977.0000	1266.0000	304.8000	9.5000	653.0000	1.5000	56.0000	78.0400
Mean	381.4857	716.2571	821.0563	194.3686	2.8863	74.8286	0.7931	25.8000	39.4977
Standard deviation	78.2170	183.0864	360.6145	53.3434	3.6948	155.4018	0.5515	13.9765	15.0914
Kurtosis	−0.0783	−1.5975	−0.4688	−1.0423	−1.1518	10.2118	−1.3095	0.6341	0.2336
Skewness	0.6454	0.0187	−0.8930	0.5658	0.7440	3.1509	−0.1092	0.6736	0.3253

TABLE 8.3
Combination of Various Variable Subset

Sr. No.	Input Combinations	Removed Parameter	Output	Model No.
1	C + FA + CA + W + SP/HRWR + Fly ash + Steel% + Curing Days	-	Compressive strength	M1
2	FA + CA + W + SP/HRWR + Fly ash + Steel% + Curing days	C	Compressive strength	M2
3	C + CA + W + SP/HRWR + Fly ash + Steel% + Curing days	FA	Compressive strength	M3
4	C + FA + W + SP/HRWR + Fly ash + Steel% + Curing days	CA	Compressive strength	M4
5	C + FA + CA + SP/HRWR + Fly ash + Steel% + Curing days	W	Compressive strength	M5
6	C + FA + CA + W + fly ash + Steel% + Curing days	SP/HRWR	Compressive strength	M6
7	C + FA + CA + W + SP/HRWR + Steel% + Curing days	Fly Ash	Compressive strength	M7
8	C + FA + CA + W + SP/HRWR + Fly ash + Curing days	Steel (%)	Compressive strength	M8
9	C + FA + CA + W + SP/HRWR + Fly ash + Steel%	Curing Days	Compressive strength	M9
10	CA + W + SP/HRWR + Fly ash + Steel% + Curing days	C + FA	Compressive strength	M10
11	C + W + SP/HRWR + Fly ash + Steel% + Curing days	FA + CA	Compressive strength	M11
12	C + FA + SP/HRWR + Fly ash + Steel% + Curing days	CA + W	Compressive strength	M12
13	C + FA + CA + Fly ash + Steel% + Curing days	SP/HRWR + W	Compressive strength	M13
14	C + FA + CA + W + Steel% + Curing days	SP/HRWR + Fly ash	Compressive strength	M14
15	C + FA + CA + W + SP/HRWR + Curing days	Fly ash + Steel%	Compressive strength	M15
16	C + FA + CA + W + SP/HRWR + Fly ash	Steel% + Curing days	Compressive strength	M16
17	C + CA + SP/HRWR + Fly ash + Steel% + Curing days	FA + W	Compressive strength	M17
18	FA + W + SP/HRWR + Fly ash + Steel% + Curing days	C + CA	Compressive strength	M18
19	C + FA + CA + SP/HRWR + Steel% + Curing days	W + Fly ash	Compressive strength	M19
20	C + FA + CA + W + Fly ash + Curing days	SP/HRWR + Steel%	Compressive strength	M20

8.4 RESULT ANALYSIS

8.4.1 Assessment of ANN-Based Model

The performance of 20 models with variations in input parameters was evaluated by applying statistic measures on training and testing datasets such as CC, RMSE, and MAE. These measures help to compare the accuracy of the predicted results. Performances of models are listed in Table 8.4. The result shows that the M7 model predicts better results as compared to other models with CC (0.9788), RMSE (3.9719), and MAE (2.9744). Figure 8.3 shows the graph between observed and predicted compressive strength using the given models. Scatter graphs show the observed and predicted compressive strength of the best model by applying the ANN technique in training and testing datasets. It also shows that most of the predicted values of the novel model lie under the range of ±25% error band from the perfect agreement.

8.4.2 Assessment of ANN–Cross-Validation (Ten-Fold)-Based Model

To check the performance of the given models, statistic measures were applied on the datasets like CC, RMSE, and MAE. Performances of models are listed in Table 8.5. The result shows that the M15 model predicts better results as compared to other models

TABLE 8.4
Statistic Quantitative Results of ANN

	ANN					
	Training			Testing		
Model No.	CC	MAE	RMSE	CC	MAE	RMSE
M1	0.9387	3.5249	5.6711	0.9715	3.4234	4.456
M2	0.9366	3.7099	5.7438	0.969	3.4526	4.4046
M3	0.9375	3.6163	5.7732	0.9541	3.9426	5.0251
M4	0.9356	3.4932	5.6877	0.966	3.636	4.497
M5	0.9394	3.5294	5.6702	0.9634	3.387	4.5631
M6	0.9396	3.4581	5.6575	0.9606	3.4377	4.5891
M7	0.9402	3.3036	5.5591	0.9788	2.9744	3.9719
M8	0.9367	3.5323	5.7336	0.9782	3.6178	4.3559
M9	0.9241	3.928	6.1298	0.9515	3.9991	5.0842
M10	0.9107	5.1751	7.2742	0.9007	5.8702	7.4624
M11	0.929	4.1397	6.2016	0.9573	4.4807	5.5174
M12	0.9378	3.415	5.6115	0.9583	3.9151	4.7997
M13	0.9364	3.3779	5.6145	0.9551	3.5267	4.4376
M14	0.9365	3.4058	5.6328	0.9565	3.2572	4.4896
M15	0.9391	3.3059	5.6497	0.959	3.5725	4.6614
M16	0.9281	3.8181	6.0047	0.957	4.1791	5.1015
M17	0.9346	4.2787	6.2784	0.955	4.5906	5.834
M18	0.9382	3.3383	5.5691	0.9782	3.3459	4.2673
M19	0.9389	4.1121	6.1788	0.942	4.9245	6.2334
M20	0.9355	3.4862	5.7622	0.969	3.4715	4.2678

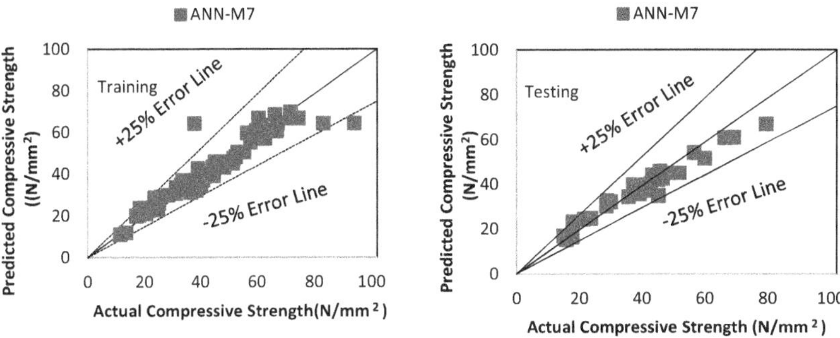

FIGURE 8.3 Scatter graphs of observed and predicted compressive strength of the best model in the training and testing by ANN technique.

TABLE 8.5
Statistic Quantitative Results of Cross-Validation

	ANN–Cross-Validation (Ten-Fold)		
	Training & Testing		
Model No.	**CC**	**MAE**	**RMSE**
M1	0.872	4.7811	8.1018
M2	0.8618	5.0713	8.3795
M3	0.8473	5.6948	8.949
M4	0.8449	5.4094	8.9861
M5	0.8578	5.4466	8.5514
M6	0.8519	5.5852	8.8576
M7	0.8668	5.2835	8.4136
M8	0.9031	4.2083	6.8714
M9	0.8503	5.7073	8.6637
M10	0.7491	8.0894	11.6736
M11	0.8471	5.5659	8.7158
M12	0.862	5.5404	8.6257
M13	0.8557	5.4611	8.5863
M14	0.858	5.3663	8.5127
M15	0.9059	4.2074	6.7505
M16	0.8816	5.2197	7.5117
M17	0.8426	5.8083	8.9416
M18	0.8673	5.1707	8.3475
M19	0.8614	5.6603	8.6715
M20	0.899	4.3961	7.0499

with CC (0.9059), RMSE (6.7505), and MAE (4.2074). Figure 8.4 shows the graph between observed and predicted compressive strength using cross-validation to evaluate the dataset. Scatter graphs show the observed and predicted compressive strength of the best model (M15) using cross-validation by applying the ANN technique. It also shows that most of the predicted values of the novel model lie under the range of ±25% error band from the perfect agreement.

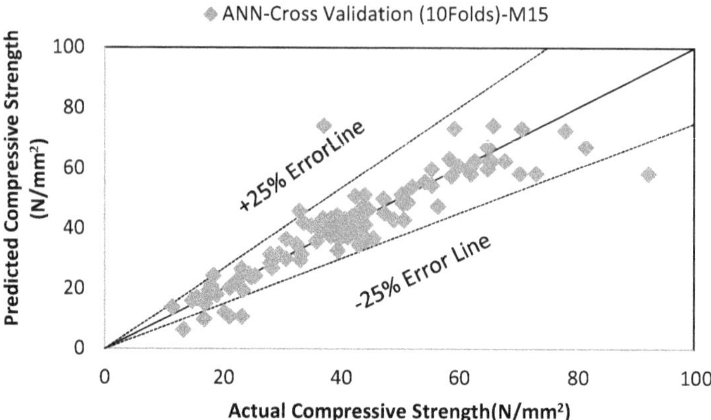

FIGURE 8.4 Scatter graphs of observed and predicted compressive strength of the best model in the training and testing by cross-validation ANN technique.

8.4.3 Assessment of LR-Based Model

LR-based models predict the relationship between two variables or factors. Table 8.6 shows the equations generated by the LR for every model, whereas Table 8.7 shows that model 5 (M5) predicts better results for the compressive strength for the testing dataset. The scatter plot (Figure 8.5) shows the graphical representation of observed versus predicted compressive strength. Table 8.7 listed the performance parameters with CC being 0.8391, MAE being 6.4804, and RMSE being 8.1961 for the testing dataset. It also shows that most of the predicted values of the novel model lie under the range of ±25% error band.

8.4.4 Comparison among Best Developed Models

In this study, various soft computing techniques were used to estimate concrete compressive strength. Variations in input parameters were done in different models to check the performance of the individual model. Comparison of results was done by applying statistics measures, as listed in Table 8.8. An agreement plot was also prepared between actual and the predicted results for the best models obtained from ANN, ANN–cross-validation, and LR algorithms (Figure 8.6). Relative error was minimum in ANN-M7 for the testing subset (Figure 8.6). Statistical analysis showed that ANN-M7 predicts the highest CC value (0.9788) and minimum RMSE and MAE values (3.9719 and 2.9744), followed by ANN–cross-validation M15 where CC, RMSE, and MAE values are 0.9059, 6.7505, and 4.2074, respectively.

For the graphical representation of the compressive strength of a concrete mix containing steel fibre boxplot is made. Table 8.9 predicts quartile values of 25%, 50%, and 75% in addition to the actual amount. The inter quartile range (IQR) of ANN-M7 is closer to that of actual data, as shown in Figure 8.7 and Table 8.9.

TABLE 8.6
Equation of LR Models

Model No.	LR
M1	CS (N/mm^2) = 0.1188 * Cement (kg/m^3) + 0.0158 * Fine aggregate (kg/m^3) – 0.2103 * Water (kg/m^3) + 0.4542 * SP/HRWR + 0.039 * Fly ash (kg/m^3) + 3.7009 * Steel % + 0.277 * Curing days + 10.4919
M2	CS (N/mm^2) = –0.0392 * Fine aggregate (kg/m^3) + 4.4227 * Steel % + 0.2681 * Curing days + 61.1572
M3	CS (N/mm^2) = 0.0875 * Cement (kg/m^3) + 0.0615 * Coarse aggregate (kg/m^3) + 0.0965 * Water (kg/m^3) + 0.7509 * SP/HRWR + 0.0616 * Fly ash (kg/m^3) + 3.4792 * Steel % + 0.162 * Curing days – 76.1472
M4	CS (N/mm^2) = 0.1188 * Cement (kg/m^3) + 0.0158 * Fine aggregate (kg/m^3) – 0.2103 * Water (kg/m^3) + 0.4542 * SP/HRWR + 0.039 * Fly ash (kg/m^3) + 3.7009 * Steel % + 0.277 * Curing days + 10.4919
M5	CS (N/mm^2) = 0.1191 * Cement (kg/m^3) + 0.0312 * Fine aggregate (kg/m^3) + 0.0571 * Coarse aggregate (kg/m^3) + 0.6431 * SP/HRWR + 0.0779 * Fly ash (kg/m^3) + 3.8143 * Steel % + 0.2081 * Curing days –91.5239
M6	CS (N/mm^2) = 0.1013 * Cement (kg/m^3) + 0.0304 * Fine aggregate (kg/m^3) + 0.0658 * Coarse aggregate (kg/m^3) + 0.0537 * Water (kg/m^3) + 0.0861 * Fly ash (kg/m^3) + 3.3824 * Steel % + 0.1818 * Curing days –99.3141
M7	CS (N/mm^2) = 0.0846 * Cement (kg/m^3) + –0.0199 * Fine aggregate (kg/m^3) + 0.0129 * Coarse aggregate (kg/m^3) –0.0536 * Water (kg/m^3) + 1.1004 * SP/HRWR + 0.2792 * Curing days + 13.3575
M8	CS (N/mm^2) =0.1115 * Cement (kg/m^3) + 0.0272 * Fine aggregate (kg/m^3) + 0.0681 * Coarse aggregate (kg/m^3) + 0.0589 * Water (kg/m^3) + 0.6001 * SP/HRWR + 0.0801 * Fly ash (kg/m^3) + 0.2002 * Curing days –103.5341
M9	CS (N/mm^2) =0.1058 * Cement (kg/m^3) –0.2009 * Water (kg/m^3) + 0.036 * Fly ash (kg/m^3) + 36.2307
M10	CS (N/mm^2) = –0.0492 * Water (kg/m^3) – 0.8844 * SP/HRWR + 0.0284 * Fly ash (kg/m^3) + 5.5936 * Steel % + 0.3387 * Curing days + 39.5097
M11	CS (N/mm^2) = 0.0998 * Cement (kg/m^3) – 0.1805 * Water (kg/m^3) + 0.0332 * Fly ash (kg/m^3) + 0.0332 * Fly ash (kg/m^3) + 0.2356 * Curing days + 29.1613
M12	CS (N/mm^2) = 0.0333 * Cement (kg/m^3) – 0.0361 * Fine aggregate (kg/m^3) + 0.3144 * Curing days + 48.1452
M13	CS (N/mm^2) = 0.1089 * Cement (kg/m^3) + 0.0322 * Fine aggregate (kg/m^3) + 0.0565 * Coarse aggregate (kg/m^3) + 0.0829 * Fly ash (kg/m^3) + 3.2708 * Steel % + 0.1892 * Curing days –85.3515
M14	CS (N/mm^2) = 0.0684 * Cement (kg/m^3) – 0.0208 * Fine aggregate (kg/m^3) – 0.0992 * Water (kg/m^3) + 0.2591 * Curing days + 43.8698
M15	CS (N/mm^2) = 0.0846 * Cement (kg/m^3) – 0.0199 * Fine aggregate (kg/m^3) + 0.0129 * Coarse aggregate (kg/m^3) –0.0536 * Water (kg/m^3) + 1.1004 * SP/HRWR + 0.2792 * Curing days + 13.3575
M16	CS (N/mm^2) = 0.1058 * Cement (kg/m^3) – 0.2009 * Water (kg/m^3) + 0.036 * Fly ash (kg/m^3) + 36.2307
M17	CS (N/mm^2) =0.0992 * Cement (kg/m^3) + 0.0444 * Coarse aggregate (kg/m^3) + 0.6282 * SP/HRWR + 0.0538 * Fly ash (kg/m^3) + 0.1705 * Curing days –44.7239

(Continued)

TABLE 8.6 (*Continued*)
Equation of LR Models

Model No.	LR
M18	CS (N/mm²) = −0.0392 * Fine aggregate (kg/m³) + 4.4227 * Steel % + 0.2681 * Curing days + 61.1572
M19	CS (N/mm²) =0.0759 * Cement (kg/m³) − 0.0239 * Fine aggregate (kg/m³) + 0.0209 * Coarse aggregate (kg/m³) + 1.1541 * SP/HRWR + 0.2754 * Curing Days + 2.7046
M20	CS (N/mm²) = 0.1033 * Cement (kg/m³) + 0.0288 * Fine aggregate (kg/m³) + 0.0657 * Coarse aggregate (kg/m³) + 0.0488 * Water (kg/m³) + 0.0843 * Fly ash (kg/m³) + 0.1837 * Curing days −95.5426

TABLE 8.7
Statistic Quantitative Results of LR

	LR					
	Training			Testing		
Model No.	CC	MAE	RMSE	CC	MAE	RMSE
M1	0.7679	7.7382	10.2437	0.7837	7.4329	9.3442
M2	0.5686	10.4941	13.1561	0.6322	10.0327	12.1957
M3	0.8318	6.324	8.8778	0.822	6.5942	8.661
M4	0.7679	7.7382	10.2437	0.7837	7.4329	9.3442
M5	0.8467	6.2406	8.5094	0.8391	6.4804	8.1961
M6	0.8402	6.3352	8.673	0.8201	7.0502	8.763
M7	0.7127	8.5961	11.2181	0.735	8.7135	10.5226
M8	0.8419	6.0946	8.6291	0.8336	6.5241	8.3761
M9	0.7311	8.1626	10.9116	0.7711	7.3225	9.5317
M10	0.5166	10.815	13.6932	0.5922	10.0256	12.4062
M11	0.7507	7.6223	10.5652	0.7855	6.9791	9.3427
M12	0.6169	9.4194	12.587	0.6836	9.1693	11.4445
M13	0.8381	6.3749	8.7253	0.8263	6.888	8.5794
M14	0.682	8.9651	11.7226	0.7264	8.7866	10.7266
M15	0.7127	8.5961	11.2181	0.7305	8.7135	10.5226
M16	0.7311	8.1626	10.9116	0.7711	7.3225	9.5317
M17	0.8195	6.5595	9.1652	0.8248	6.3965	8.5525
M18	0.5686	10.4941	13.1561	0.6322	10.0327	12.1957
M19	0.71	8.7735	11.2615	0.7281	8.8049	10.5909
M20	0.8343	6.2968	8.8162	0.8208	6.907	8.7523

The results from Table 8.10 indicate that cement and fine aggregate has a significant effect of an ANN-based model to estimate the compressive strength of a concrete mix in comparison to other input parameters for this dataset.

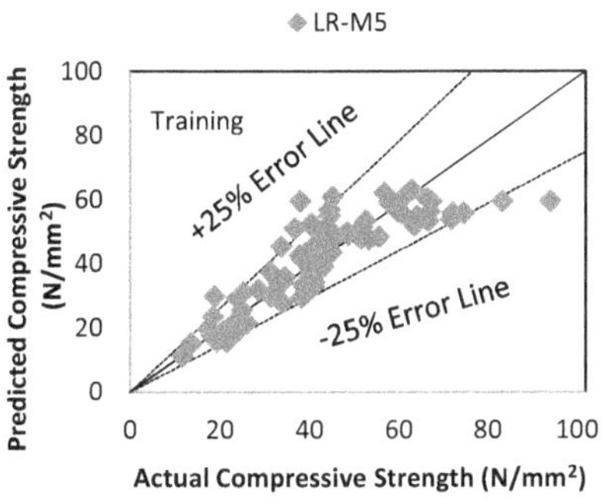

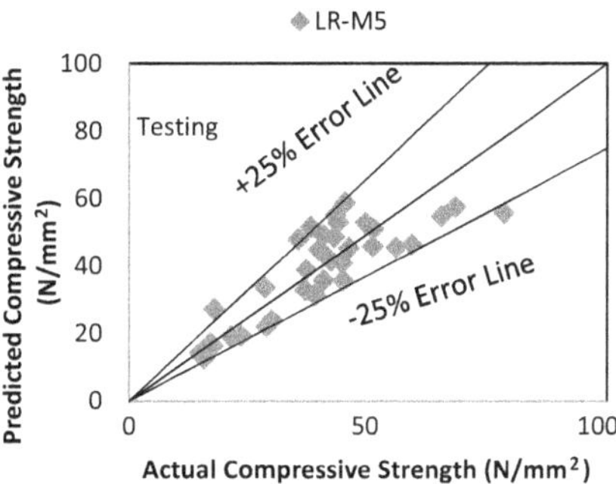

FIGURE 8.5 Scatter graphs of observed and predicted compressive strength of the best model in the training and testing by LR technique.

TABLE 8.8
Comparison of Model Prediction

Models	CC	MAE (MPa)	RMSE (MPa)	Order
ANN-M7	0.9788	2.9744	3.9719	1
ANN–cross-validation Ten-Fold M15	0.9059	4.2074	6.7505	2
LR-M5	0.8391	6.4804	8.1961	3

8.5 CONCLUSION

This paper consists of 20 models based on variations in input parameters. Soft computing techniques such as ANN, ANN–cross-validation (ten-fold), and LR were applied on every model for the prediction of concrete compressive strength. Evaluation of models was checked by applying performance measures like CC, RMSE, and MAE. The results acquired in this examination are summed up as follows:

1. The results obtained from ANN-M7 were found to be most suitable to predict the concrete compressive strength. Cement, fine aggregate, coarse aggregate, water, SP/HRWR, and curing days were considered as input parameters in ANN-M7.
2. Among all three algorithms, ANN-M7 predicts better results followed by ANN–cross-validation (ten-fold) with the highest CC values (0.9788 and 0.9059), lower MAE values (2.9744 and 4.2074), and lower RMSE values (3.9719 and 6.7505).
3. The scatter diagram shows that the ANN-M7 has minimum error bandwidth, and it is a good fit for predicting the output.
4. The box plot likewise confirms that the performance of the ANN-M7-based model is superior to other applied models for a given dataset.
5. The results of the study also conclude that cement and fine aggregates have a major impact on the prediction of compressive strength concrete with the ANN model in comparison to other input variables for this dataset.

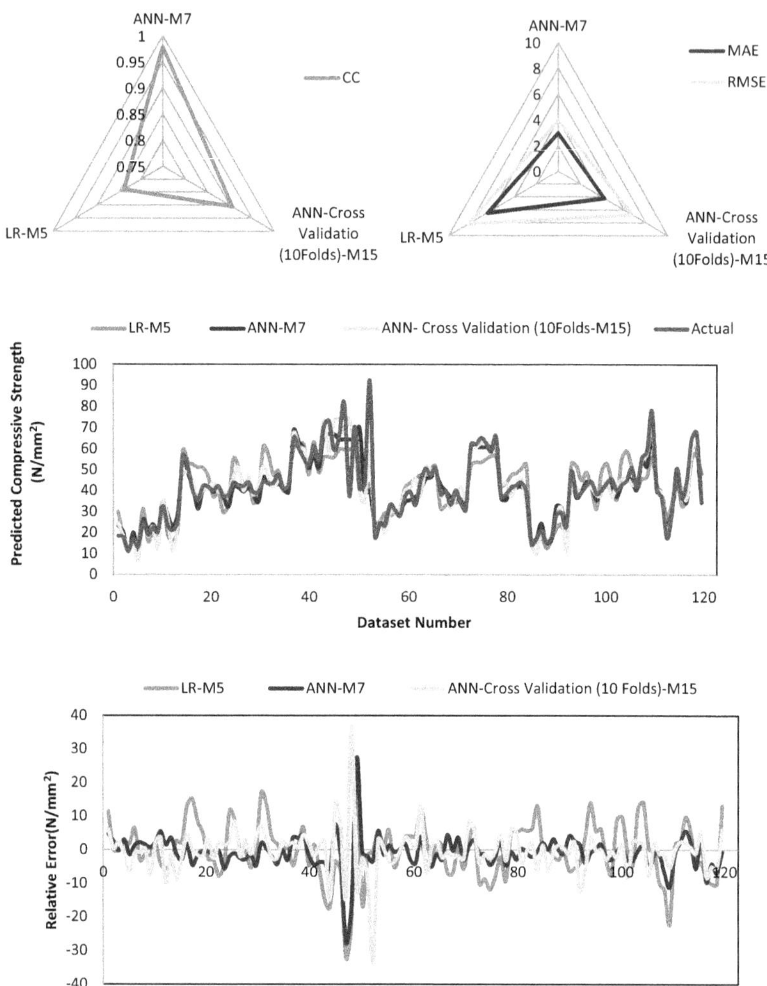

FIGURE 8.6 Comparison between predicted and actual values of LR-M5, ANN-M7, and ANN–cross-validation ten-fold M15.

8.6 THE INTEREST OF CONFLICT STATEMENT

There is no interest in conflict with anybody who is related to this research article by the authors.

TABLE 8.9
Statistics of Actual and Predicted Values of Testing Dataset Using Various Soft Computing Techniques

Statistic	LR-M5	ANN-M7	ANN–Cross-Validation (Ten-Fold)	Actual
Minimum	12.7	15.396	9.929	14.71
Maximum	58.878	66.904	72.578	78.04
First Quartile	29.3825	32.3365	30.1635	28.91
Mean	39.20183	38.30451	38.33469	39.49771
Third Quartile	50.2145	44.6	44.5495	45.2
IQR	20.832	12.2635	14.386	16.29
Median	42.75	39.728	41.043	40.3

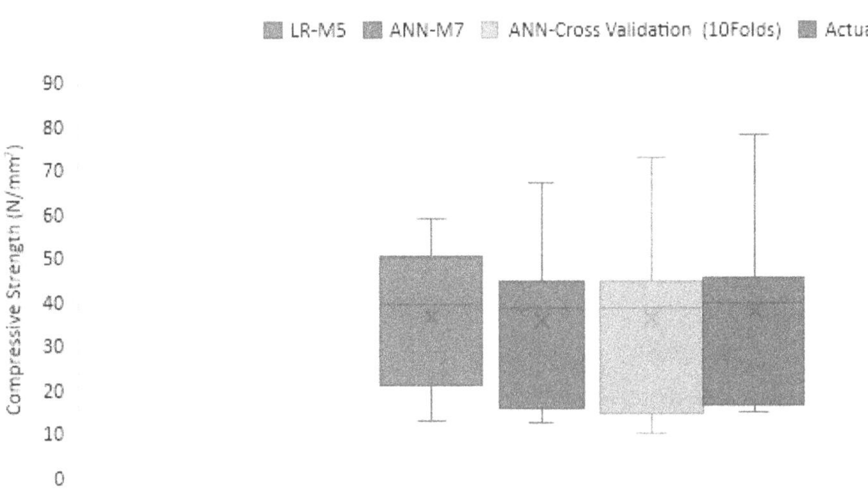

FIGURE 8.7 Box plot diagram for all applicable models using testing dataset.

TABLE 8.10
Sensitivity Analysis Results with an ANN-Based Model

Input Combination								Output		ANN-Based Model		
Cement (kg/m³)	Fine Aggregate (kg/m³)	Coarse Aggregate (kg/m³)	Water (kg/m³)	SP/HRWR	Fly Ash (kg/m³)	Steel %	Curing Days	CS (N/mm²)	Removed Parameter	CC	MAE	RMSE
									-	0.9715	3.4234	4.456
									C	0.969	3.4526	4.4046
									FA	0.9541	3.9426	5.0251
									CA	0.966	3.636	4.497
									W	0.9634	3.387	4.5631
									SP/HRWR	0.9606	3.4377	4.5891
									Fly ash	0.9788	2.9744	3.9719
									Steel (%)	0.9782	3.6178	4.3559
									Curing days	0.9515	3.9991	5.0842
									C + FA	0.9007	5.8702	7.4624
									FA + CA	0.9573	4.4807	5.5174
									CA + W	0.9583	3.9151	4.7997
									SP/HRWR + W	0.9551	3.5267	4.4376
									SP/HRWR + Fly ash	0.9565	3.2572	4.4896
									Fly ash + Steel%	0.959	3.5725	4.6614
									Steel% + Curing days	0.957	4.1791	5.1015
									FA + W	0.955	4.5906	5.834
									C + CA	0.9782	3.3459	4.2673
									W + Fly ash	0.942	4.9245	6.2334
									SP/HRWR + Steel%	0.969	3.4715	4.2678

REFERENCES

Al-Gemeel A.N., Zhuge Y., and Youssf O., Use of hollow glass microspheres and hybrid fibres to improve the mechanical properties of engineered cementitious composite. *Construction and Building Materials.* 2018; 171(2018): 858–870. Doi: 10.1016/j.conbuildmat.2018.03.172.

Chopra P., Sharma R.K., Kumar M., and Chopra T. Comparison of machine learning techniques for the prediction of compressive strength of concrete. *Advances in Civil Engineering.* 2018. Doi: 10.1155/2018/5481705.2.

Deepa C., SathiyaKumari K., and Pream Sudha V. Prediction of the compressive strength of high performance concrete mix using tree based modeling. *International Journal of Computer Application.* 2010; 6(5): 18–24.

Ganesan N., Abraham R., and Raj S.D. Durability characteristics of steel fibre reinforced geopolymer concrete. *Construction and Building Materials.* 2015; 93(2015): 471–476. Doi: 10.1016/j.conbuildmat.2015.06.014.

Goldberg Y. Neural network methods for natural language processing. *Synthesis Lectures on Human Language Technologies.* 2017; 10(1): 1–309. Doi: 10.2200/S00762ED1V01Y201703HLT037.

Guo Y.C., Zhang J.H., Chen G.M., and Xie Z.H. Compressive behavior of concrete structures incorporating recycled concrete aggregates, rubber crumb and reinforced with steel fibre, subjected to elevated temperatures. *Journal of Cleaner Production.* 2014; 72(2014): 193–203. Doi: 10.1016/j.jclepro.2014.02.036.

Haddadou N., Chaid R., and Ghernouti Y. Experimental study on steel fibre reinforced self-compacting concrete incorporating high volume of marble powder. *European Journal of Environmental and Civil Engineering.* 2015; 19(1): 48–64. Doi:10.1080/19648189.2014.929537.

Han Q., Gui C., Xu J., and Lacidogna G. A generalized method to predict the compressive strength of high-performance concrete by improved random forest algorithm. *Construction and Building Materials.* 2019; 226(0): 734–742. Doi: 10.1016/j.conbuildmat.2019.07.315.

Laghari R.A., Li J., Laghari A.A., and Wang S. A review on application of soft computing techniques in machining of particle reinforcement metal matrix composites. *Archives of Computational Methods in Engineering.* 2019. Doi: 10.1007/s11831-019-09340-0

Lau A. and Anson M. Effect of high temperatures on high performance steel fibre reinforced concrete. *Cement and Concrete Research.* 2006; 36 (2006): 1698–1707. Doi: 10.1016/j.cemconres.2006.03.024.

Nalanth N., Venkatesan P.V., and Ravikumar M.S. Evaluation of the fresh and hardened properties of steel fibre reinforced self-compacting concrete using recycled aggregates as a replacement material. *Advances in Civil Engineering.* 2014. Doi: 10.1155/2014/671547.

Nataraja M.C., Nagaraj T.S., and Basavaraja S.B., Reproportioning of steel fibre reinforced concrete mixes and their impact resistance. *Cement and Concrete Research.* 2005; 35(12): 2350–2359. Doi: 10.1016/j.cemconres.2005.06.011.

Neves R.D. and Almeida O.C.O.F.D. Compressive behaviour of steel fibre reinforced concrete. *Structural Concrete.* 2005; 6(1): 1–8. DOI: 10.1680/stco.2005.6.1.1.

Nhu V.H., Shahabi H., Nohani E., Shirzadi A., Ansari N.A., Bahrami S., Miraki S., Geertsema M., and Nguyen H. Daily water level prediction of zrebar lake(Iran): A Comparison between M5P, Random Forest, Random Tree and Reduced Error Pruning Trees Algorithms. *International Journal of Geo-Information.* 2020; 9: 479. doi:10.3390/ijgi9080479.

Sepahvand A., Singh B., Sihag P., Samani A.N., Ahmadi H., and Nia S.F. Assessment of the various soft computing techniques to predict sodium absorption ratio (SAR). *ISH Journal of Hydraulic Engineering.* 2019. Doi: 10.1080/09715010.2019.1595185.

Sharma N., Thakur M.S., Goel P.L., and Sihag P. A review: Sustainable compressive strength properties of concrete mix with replacement by marble powder. *Journal of Achievements in Materials and Manufacturing Engineering*. 2020; 98(1): 11–23. Doi: 10.5604/01.3001.0014.0813.

Siddique R., Singh M., and Jain M. Recycling copper slag in steel fibre concrete for sustainable construction. *Journal of Cleaner Production*. 2020. Doi: 10.1016/j.jclepro.2020.122559.

Singh B., Sihag P., Tomar A., and Sehgad A. Estimation of compressive strength of high-strength concrete by random forest and M5P model tree approaches. *Journal of Materials and Engineering Structures*. 2019; 6(2019): 583–592.

Sobhani J., Najimi M., Pourkhorshidi A.R., Parhizkar T. Prediction of the compressive strength of no-slump concrete: A comparative study of regression, neural network and ANFIS models. *Construction and Building Materials*. 2010; 24(2010): 709–718. Doi: 10.1016/j.conbuildmat.2009.10.037.

Sounthararajan V.M. and Sivakumar A. Accelerated properties of steel fibre reinforced concrete containing finer sand. *ARPN Journal of Engineering and Applied Sciences*. 2013; 8(1): 57–63.

Tadepalli P.R., Mo Y.L. and Hsu T.T.C. Mechanical property of steel fibre concrete. *Magazine of Concrete Research*. 2015; 65(8). Doi: 10.1680/macr.12.00077.

Thakur M.S., Pandhiani S.M., Kashyap V., Upadhya A. and Sihag P. Predicting Bond Strength of FRP Bars in Concrete Using Soft Computing Techniques. *Arabian Journal of Science and Engineering*. 2021. Doi: 10.1007/s13369-020-05314-8.

Vakharia V., and Gujar R., Prediction of compressive strength and portland cement composition using cross-validation and feature ranking techniques, *Construction and Building Materials*. 2019; 225(2019): 292–301. Doi: 10.1016/j.conbuildmat.2019.07.224.

Zhang J., Wang Q., and Wang Z. Properties of polyvinyl alcohol-steel hybrid fiber-reinforced composite with high-strength cement matrix. *Journal of Materials in Civil Engineering*. 2017; 29(7). Doi: 10.1061/(ASCE)MT.1943–5533.0001868.

Zongjin L. *Advanced Concrete Technology*. John Wiley and Sons 2011, Hoboken, New Jersey.

9 Using Regression Model to Estimate the Splitting Tensile Strength for the Concrete with Basalt Fiber Reinforced Concrete

Fadi Hamzeh Almohammed, Ahmad Alyaseen, and Ankita Upadhya
Shoolini University

CONTENTS

9.1 INTRODUCTION

One of the most commonly used construction materials is concrete. Concrete has various advantages over other common construction materials, such as durability, formability, and required mechanical strength, but it also has a few limitations, such

DOI: 10.1201/9781003184331-9

as low tensile strength and strain capacity (Tassew and Lubell, 2014; Shaikh, 2013; Jiang et al., 2014).

Because of its ease of preparation, low cost, and durability, concrete is the favored building material for engineering structures. Concrete, on the other hand, has flaws such as low tensile strength and cracking susceptibility. The impermeability of the structure is lowered as a result of fracture growth, and the structure's durability is compromised (Tassew and Lubell, 2014). Many studies have demonstrated that adding fibers to concrete can significantly improve the mechanical qualities and durability of the concrete by overcoming internal faults. As a result, fiber concrete is a type of engineering material that has a wide range of applications (Afroughsabet et al., 2016).

Fiber-reinforced concrete's qualities are determined by the type, quantity, and size of fibers used, as well as how the fibers are combined. The performance of a single aspect of the concrete can be increased by using a single type of fiber, whereas the performance of the concrete can be improved in numerous ways by adding fibers with diverse qualities and sizes to the concrete.

Different data-driven models have been successfully implemented in a variety of fields in recent years, including fiber-reinforced polymer (FRP) bond strength (Koroglu, 2018; Bashir and Ashour, 2012; Coelho et al., 2016; Bolandi et al., 2019), concrete compressive strength (Kumar et al., 2019; Singh et al., 2019b; Aggarwal et al., 2019), hydrology (Kwin et al., 2016; Zaji et al., 2018), hydraulic conductivity (Sihag et al., 2019a; Singh et al., 2019b), soil temperature (Kisi et al., 2016; Sihag et al., 2019a), water quality (Heddam, 2014; Sepahvand et al., 2019) and air quality (Mehdipour et al., 2018), and other civil engineering-related problems (Golafshani et al., 2012).

Basalt fiber (BF) is a type of fiber manufactured from basalt. It possesses a high modulus of elasticity, interfacial shear strength, adequate heat resistance, acceptable chemical resistance, and a less expensive manufacturing procedure when compared to other fibers (Mohammadyan-Yasouj and Ghaderi, 2020). Basalt fiber has been examined for its mechanical qualities and endurance in cementitious composites in the form of a single type, bundled mesh, or composite fibers (Mohammadyan-Yasouj and Ghaderi, 2020).

Basalt fiber beats other fibers in terms of tensile and flexural properties resistance, despite its low compressive strength. The majority of tensile properties research employs split tensile testing. There is a shortage of research on the direct tensile test due to the test's stringent requirements.

One of the problems with traditional control systems is that complex plants are difficult to regulate using traditional methods since they cannot be accurately represented by mathematical models. Soft computing, on the other hand, tackles difficult issues by using partial truth, ambiguity, and approximation (Ibrahim, 2016). "Soft computing's guiding notion is to make use of tolerance for imprecision, uncertainty, and imprecise information in order to obtain predictability, robustness, lower costs, and a stronger relationship with reality," says Dr. Zadeh, a pioneer of fuzzy logic (Ibrahim, 2016). Soft computing has become popular and has attracted academic attention from people of many backgrounds due to features such as nonlinear programming, optimization, intelligent control, and decision-making help (Jang et al., 1997).

There are numerous approaches for modeling material properties, including statistical techniques, computational modeling, and more recently, regression analysis and artificial neural networks (ANN) (Mohammed et al., 2020). Soft computing approaches are equally appealing to civil engineering teachers. Adaptive neuro-fuzzy inference systems include the neuro-fuzzy inference system (ANFIS) and genetic programming (Jalal, 2015). Predicting compressive strength is significant in concrete construction since it gives a notion of project scheduling, quality control, and concrete form removal time. Several regression-based algorithms for estimating concrete strength have been presented (Deepa et al., 2010).

In civil engineering problems in general and structural engineering in particular, techniques such as linear regression analysis, M5P modal trees, and MLP (multilayer perceptron) are applied (Deepa et al., 2010).

This book chapter is trying to implement the Gaussian process (GP) model, support vector machine (SVM) model, and linear regression (LR) to estimate the splitting tensile strength for the concrete with basalt fiber reinforced concrete (BFRC).

9.2 SOFT COMPUTING TECHNIQUE REVIEW

9.2.1 Gaussian Process Regression

GP is a typical Gaussian distribution concept with mean and covariance represented as a vector and matrix, respectively (Sihag et al., 2017; Thakur et al., 2021). A regression approach known as GP regression. It is a fairly simple technique with excellent target prediction ability (Thakur et al., 2021).

The GP regression model is based on Rasmussen's (2006) notion that adjoining observations should express information about each other; it is a means of describing a prior directly over function space. The Gaussian distribution's mean and covariance are vector and matrix, respectively, but the Gaussian process is an over function. The GP regression model can recognize the predicted distribution that is similar to the test input (Sihag et al., 2017).

A GP is a set of random variables with a joint multivariate Gaussian distribution for any finite number of them. Let x and y represent the input and output domains, respectively, from which n pairs(xi,yi) are dispersed independently and identically. Let $y \subseteq R$ be the regression coefficient; then, a GP on is defined by a mean function: R and a covariance function $k \rightarrow R$. Kuss (2006) is a good resource for more information on GP regression and other covariance functions.

9.2.2 Support Vector Machines (SVM)

SVM is derived from statistical learning theory and relies on optimal class division in the case of classification problems, according to Vapnik (1995). SVM picks a classifier with the least generalization error from an infinite number of linear classifiers or sets an upper limit on the error derived via structural risk minimization for two-class classification problems. Thus, the selected hyperplane and the addition of distances of the hyperplane from the nearest point of two classes might be used to achieve the largest margin between the two classes.

By offering an option-insensitive loss function, Cortes and Vapnik (1995) created support vector regression (SVR). This loss function allows the idea of edge to be used for regression problems, where the margin is defined as the sum of the hyperplane separations from the two classes' nearest point. The goal of the SVR is to find a function that has the least amount of divergence from the real target vectors for all available information and is as level as possible (Smola 1996). For non-linear support vector regression, Vapnik (1995) proposed the kernel function. Readers are directed to Vapnik and Vapnik (1998) for further information on support vector regression.

9.2.3 Multiple Linear Regression (MLR)

Using observed data, multiple linear regression is utilized to discover the link between input and output variables. For nonlinear and difficult issues, MLR is frequently used (Thakur et al., 2021). MLRs are a multivariate statistical approach that fits a linear equation to observed data to represent the linear correlations between the dependent variable y and two or more independent variables. Each response of independent variables is related to the value of the dependent variable y. The equation for y's regression can be written as follows:

$$y = m_1x_1 + m_2x_2 + m_3x_3 + \cdots + m_nx_n + C \tag{9.1}$$

where y is the dependent variable, x_1, x_2... x_n are the independent variables, m_1, m_2... m_n are the regression coefficients, and C is constant.

MLR models were the usual method for estimating responses between a dependent variable and various independent factors where the dependent variable and independent variables had a linear connection (Singh et al., 2019).

9.2.4 Performance Evaluation Indices

9.2.4.1 Correlation Coefficient (CC)

The correlation reveals the strength of a linear relationship between two variables. CC is expressed by equation (9.3) ($O1$, $P1$), ($O2$, $P2$),... (On, Pn) given a set of data ($O1$, $P1$).

The correlation coefficient is always in the range of −1 to 1 with 1 and −1 indicating total correlation (in this case, all points are on a straight line). Positive correlation denotes a positive relationship between variables (a rise in one variable is accompanied by a rise in another), whereas negative correlation denotes a negative relationship between variables (the increase in value is that one variable resembles the decrease in another variable). Small worth (changeable). A correlation value near 0 indicates that the variables are unrelated.

Because CC is calculated using a method that standardizes the variables, changes in the scale or measuring unit have no effect on its value. As a result, the correlation coefficient is typically more helpful than a graphical description when examining the intensity of the link between two variables.

$$\text{Correlation coefficient CC} = \frac{\sum_{i=1}^{N}(O_i - \bar{O})(P_i - \bar{P})}{\sqrt{\sum_{i=1}^{N}(O_i - \bar{O})^2}\sqrt{\sum_{i=1}^{N}(P_i - \bar{P})^2}} \tag{9.2}$$

9.2.4.2 Root Mean Square Error (RMSE)

The root mean square deviation or root mean square error (RMSE) is a widely used statistical measure metric for comparing a model or estimate's expected and actual values (sample or population value). The root mean square deviation is the square root of the root mean square average.

$$\text{RMSE} = \sqrt{\frac{1}{N}\sum_{i=1}^{N}(P_i - O_i)^2} \tag{9.3}$$

9.2.4.3 Mean Absolute Error

The mean absolute error (MAE) is a statistic that compares the errors of two observations that describe the same phenomena. Equation (9.4) represents the MAE calculation formula.

$$\text{MAE} = \frac{1}{N}\sum_{i=1}^{N}|P_i - O_i| \tag{9.4}$$

9.3 DATA SET

Table 9.1 shows the total data set collected from lateral review. The data collection consists of 78 observations gathered from prior research studies. Out of 78 observations, 54 were chosen at random as the training data set and 24 were chosen for model validation and testing.

Table 9.2 contains a description of the feature of all overall data.

9.4 MATERIAL METHODOLOGY

In this work, data for compressive strength of concrete with basalt fiber-reinforced polymer (BFRP) is collected, and soft computing approaches are used to predict compressive strength, with the goal of determining which technique is the best. Cement, fine aggregate or crushed sand, coarse aggregate, water, superplasticizer, fly ash, basalt fiber reinforced polymer, fiber diameter, fiber length, curing time, and tensile strength are among the ten inputs that were collected. One of these strategies will produce the greatest values and coefficient performance if the data is divided into training and testing data sets and soft computing techniques are used. For the following procedures, the analysis will rely on WEKA software. GP, SVN_PUK, SVM_Poly, and LR, the data set that was used in the analysis.

TABLE 9.1
The Total Data Collected from Lateral Review[a]

Cement (kg/m^3)	Fine Aggregate/ Crushed Sand (kg/m^3)	Coarse Aggregate (kg/m^3)	Water (kg/m^3)	SP (kg/m^3)	Fly Ash (kg/m^3)	BFRP	Diameter (µm)	Length (mm)	Curing Time (Day)	Splitting Tensile Strength
514.5	942	824.4	343.5	0	0	0	0	0	28	2.34
514.5	942	824.4	343.5	0	0	0.1	16	18	28	2.42
514.5	942	824.4	343.5	0	0	0.3	16	18	28	2.63
514.5	942	824.4	343.5	0	0	0.5	16	18	28	2.81
514.5	942	824.4	343.5	0	0	1	16	18	28	2.93
514.5	942	824.4	343.5	0	0	1.5	16	18	28	3.28
350	1100	740	0.5	3.5	0	0	17	24	28	3.65
350	1100	740	0.5	3.5	0	0.5	17	24	28	3.99
350	1100	740	0.5	3.5	0	1	17	24	28	3.92
350	1100	740	0.5	3.5	0	2	17	24	28	4.02
350	1100	740	0.5	3.5	0	3	17	24	28	3.85
380	720	1215.60	171	1.9	0	0	0	0	28	2.95
380	720	1215.60	171	1.9	0	0.5	16	24	28	4.21
380	720	1215.60	171	1.9	0	0.75	16	24	28	4.52
380	720	1215.60	171	1.9	0	1	16	24	28	3.98
217	694	1416	130	2.71	54	0	0	0	28	2.29
217	694	1416	130	2.71	54	0.1	17.4	6	28	2.36
217	694	1416	130	2.71	54	0.2	17.4	6	28	2.38
217	694	1416	130	2.71	54	0.3	17.4	6	28	2.5
217	694	1416	130	2.71	54	0.4	17.4	6	28	2.51
217	694	1416	130	2.71	54	0.5	17.4	6	28	2.4
217	694	1416	130	2.71	54	0.1	17.4	12	28	2.28

(Continued)

TABLE 9.1 (*Continued*)
The Total Data Collected from Lateral Review[a]

Cement (kg/m³)	Fine Aggregate/ Crushed Sand (kg/m³)	Coarse Aggregate (kg/m³)	Water (kg/m³)	SP (kg/m³)	Fly Ash (kg/m³)	BFRP	Diameter (µm)	Length (mm)	Curing Time (Day)	Splitting Tensile Strength
217	694	1416	130	2.71	54	0.2	17.4	12	28	2.27
217	694	1416	130	2.71	54	0.3	17.4	12	28	2.26
217	694	1416	130	2.71	54	0.4	17.4	12	28	2.28
217	694	1416	130	2.71	54	0.5	17.4	12	28	2.2
375	692	1207	169	3.75	0	0	0	0	28	2.9
375	692	1207	169	3.75	0	0.5	13	12.7	28	3
375	692	1207	169	4.87	0	1	13	12.7	28	3.3
375	692	1207	169	5.62	0	1.5	13	12.7	28	3.1
442	678	984	156	4.16	78	0	0	0	3	2.82
442	678	984	156	4.16	78	0.5	15	18	3	2.84
442	678	984	156	4.16	78	1	15	18	3	2.78
442	678	984	156	4.16	78	1.5	15	18	3	2.77
442	678	984	156	4.16	78	2	15	18	3	2.51
442	678	984	156	4.16	78	0	0	0	7	3.01
442	678	984	156	4.16	78	0.5	15	18	7	3.07
442	678	984	156	4.16	78	1	15	18	7	3.02
442	678	984	156	4.16	78	1.5	15	18	7	2.91
442	678	984	156	4.16	78	2	15	18	7	2.87
442	678	984	156	4.16	78	0	0	0	28	3.69
442	678	984	156	4.16	78	0.5	15	18	28	3.74
442	678	984	156	4.16	78	1	15	18	28	3.33
442	678	984	156	4.16	78	1.5	15	18	28	3.28

(Continued)

TABLE 9.1 (*Continued*)
The Total Data Collected from Lateral Review[a]

Cement (kg/m^3)	Fine Aggregate/ Crushed Sand (kg/m^3)	Coarse Aggregate (kg/m^3)	Water (kg/m^3)	SP (kg/m^3)	Fly Ash (kg/m^3)	BFRP	Diameter (μm)	Length (mm)	Curing Time (Day)	Splitting Tensile Strength
442	678	984	156	4.16	78	2	15	18	28	3.25
442	678	984	156	4.16	78	0	0	0	60	3.71
442	678	984	156	4.16	78	0.5	15	18	60	3.84
442	678	984	156	4.16	78	1	15	18	60	3.78
442	678	984	156	4.16	78	1.5	15	18	60	3.54
442	678	984	156	4.16	78	2	15	18	60	3.48
442	678	984	156	4.16	78	0	0	0	90	3.85
442	678	984	156	4.16	78	0.5	15	18	90	4.21
442	678	984	156	4.16	78	1	15	18	90	4.02
442	678	984	156	4.16	78	1.5	15	18	90	4.02
442	678	984	156	4.16	78	2	15	18	90	3.8
420	656	1069	210	2.4	60	0	0	0	7	3.5
420	656	1069	210	2.4	60	1	15	30	7	3.4
420	656	1069	210	2.4	60	1.5	15	30	7	3.7
420	656	1069	210	2.4	60	2	15	30	7	3.4
420	656	1069	210	2.4	60	2.5	15	30	7	3.6
420	656	1069	210	2.4	60	3	15	30	7	3.3
420	656	1069	210	2.4	60	3.5	15	30	7	4.4
420	656	1069	210	2.4	60	0	0	0	28	4
420	656	1069	210	2.4	60	1	15	30	28	4.7
420	656	1069	210	2.4	60	1.5	15	30	28	4.4

(*Continued*)

TABLE 9.1 (*Continued*)
The Total Data Collected from Lateral Review[a]

Cement (kg/m^3)	Fine Aggregate/ Crushed Sand (kg/m^3)	Coarse Aggregate (kg/m^3)	Water (kg/m^3)	SP (kg/m^3)	Fly Ash (kg/m^3)	BFRP	Diameter (μm)	Length (mm)	Curing Time (Day)	Splitting Tensile Strength
420	656	1069	210	2.4	60	2	15	30	28	3.9
420	656	1069	210	2.4	60	2.5	15	30	28	4.8
420	656	1069	210	2.4	60	3	15	30	28	4.9
420	656	1069	210	2.4	60	3.5	15	30	28	4.6
450	670	1100	180	0	0	0	0	0	28	5.16
450	670	1100	180	0	0	1	18	25	28	5.16
450	670	1100	180	0	0	2	18	25	28	5.4
450	670	1100	180	0	0	3	18	25	28	6
425	846.72	705.99	225	1.8	75	0	16	24	28	3.6
425	846.72	705.99	225	1.8	75	0.15	16	24	28	3.91
425	846.72	705.99	225	1.8	75	0.2	16	24	28	3.95
425	846.72	705.99	225	1.8	75	0.25	16	24	28	4.02
425	846.72	705.99	225	1.8	75	0.3	16	24	28	4.06

[a] Cement = C, Fine Aggregate = FA, Coarse Aggregate = CA, W = Water, Sp = Superplasticizer

TABLE 9.2
Features of Data Set

	Cement (kg/m³)	Fine Aggregate/ Crushed Sand (kg/m³)	Coarse Aggregate (kg/m³)	Water (kg/m³)	SP (kg/m³)	Fly Ash (kg/m³)	BFRP
Training Data Set							
Mean	385.825	751.85	1062.595	158.425	3.231	46.95	0.79375
Median	442	692	984	156	3.625	54	0.5
Mode	442	678	984	156	4.16	78	0.5
Standard deviation	95.51195	141.4303	218.2008	79.95316	1.363229	34.2629	0.742845
Sample variance	9122.533	20,002.54	47,611.58	6392.507	1.858394	1173.946	0.551819
Range	297.5	444	676	343	5.62	78	3
Minimum	217	656	740	0.5	0	0	0
Maximum	514.5	1100	1416	343.5	5.62	78	3
Largest(1)	514.5	1100	1416	343.5	5.62	78	3
Smallest(1)	217	656	740	0.5	0	0	0
Confidence level(95.0%)	30.5462	45.23162	69.784	25.57026	0.435982	10.95781	0.237573
Testing Data Set							
Mean	403.1667	725.78	1054.083	178.8125	2.760833	48.375	1.002083
Median	422.5	678	1026.5	162.5	2.71	60	1
Mode	442	678	984	156	4.16	78	1
Standard deviation	80.65192	113.8512	190.691	67.4269	1.369176	32.89088	0.948967
Sample variance	6504.732	12,962.09	36,363.05	4546.387	1.874643	1081.81	0.900539
Range	297.5	444	710.01	343	4.16	78	3.5
Minimum	217	656	705.99	0.5	0	0	0
Maximum	514.5	1100	1416	343.5	4.16	78	3.5
Largest(1)	514.5	1100	1416	343.5	4.16	78	3.5
Smallest(1)	217	656	705.99	0.5	0	0	0
Confidence level (95.0%)	34.05632	48.07514	80.52174	28.47188	0.578152	13.8886	0.400714

9.5 RESULTS AND DISCUSSION

In this chapter GP, SVM_PUK, SVM_Poly, and LR based models have been developed using WEKA software to predict the values of tensile strength of concrete with BFRP. Model development is a trial-and-error process. Many trials have been carried out to find the optimum value user-defined parameters of various tree and forest-based models. Figures 9.9 and 9.10 represent the agreement plot between observed and predicted values using GP, SVM_PUK, SVM_Poly, and LR and For the training and testing stages, GP regression model algorithms are used. Most of the predicted values using GP, SVM_PUK, SVM_Poly, and LR regression models lie within ± 25 % error band. Table 9.3 includes the training and testing data sets as well as the performance assessment index values for various regression models. Outcomes of Table 9.3 conclude that the SVM_PUK model outperforms the GP, SVM_Poly, and LR models for training and testing stages with higher values of CC (0.9942, 0.9170) and lower values of MAE (0.0337, 0.2724) and RMSE (0.0934, 0.3505).

Figures 9.1–9.8 show the results for an estimate the splitting tensile strength after using every technique individually for training and testing data set. Figures 9.9 and 9.10 show all actual and predicted values for all the techniques. Figure 9.11 shows the comparison between the actual and predicted values for all techniques. Figure 9.12 shows the graph error for all the techniques.

9.5.1 Sensitivity Analysis

Use the SVM_PUK model for sensitivity testing to determine the key input variables in compressive strength. Create a set of different test data by removing one input parameter at a time, as shown in Table 9.4. According to CC, the impact of each input variable on compressive strength is reported using CC, MAE, and RMSE. When compared to the other input factors in this data set, Table 9.4 reveals that curing time has a considerable impact on concrete tensile strength predicting.

TABLE 9.3
Performance Evaluation Parameters for GP, SVM_PUK, SVM_Poly and LR

	CC	MAE	RMSE
	Training		
GP	0.9561	0.2647	0.3498
SVM_PUK	0.9942	0.0337	0.0934
SVM_Poly	0.9084	0.2589	0.3617
LR	0.8679	0.3555	0.4287
	Testing		
GP	0.9029	0.2902	0.3452
SVM_PUK	0.917	0.2724	0.3505
SVM_Poly	0.8605	0.3346	0.4082
LR	0.8481	0.3357	0.4164

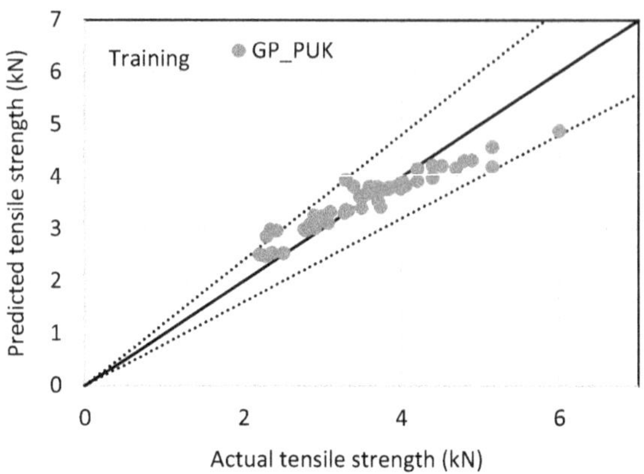

FIGURE 9.1 Actual versus predicted values using GP using training data set.

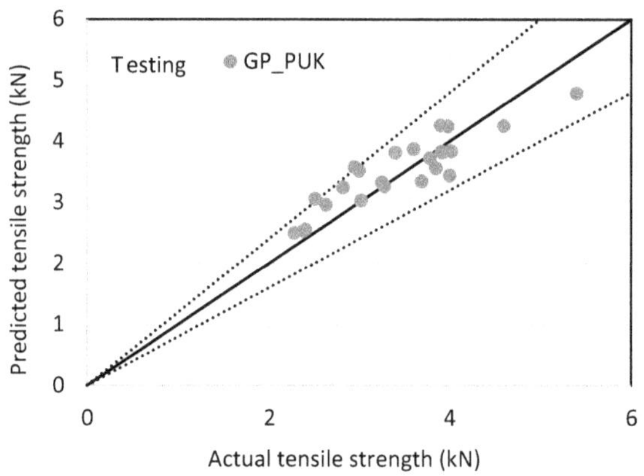

FIGURE 9.2 Actual versus predicted values using GP using testing data set.

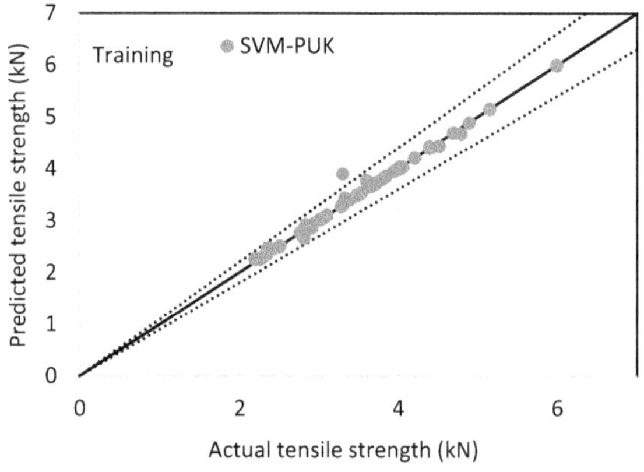

FIGURE 9.3 Actual versus predicted values using SVM_PUK using training data set.

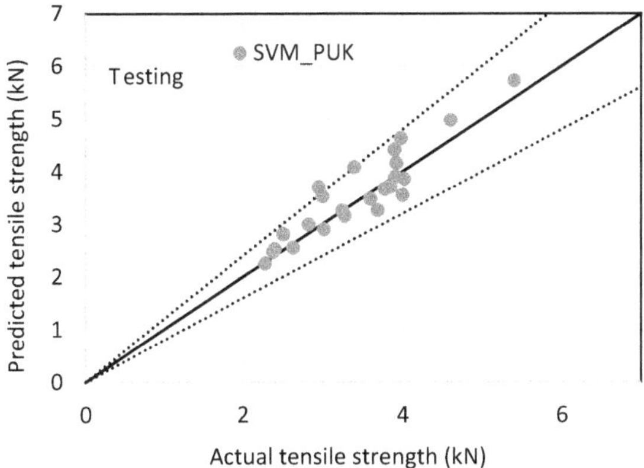

FIGURE 9.4 Actual versus predicted values using SVM_PUK using testing data set.

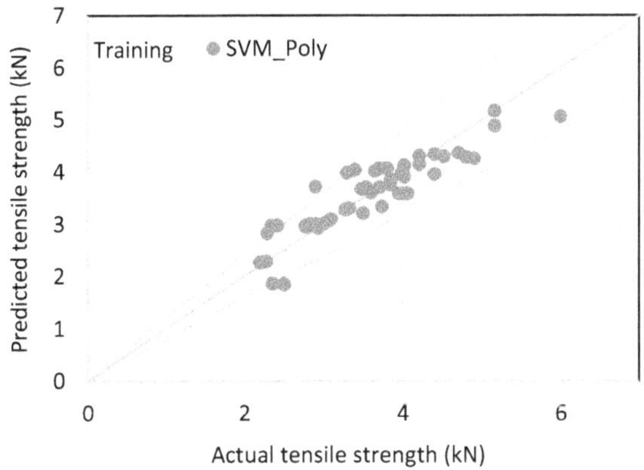

FIGURE 9.5 Actual versus predicted values using SVM_Poly using training data set.

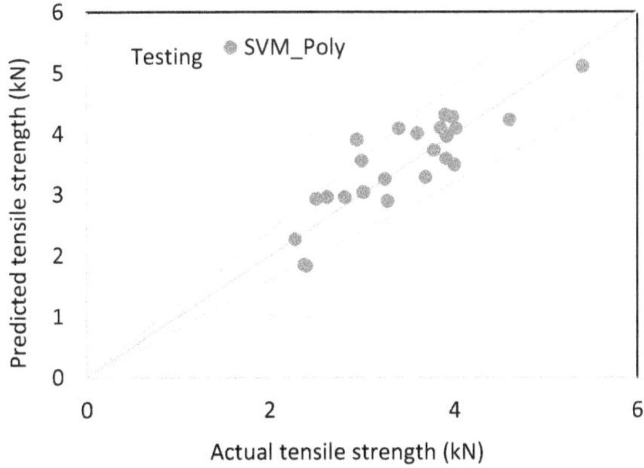

FIGURE 9.6 Actual versus predicted values using SVM_Poly using testing data set.

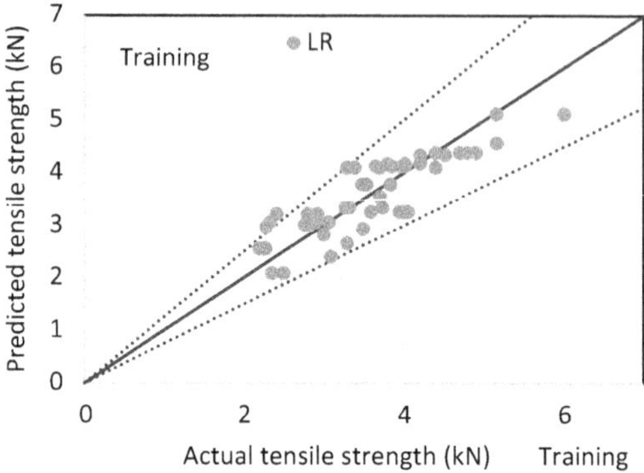

FIGURE 9.7 Actual versus predicted values using LR using training data set.

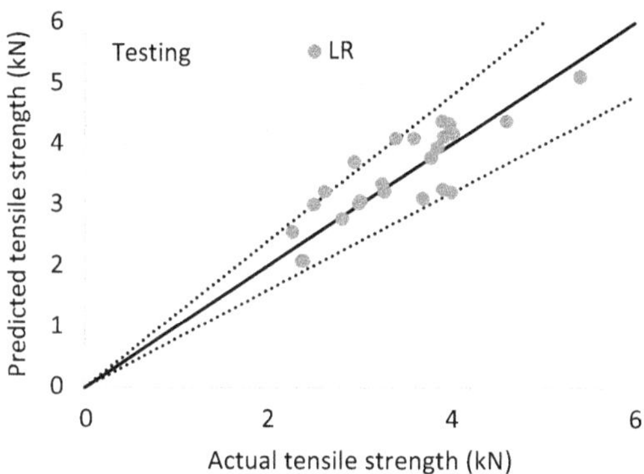

FIGURE 9.8 Actual versus predicted values using LR using testing data set.

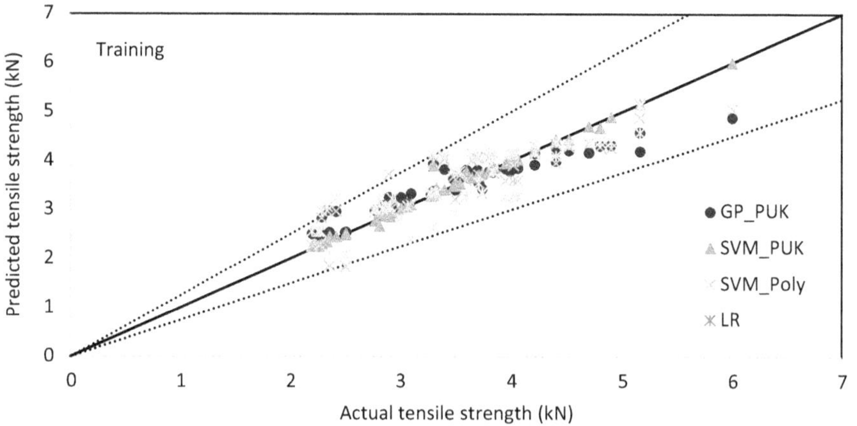

FIGURE 9.9 Actual versus predicted values using GP, SVM_PUK, SVM_Poly and LR using training data set.

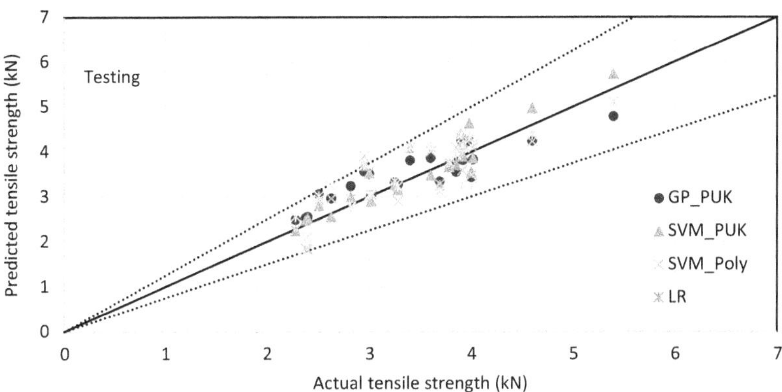

FIGURE 9.10 Actual versus predicted values using GP, SVM_PUK, SVM_Poly and LR using testing data set.

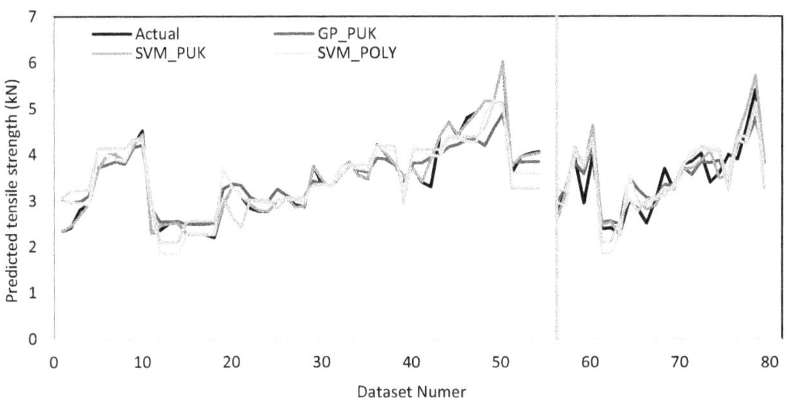

FIGURE 9.11 Comparative between actual and predicted values for GP, SVM_PUK, SVM_Poly and LR.

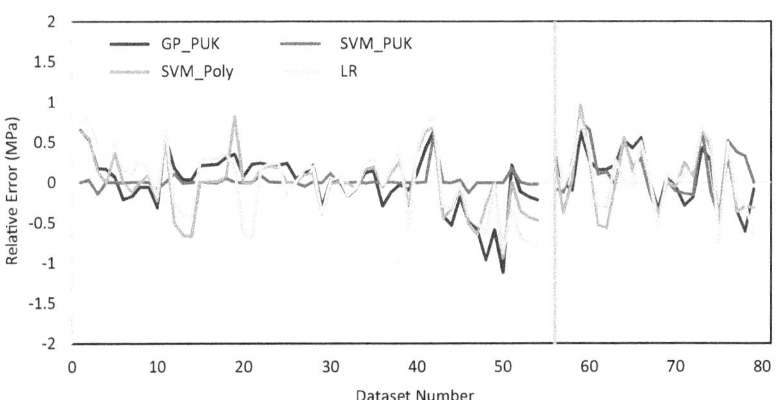

FIGURE 9.12 Error graph for GP, SVM_PUK, SVM_Poly and LR.

TABLE 9.4
Sensitivity Results Using SVM_PUK

Input										Output	SVM_PUK		
Cement (kg/m³)	Fine Aggregate/ Crushed Sand (kg/m³)	Coarse Aggregate (kg/m³)	Water (kg/m³)	SP (kg/m³)	Fly Ash (kg/m³)	BFRP	Diameter (μm)	Length (mm)	Curing Time (Day)	Splitting Tensile Strength	CC	MAE	RMSE
											0.917	0.2724	0.3505
											0.9086	0.2809	0.3675
											0.914	0.2747	0.3587
											0.9149	0.2743	0.3563
											0.9182	0.2696	0.3488
											0.9363	0.2358	0.3008
											0.9299	0.2503	0.3191
											0.9022	0.2529	0.3327
											0.9213	0.2623	0.3457
											0.9218	0.2702	0.3481
											0.791	**0.4273**	**0.5345**

9.6 CONCLUSION

To forecast the compressive strength of concrete with BFRP, GP, SVM_PUK, SVM_Poly, and LR models were constructed in this study. The information was gathered from a variety of articles, each with a different percentage of BFRP and a varied length of BFRP. In a comparative study using performance assessment indices (GP, SVM_PUK, SVM_Poly and LR), the proposed SVM_PUK method outperformed the other algorithms with CC (0.9942, 0.9170), RMSE (0.0934, 0.3505), and MAE (0.0337, 0.2724), and LR gives the lower values of CC (0.8679, 0.8481), RMSE (0.4287, 0.4164), and MAE (0.3555, 0.3357). Another important finding of this study is that GP outperforms SVM_Poly and LR model. The results of the sensitivity investigation reveal that the curing time of bacterial concrete is critical in determining its splitting tensile strength, using this data set and the SVM_PUK.

REFERENCES

Afroughsabet, V., Biolzi, L. and Ozbakkaloglu, T., 2016. High-performance fiber-reinforced concrete: A review. *Journal of Materials Science*, *51*(14), pp. 6517–6551. Doi: 10.1007/s10853-016-9917-4.

Aggarwal, Y., Aggarwal, P., Sihag, P., Pal, M. and Kumar, A., 2019. Estimation of punching shear capacity of concrete slabs using data mining techniques. *International Journal of Engineering*, *32*(7), pp. 908–914. Doi: 10.5829/ije.2019.32.07a.02.

Bashir, R. and Ashour, A., 2012. Neural network modelling for shear strength of concrete members reinforced with FRP bars. *Composites Part B: Engineering*, *43*(8), pp. 3198–3207. Doi: 10.1016/j.compositesb.2012.04.011.

Bolandi, H., Banzhaf, W., Lajnef, N., Barri, K. and Alavi, A.H., 2019. An intelligent model for the prediction of bond strength of FRP bars in concrete: A soft computing approach. *Technologies*, *7*(2), p. 42. Doi: 10.3390/technologies7020042.

Coelho, M.R., Sena-Cruz, J.M., Neves, L.A., Pereira, M., Cortez, P. and Miranda, T., 2016. Using data mining algorithms to predict the bond strength of NSM FRP systems in concrete. *Construction and Building Materials*, *126*, pp. 484–495. Doi: 10.1016/j.conbuildmat.2016.09.048.

Cortes, C. and Vapnik, V., 1995. Support-vector networks. *Machine Learning*, *20*(3), pp. 273–297. Doi: 10.1007/BF00994018.

Deepa, C., SathiyaKumari, K. and Sudha, V.P., 2010. Prediction of the compressive strength of high performance concrete mix using tree based modeling. *International Journal of Computer Applications*, *6*(5), pp.18–24.

Golafshani, E.M., Rahai, A., Sebt, M.H. and Akbarpour, H., 2012. Prediction of bond strength of spliced steel bars in concrete using artificial neural network and fuzzy logic. *Construction and Building Materials*, *36*, pp. 411–418. Doi: 10.1016/j.conbuildmat.2012.04.046.

Heddam, S., 2014. Modelling hourly dissolved oxygen concentration (DO) using dynamic evolving neural-fuzzy inference system (DENFIS)-based approach: Case study of Klamath River at Miller Island Boat Ramp, OR, USA. *Environmental Science and Pollution Research*, *21*(15), pp. 9212–9227. Doi: 10.1007/s11356-014-2842-7.

Ibrahim, D., 2016. An overview of soft computing. *Procedia Computer Science*, *102*, pp. 34–38. Doi: 10.1016/j.procs.2016.09.366.

Jalal, M., 2015. Soft computing techniques for compressive strength prediction of concrete cylinders strengthened by CFRP composites. *Science and Engineering of Composite Materials*, *22*(1), pp. 97–112. Doi: 10.1515/secm-2013-0240.

Jang, J.S.R., Sun, C.T. and Mizutani, E., 1997. Neuro-fuzzy and soft computing-a computational approach to learning and machine intelligence [Book Review]. *IEEE Transactions on automatic control*, *42*(10), pp.1482–1484.

Jiang, C., Fan, K., Wu, F. and Chen, D., 2014. Experimental study on the mechanical properties and microstructure of chopped basalt fibre reinforced concrete. *Materials & Design*, *58*, pp.187–193. Doi: 10.1016/j.matdes.2014.01.056.

Kisi, O., Sanikhani, H. and Cobaner, M., 2017. Soil temperature modeling at different depths using neuro-fuzzy, neural network, and genetic programming techniques. *Theoretical and Applied Climatology*, *129*(3), pp. 833–848. Doi: 10.1007/s00704-016-1810-1.

Köroğlu, M.A., 2019. Artificial neural network for predicting the flexural bond strength of FRP bars in concrete. *Science and Engineering of Composite Materials*, *26*(1), pp. 12–29. Doi: 10.1515/secm-2017-0155.

Kumar, M., Sihag, P. and Singh, V., 2019. Enhanced soft computing for ensemble approach to estimate the compressive strength of high strength concrete. *Journal of Materials and Engineering Structures*, *6*(1), pp. 93–103.

Kuss, M., 2006. *Gaussian Process Models for Robust Regression, Classification, and Reinforcement Learning* (Doctoral dissertation, echnische Universität Darmstadt Darmstadt, Germany).

Kwin, C.T., Talei, A., Alaghmand, S. and Chua, L.H., 2016. Rainfall-runoff modeling using dynamic evolving neural fuzzy inference system with online learning. *Procedia engineering*, *154*, pp. 1103–1109. Doi: 10.1016/j.proeng.2016.07.518.

Mehdipour, V., Stevenson, D.S., Memarianfard, M. and Sihag, P., 2018. Comparing different methods for statistical modeling of particulate matter in Tehran, Iran. *Air Quality, Atmosphere & Health*, *11*(10), pp. 1155–1165. Doi: 10.1007/s11869-018-0615-z.

Mohammadyan-Yasouj, S.E. and Ghaderi, A., 2020. Experimental investigation of waste glass powder, basalt fibre, and carbon nanotube on the mechanical properties of concrete. *Construction and Building Materials*, *252*, p. 119115. Doi: 10.1016/j.conbuildmat.2020.119115.

Mohammed, A., Rafiq, S., Sihag, P., Kurda, R. and Mahmood, W., 2020. Soft computing techniques: Systematic multiscale models to predict the compressive strength of HVFA concrete based on mix proportions and curing times. *Journal of Building Engineering*, p. 101851. Doi: 10.1016/j.jobe.2020.101851

Rasmussen, C.E. and Williams, C.K.I., 2006. *Gaussian Processes for Machine Learning*. The MIT Press, Cambridge, MA.

Sepahvand, A., Singh, B., Sihag, P., Nazari Samani, A., Ahmadi, H. and Fiz Nia, S., 2019. Assessment of the various soft computing techniques to predict sodium absorption ratio (SAR). *ISH Journal of Hydraulic Engineering*, pp. 1–12. Doi: 10.1080/09715010.2019.1595185.

Shaikh, F.U.A., 2013. Review of mechanical properties of short fibre reinforced geopolymer composites. *Construction and building materials*, *43*, pp. 37–49. Doi: 10.1016/j.conbuildmat.2013.01.026.

Sihag, P., Esmaeilbeiki, F., Singh, B., Ebtehaj, I. and Bonakdari, H., 2019. Modeling unsaturated hydraulic conductivity by hybrid soft computing techniques. *Soft Computing*, *23*(23), pp. 12897–12910. Doi: 10.1007/s00500-019-03847-1.

Sihag, P., Tiwari, N.K. and Ranjan, S., 2017. Modelling of infiltration of sandy soil using gaussian process regression. *Modeling Earth Systems and Environment*, *3*(3), pp. 1091–1100. Doi: 10.1007/s40808-017-0357-1.

Singh, B., Sihag, P., Pandhiani, S.M., Debnath, S. and Gautam, S., 2019a. Estimation of permeability of soil using easy measured soil parameters: Assessing the artificial intelligence-based models. *ISH Journal of Hydraulic Engineering*, pp. 1–11. Doi: 10.1080/09715010.2019.1574615.

Singh, B., Sihag, P., Tomar, A. and Sehgal, A., 2019b. Estimation of compressive strength of high-strength concrete by random forest and M5P model tree approaches. *Journal of Materials and Engineering Structures*, *6*(4), pp. 583–592.

Smola, A.J., 1996. *Regression Estimation with Support Vector Learning Machines* (Doctoral dissertation, Master's thesis, Technische Universität München).

Tassew, S.T. and Lubell, A.S., 2014. Mechanical properties of glass fiber reinforced ceramic concrete. *Construction and Building Materials*, *51*, pp. 215–224. Doi: 10.1016/j.conbuildmat.2013.10.046.

Thakur, M.S., Pandhiani, S.M., Kashyap, V., Upadhya, A. and Sihag, P., 2021. Predicting bond strength of FRP bars in concrete using soft computing techniques. *Arabian Journal for Science and Engineering*, *46*(5), pp. 4951–4969. Doi: 10.1007/s13369-020-05314-8.

Vapnik, V.N., 1995. The nature of statistical learning. *Theory*, Springer Nature Switzerland AG.

Vapnik, V.N. and Vapnik, V., 1998. *Statistical Learning Theory*. vol. 1 Wiley, New York.

Zaji, A.H., Bonakdari, H. and Gharabaghi, B., 2018. Reservoir water level forecasting using group method of data handling. *Acta Geophysica*, *66*(4), pp. 717–730. Doi: 10.1007/s11600-018-0168-4.

10 Prediction of Compressive Strength of Self-Compacting Concrete Containing Silica's Using Soft Computing Techniques

Pranjal Kumar Pandey, Paratibha Aggarwal, and Yogesh Aggarwal
NIT Kurukshetra

Sejal Aggarwal
Manipal University

CONTENTS

DOI: 10.1201/9781003184331-10

10.1 INTRODUCTION

Self-compacting concrete (SCC) is amongst most practical concrete in construction industry. Back in the 1980s when Japan construction industry was suffering from durability issues and lack of skilled labour, Prof. Hajjim Okamura came up with the concept of SCC [1]. As the name suggest it's a special type of high performance concrete. The requirements of SCC are achieved by properties in its fresh state. The main properties are like "filling ability", "passing ability" and "segregation resistance". Therefore, bleeding and segregation is avoided and stability is maintained at the same time [2,3]. It sets under its own weight and can easily occupy the confined spaces between reinforcements. It finds huge application in prefabricated industries.

Cement industry contributes 7%–9% of total global CO_2 emission which leads to global warming causing an increase in earth temperature. Thus, researchers across the globe are working towards eco-friendly concrete, as in cement production fossil fuels are burnt and huge amount of CO_2 is emitted in atmosphere. Thus, the concept of blended cement with supplementary cementitious materials is in play. Commonly used supplementary cementitious materials are fly ash [4], silica fume [5], nano-silica fume [6–9], micro-silica fume, perlite powder, ground-granulated blast-furnace slag [10], viscosity modifying admixtures and new generation of superplasticizers. Fly ash improves the rheological properties of SCC, reduces the cracks in concrete as heat of hydration is lowered and the amount of superplasticizer required to obtain slump flow as in that of ordinary Portland cement concrete is reduced.

Silica fume a by-product of powder industry being a pozzolanic material is used to overcome low early compressive strength of SCC. Silica fume reduces the water absorption and porosity, increases the electrical resistivity and highly effective in controlling the chloride ion penetration. Nano- and micro-silica particles modify the material to microscopic level to enhance its properties. According to various studies use of nano-silica is quite outstanding. Nano-silica act as a filling agent with C–S–H gel for voids and also forms a nanostructured C–S–H gel material, thereby improving the microstructure.

The compressive strength of SCC is a non-linear function of proportion of its ingredients and there is no direct relationship between mix proportioning and SCC strength, hence there is a need to use appropriate tool for their prediction based on its constituents and design. Multiple techniques of modelling the properties of material are explored by researchers including computational modelling, statistical techniques and various tools like artificial neural network (ANN) [11], support vector machine algorithm [12], random forest (RF) [13], multivariate adaptive regression splines (MARS) and gene expression programming (GEP) methods [14] in predicting 28 days compressive strength of concrete. These machine learning tools have ability to learn from input–output relation of any complex problem thus eliminating use of any specific equation form. Padmini et al. [15] have successfully modelled neuro fuzzy technique to find bearing capacity of shallow foundation. ANN finds its application in structural engineering [16].

The objective of this paper is to create the model and compare the potential of artificial intelligence technique (ANN, GPR, RF, SVM, and use ensemble, i.e., in form of bagging with ANN and RF) in estimating the 28 days compressive strength of SCC incorporating silica using data obtained from literature. To build the model a total of 99 data were collected. These data are arranged in form of eight input parameters and machine learning tools are used to predict the strength as output parameter. It was observed that, in spite of intricate data these tools can be effectively used to predict the compressive strength and help in decision making. A sensitivity analysis was also performed to suggest the most influencing input parameter for prediction of compressive strength. The proposed model was validated with help of experimental results.

10.2 SOFT COMPUTING TECHNIQUES

10.2.1 Artificial Neural Network (ANN)

ANN is computational tool inspired by biological neural network for modelling complex non-linear relationship. It does not need a specific equation. It's a multilayer feed forward network with input layer, hidden layer and output layer programmed with eight input parameters and compressive strength as output. Based on lowest average square error criteria maximum number of hidden layers and neurons are calculated by trial and error. Briefly, for a given input and the linkages, activation rate is found and forwarded from the input layer to the output layer via the hidden layers. Then the errors in the output are initiated. The ANN technique offers an advantage since it can use an unlimited number of characteristic parameters of the phenomenon, compared to statistical methods. Figure 10.1 shows the architecture of ANN model.

The optimum number of epochs is chosen during the training that gives minimum mean absolute error (MAE) and root mean square error (RMSE) value and higher R^2

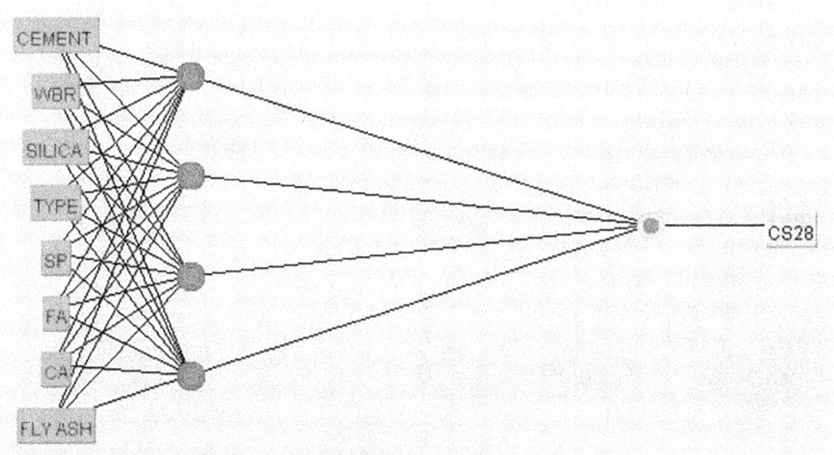

FIGURE 10.1 Architecture of ANN model.

value. Cross-validation technique is used for training and testing data set. Multiple coefficient of determination (R^2) values compare model accuracy with basic benchmark model wherein prediction is samples mean. $R^2 = 1$ results in perfect fit model. Further for selected hidden layers and nodes suitable value of parameters like learning rate and momentum is also required.

10.2.2 Linear Regression (LR)

LR is a statistical method used for predictive analysis. Statistical methods perform predictions quickly and are simpler to implement. Apart from its speed, statistical modelling has advantages over other techniques that it is rigorous and can be used to define confidence interval for the prediction [17]. Linear regression equation van be represented by Equation 10.1

$$y = a_0 + a_1 x + \varepsilon \tag{10.1}$$

where, y = Dependent variable, x = Independent variable, a_0 = Intercept of a line (gives an additional degree of freedom), a_1 = Linear Regression coefficient and ε = Random error. The x and y variables are the training data set for LR model representation.

Best fit line is obtained to minimise the error. Different values of weights give different regression lines and best fit line is obtained with help of cost function which is mean square error in LR. Mean square error is minimised using gradient descent technique by calculating the gradient of cost function. A regression model uses gradient descent to update the line coefficient by reducing the cost function. Goodness of fit is determined by R-squared method. High value of R^2 determines less difference between predicted and actual value, hence a good model

10.2.3 Support Vector Machine (SVM)

The support vector machine (SVM) was developed by [18,19] and gained popularity due to high classification and prediction performance [20,21]. SVM create a decision boundary or the best fit line that can segregate n-dimensional spaces in to classes so that a new data point can be applied easily in the correct category. This best decision boundary is called hyperplane. The hyperplane is derived by selecting training data points using selected data points called support vectors. The distance of vector from the hyperplane are called margin and optimal hyperplane maximised the margin.

Unlike other logistics regression in which output lies in range of [0, 1], the threshold values of SVM lies between [−1, 1] which act as a margin. SVM algorithms use a set of mathematical functions that are defined as the kernel. The kernel function calculates dot product in high-dimensional variables space using low dimensional space data input without making any assumptions about data features.

Standard Kernel Function Equation 10.2 and 10.3

$$K(\overline{x}) = 1, \text{ if} \|\overline{x}\| \le 1 \tag{10.2}$$

$$K(\overline{x}) = 0, \text{ otherwise} \tag{10.3}$$

Among the different types functions (linear, nonlinear, polynomial, radial basis function, and sigmoid), polynomial kernel function is used in this study.

10.2.4 Random Forest (RF)

RF regression is proposed by Breiman in 2001[22] and is considered an improved classification regression method. It is based on the concept of ensemble learning which involve combining multiple classifiers for solving complex problem. RF contain number of decision trees on various subset of a given data set and take average value of these data set to improve the predictive accuracy. The greater number of trees leads to higher accuracy and avoids over fitting. The RF algorithm is simple, independent of training set characteristics, and high prediction precision capability is high [23,24]. Two user-defined parameters were used: the number of trees grown (k) and the number of input parameters (m). A trial-and-error process is used for model development.

10.2.5 Bagging

The bagging classifier is an ensemble meta estimator that fits base classifier each on random subsets of the original data set and then aggregate their individual predictions to form a final prediction. Such a meta estimator are very useful in reducing the variance of classifier like decision tree by introducing randomization in its construction procedure and then making an ensemble out of it. Every model is generated with the help of bootstrap [25] independent subset of the observed data set. In this study all the four techniques (ANN, LR, SVM, and RF) discussed above are employed as the base learning algorithm of bagging.

10.3 DATA AND ANALYSIS

10.3.1 Data Set

A total of 99 data's (Table A.1) were collected from literature [26–30] to develop and validate the model. The data used for training purpose consist of seven input variables and 28 days compressive strength as output. Table 10.1 provides the statistics of the data used for development of model.

10.3.2 Evaluation Parameters

The model is accepted or rejected based on its ability to predict the compressive strength based upon the data it was trained on. To minimize the influence of variability of the training set, the ten-fold cross-validation was performed. The data set was randomly split into ten sets $g1$, $g2$... $g10$, of which the size and distribution were equal. The simulation tested on the data set $g1$ and trained on $g2$, $g3$... $g10$, followed by testing on the data set $g2$ and trained on $g1$, $g3$... $g10$. Finally, the mean value of all the ten data set gives the final prediction result.

The performance of the proposed model was validated with MAE, RMSE, R^2 values. A higher R^2 value mean a higher accuracy.

TABLE 10.1
Statistics of the Data

Parameters	Data Base Range		Mean	Standard Deviation	Kurtosis	Skewness
	Minimum	Maximum				
Cement (kg/m^3)	135	750	341.5697	104.8351	0.0738	−0.561
Fine aggregate(kg/m^3)	720	1166	963.4287	104.3202	0.5966	−0.9251
Coarse aggregate (kg/m^3)	595	940	704.6994	96.2075	0.7135	1.2113
Fly ash (kg/m^3)	0	420	143.4394	117.6878	−1.051	0.3114
Silica						
Silica fume (kg/m^3)[a]	0	81	28.09	20.4022	−0.3519	0.2413
Nano-silica (kg/m^3)[a]	0	14	5.2804	4.3674	−0.8312	0.6358
Superplasticizer %	0.25	2	3.6153	3.3357	−1.7268	0.4305
Water (kg/m^3)	109	285	185.3715	34.7388	1.9217	0.9012

$$\mathrm{RMSE} = \sqrt{\frac{1}{N}\sum_{n=1}^{N}(\mathrm{ACS} - \mathrm{PCS})^2}$$

$$\mathrm{MAE} = \frac{1}{N}\sum_{n=1}^{N}(\mathrm{ACS} - \mathrm{PCS})$$

$$R^2 = 1 - \frac{\sum (x - x')^2}{\sum (x - \overline{x})^2}$$

where ACS is actual compressive strength, PCS is predicted compressive strength, x is the actual value, x' is the predicted value and $(\overline{x})$ is the mean of x values. N is the number of data.

Compared to complex Statistical approach, modelling process is more direct as no mathematical relationship between input and output variable is to be setup.

10.4 RESULTS AND DISCUSSION

10.4.1 ANN and Ensemble ANN Model

Table 10.2 provides the statistics of data which was used for modelling. To examine the accuracy of result a line and scatter graph is plotted between actual and predicted value in Figure 10.2. The results of performance evaluation parameters shows that bagged ANN-based model with $R^2 = 0.9289$, MAE = 4.2069 and RMSE = 5.9004 performed better than ANN model with $R^2 = 0.9249$, MAE = 4.3658 and RMSE = 6.0664 for predicting the compressive strength of SCC containing silica.

TABLE 10.2
Summary of Network Parameters

Network Parameters	ANN
No. of hidden layers	1
Number of hidden neurons	8
Learning rate	0.04
Momentum	0.6
Iterations	500

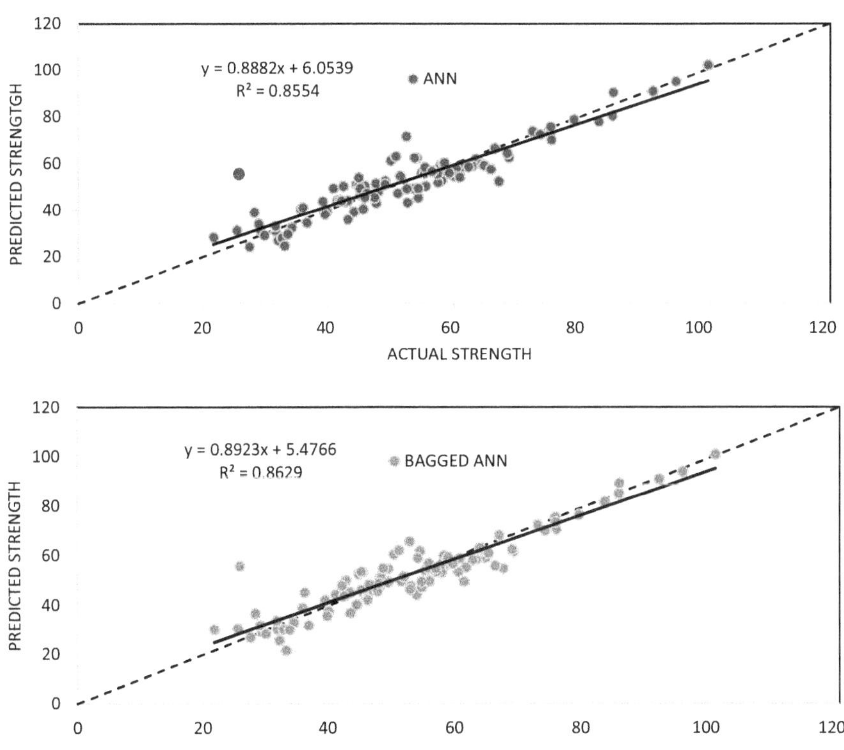

FIGURE 10.2 Actual versus predicted strength for ANN and hybrid ANN-based model.

10.4.2 LR and Ensemble LR Model

The precision of the developed models is examined with help of agreement plots between predicted and actual strength as shown in Figure 10.3. The results of performance evaluation parameters shows that bagged LR-based model with $R^2 = 0.9221$, MAE = 4.5461 and RMSE = 6.1549 achieved better results than LR model with $R^2 = 0.9192$, MAE = 4.6331 and RMSE = 6.2657.

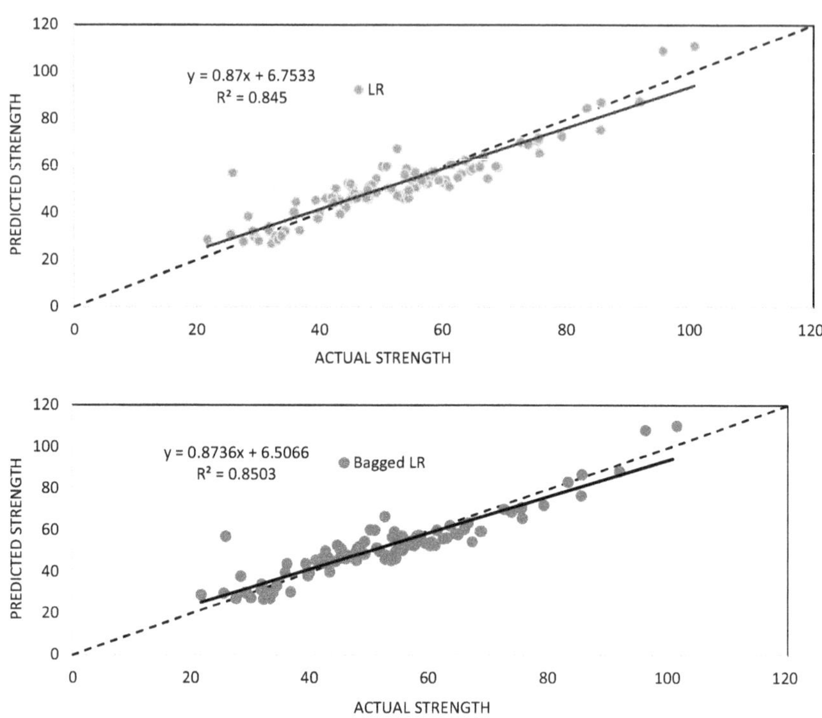

FIGURE 10.3 Actual versus predicted strength for LR and hybrid LR-based model.

10.4.3 SVM and Ensemble SVM Model

Model development in SVM is based on trial-and-error process. For the optimum development of the model large numbers of trials were conducted. In this study model development is based on polynomial kernel function. It was observed that poly kernel function performed better than radial basis function kernel. The precision of the developed models is examined with help of agreement plots between predicted and actual strength as shown in Figure 10.4. The results of performance evaluation parameters shows that bagged SVM-based model with $R^2 = 0.9204$, MAE = 4.6161 and RMSE = 6.2214 achieved better results than SVM model with $R^2 = 0.9181$, MAE = 4.71274 and RMSE = 6.2917.

10.4.4 RF and Ensemble RF Model

RF and Bagged RF are tree-based models based on trial-and-error process. The precision of the developed models is examined with help of agreement plots between predicted and actual strength as shown in Figure 10.5. The results of parameters used for performance evaluation shows that bagged RF-based model with $R^2 = 0.899$, MAE = 5.4527 and RMSE = 7.5371 achieved better results than RF model with $R^2 = 0.8953$, MAE = 5.3681 and RMSE = 7.4198 for predicting the compressive strength of SCC containing silica.

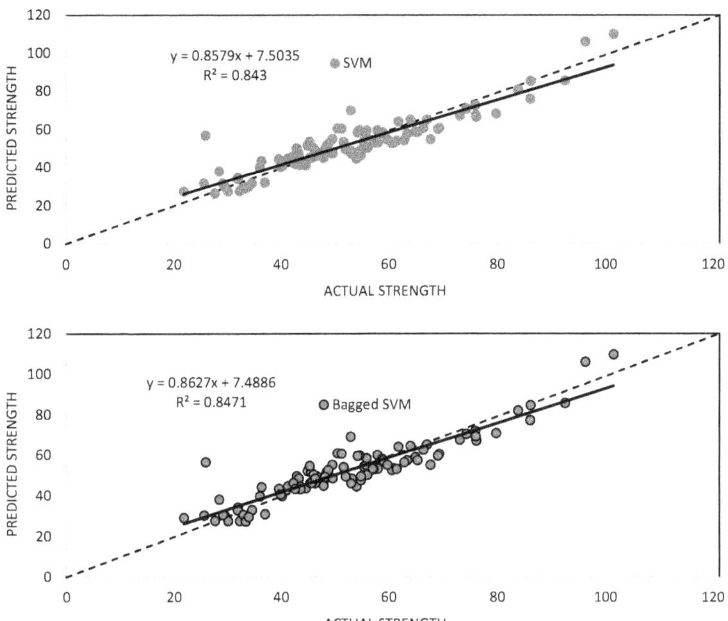

FIGURE 10.4 Actual versus predicted strength for SVM and hybrid SVM-based model.

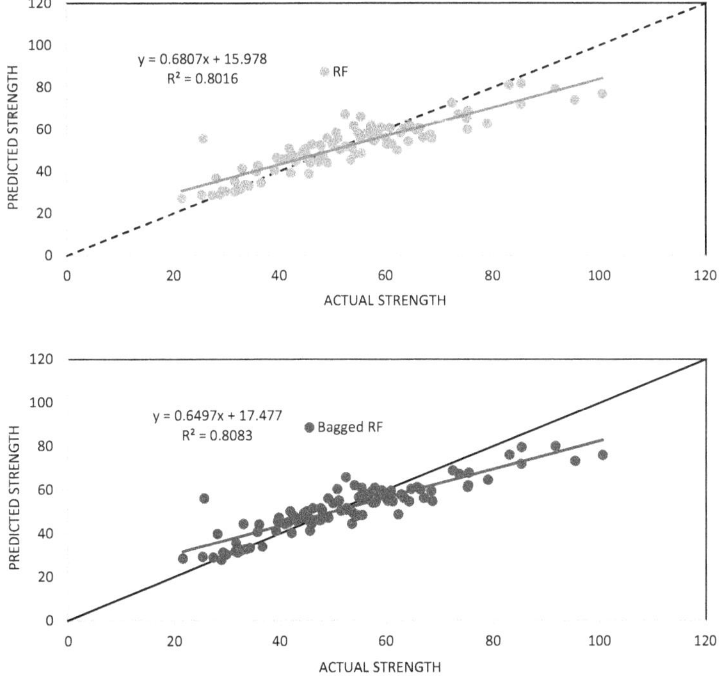

FIGURE 10.5 Actual versus predicted strength for RF and hybrid RF-based model.

10.4.5 Inter-Comparison between Applied Models

Table 10.3 provides the values of CC, RMSE and MAE obtained from training data set for all the developed models. For all the applied models a performance agreement graph and error diagram between actual and predicted values of compressive strength of SCC is indicated in Figure 10.6. From Figure 10.6, it is observed that bagged ANN model lies very close and traverse the same path as actual value

TABLE 10.3
Performance Evaluation Parameters for Applied Models

Models	CC	MAE	RSME
ANN	0.9249	4.3658	6.0664
LR	0.9192	4.6331	6.2657
SVM	0.9181	4.7127	6.2971
RF	0.8953	5.3681	7.4198
BAGGED ANN	0.9289	4.2069	5.9004
BAGGED LR	0.9221	4.5461	6.1549
BAGGED SVM	0.9204	4.6161	6.2214
BAGGED RF	0.899	5.4527	7.5371

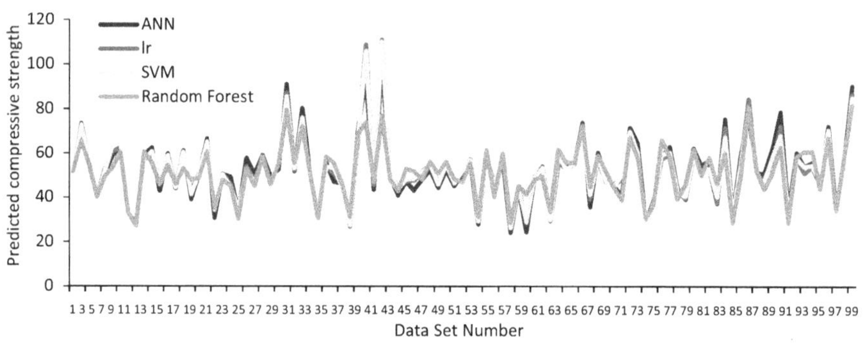

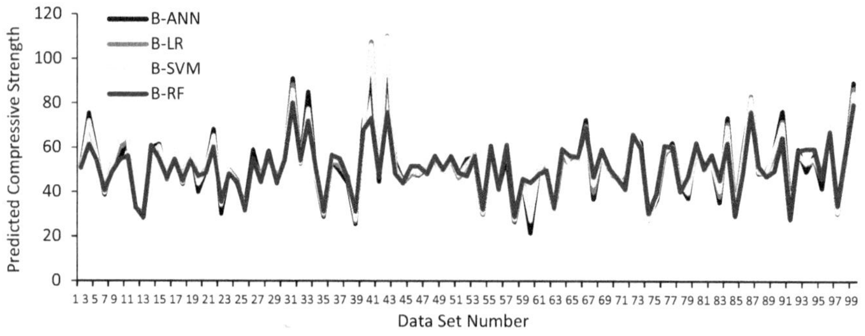

FIGURE 10.6 Performance agreement graph for all the applied models.

with minimum deviation and the error plot shows the same with bagged ANN having minimum error. Figure 10.7 clearly depicts that a higher correlation coefficient ($R^2 = 0.9289$) and lower value of RMSE (5.9004) is achieved for Bagged ANN-based model with respect to other applied models.

Relative error distribution is shown for all developed models with help of Box plot (Figure 10.8). The values of "minimum error, median, maximum error, first quartile, and third quartile" are listed in Table 10.4. The bagged-ANN model has lower quartile value of –3.913 and upper quartile (Q75) is 2.141. Its lower box width confirms of minimum error in this model.

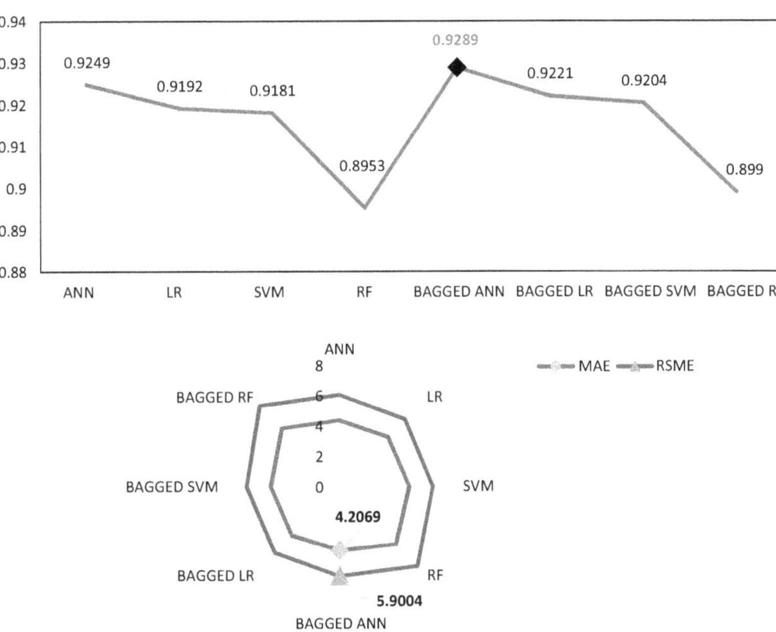

FIGURE 10.7 Inter-comparison of performance evaluation parameters for all the models.

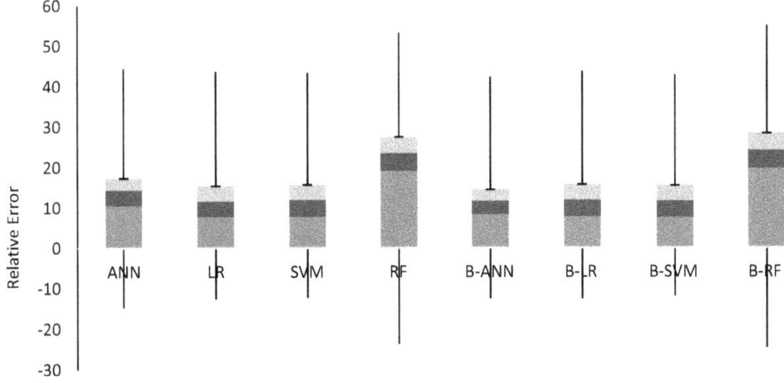

FIGURE 10.8 Box-plot graph.

TABLE 10.4
Error Distribution Statistics for All Models

Statistic	ANN	LR	SVM	RF	B-ANN	B-LR	B-SVM	B-RF
Minimum	−14.639	−12.428	−12.196	−23.558	−12.387	−12.49	−11.902	−24.567
First quartile	−3.993	−4.4555	−4.3415	−4.2795	−3.913	−4.559	−4.188	−4.7955
Median	−0.196	−0.742	−0.198	0.007	−0.704	−0.549	−0.156	−0.39
Third quartile	2.764	3.061	3.568	3.977	2.141	3.22	3.603	3.796
Maximum	29.808	31.382	31.195	29.8	29.987	31.372	30.997	30.489
Mean	0.219737	0.02734	0.091677	0.68257	0.14466	0.08822	0.325505	0.80247

10.4.6 Sensitivity Analysis

Sensitivity analysis was also done using the most fit model that is bagged ANN-based model in non-existence of each input parameter. Six different training model was created by eliminating one input data set in each model and outcome was listed as shown in Figures 10.9 and 10.10 in terms of CC and RMSE, respectively. The analysis

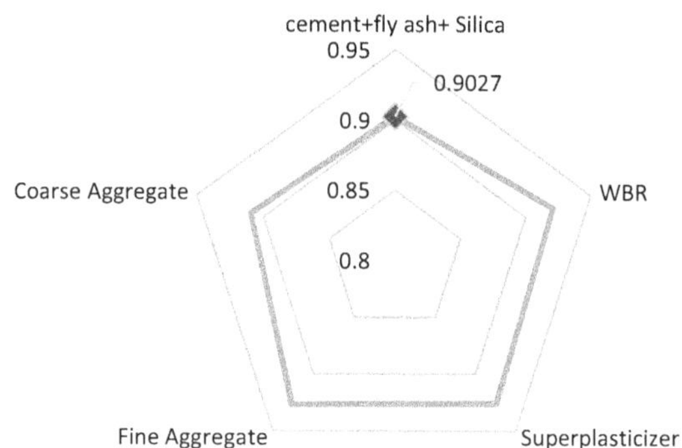

FIGURE 10.9 Sensitivity analysis result in terms of CC.

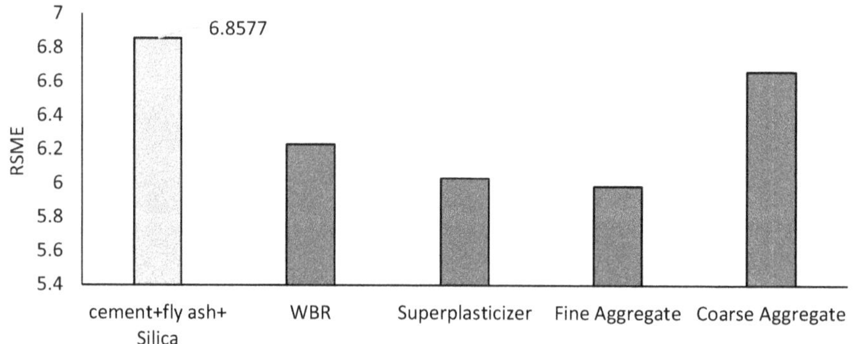

FIGURE 10.10 Sensitivity analysis result in terms of RMSE.

pointed that the most influencing parameter is cementitious material (cement + fly ash + silica) compared with other input parameters for calculating the compressive strength of SCC containing silica's.

10.4.7 Experimental Work

The predictive ability and accuracy of well-trained ANN-based bagged model was evaluated with unviewed data within the range of training input parameters. A total of five SCC mixture was prepared in laboratory with the physical properties of cement and aggregates given in Tables 10.5 and 10.6, respectively, and was tested for its compressive strength. These mix proportion were fed to ANN-based bagged model and the compressive strength associated with each mixture was predicted. The accuracy was measured in mean absolute percentage error as given in Table 10.7. For the experimental works Portland Pozzolana Cement (PPC-53 grade) procured from local market was used. Fly ash obtained from Panipat thermal power station was used as filler. Superplasticizer was acquired locally.

As shown in Table 10.7 a relatively less error was obtained between the predicted and the experimental results. From Figure 10.11 and Table 10.7, it can be deduced that for the bagged ANN model predicted results are in agreement with experimental results.

TABLE 10.5
Physical Properties of Cement

Physical Properties	Result Obtained
Fineness (retained on 90-mm sieve)	9
Normal consistency	28%
Vicat initial setting time (minutes)	75
Vicat final setting time (minutes)	215
Specific gravity	3.15
Soundness (Le-Chatelier method in mm)	9

TABLE 10.6
Physical Properties of Coarse and Fine Aggregate

Physical Test	Coarse Aggregate	Fine Aggregate
Specific gravity	2.67	2.66
Fineness modulus	6.86	2.32
Bulk density(kg/m^3)	1540	1780

TABLE 10.7
Relative Error of Predicted Results of B-ANN Model and Experimental Results

S. NO.	CEMENT	WBR	SILICA	SP	FA	CA	filler	CS28	B-ANN	Error (%)
1	154	0.35	0	0.13	980	621	360	38.8	37.31	3.840206
2	300	0.3	0	0.25	958	595	300	66.5	68.115	2.428571
3	350	0.41	0	2	755	940	219	32.55	31.64	2.795699
4	350	0.5	0	2	720	900	219	28.63	30.023	4.865526
5	400	0.33	0	7.84	929	654	160	65.4	61.421	6.084098
							Average Error			4.003%

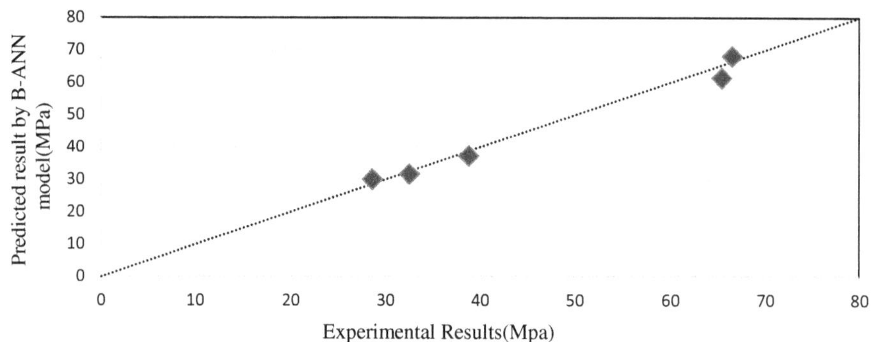

FIGURE 10.11 Comparison between experimental results and B-ANN results.

10.5 CONCLUSIONS

Based on data obtained from literatures and simulation of the compressive strength of SCC using 99 different data set under 8 input parameters, following conclusion can be drawn.

- Design mix of SCC is different from conventional concrete as it requires more workability and contains more fines. The study demonstrates the feasibility of computing technique in developing a nonlinear interaction between various parameters in solving complex civil engineering problems.
- ANN-based bagged model outperform every other model with $R^2 = 0.9289$, MAE = 4.2069 and RMSE = 5.9004 in prediction of compressive strength.
- ANN-based model performs better than techniques, i.e., linear regression, SVM and RF with $R^2 = 0.9249$, MAE = 4.3658, RMSE = 6.0064. Bagging-based RF model perform better than RF model with $R^2 = 0.899$, MAE = 5.4527, RMSE = 7.5371.

- As sensitivity analysis was performed which suggest that cementitious materials were the guiding parameter in comparison to other input parameters for prediction of compressive strength.
- The predicted results shows minimum deviation from experimental results in all the phases of validation thus clarifying the accuracy of proposed B-Artificial Neural Network (ANN) model.
- The compressive strength of SCC improved with usage of silica as partial replacement to cement based on B-ANN model. The results demonstrate the feasibility of using silica as extra cementitious material in SCC for the required strength.

TABLE A.1
Complete Data Set Used for Model Creation

S. No	Cement	WBR	Silica	Silica Type	SP	FA	CA	Filler	CS28
1	300	0.3	0	1	0.25	958	595	300	66.4
2	240	0.3	0	1	0.17	933	595	360	58
3	180	0.3	0	1	0.12	908	595	420	45.6
4	570	0.3	30	1	1.33	1072	595	0	95.3
5	540	0.3	60	1	1.43	1059	595	0	100.5
6	300	0.3	30	1	0.37	958	595	270	75.2
7	240	0.3	30	1	0.3	933	595	330	63.4
8	180	0.3	30	1	0.2	908	595	390	25.7
9	300	0.3	60	1	0.6	958	595	240	85.2
10	240	0.3	60	1	0.48	933	595	300	73.6
11	180	0.3	60	1	0.38	908	595	360	61.2
12	514	0.35	0	1	1.5	1131	621	0	83
13	257	0.35	0	1	0.26	1023	621	257	59.2
14	206	0.35	0	1	0.19	1001	621	309	52.6
15	154	0.35	0	1	0.13	980	621	360	39.8
16	489	0.35	26	1	1.6	1120	621	0	85.3
17	463	0.35	51	1	1.75	1110	621	0	91.6
18	257	0.35	26	1	0.4	1023	621	231	68.4
19	206	0.35	26	1	0.32	1001	621	283	57.4
20	154	0.35	26	1	0.22	980	621	334	45.9
21	257	0.35	51	1	0.62	1023	621	206	75.4
22	206	0.35	51	1	0.5	1001	621	257	64.7
23	154	0.35	51	1	0.4	980	621	309	51.1
24	450	0.4	0	1	1.8	1166	640	0	72.4
25	225	0.4	0	1	0.26	1072	640	225	41.9
26	180	0.4	0	1	0.21	1053	640	270	35.7
27	135	0.4	0	1	0.13	1034	640	315	31.7
28	428	0.4	23	1	1.9	1157	640	0	75.3
29	405	0.4	45	1	2.1	1147	640	0	79
30	225	0.4	23	1	0.43	1072	640	203	51.5

(Continued)

TABLE A.1 (*Continued*)
Complete Data Set Used for Model Creation

S. No	Cement	WBR	Silica	Silica Type	SP	FA	CA	Filler	CS28
31	180	0.4	23	1	0.34	1053	640	248	39.2
32	135	0.4	23	1	0.28	1034	640	293	28.2
33	225	0.4	45	1	0.64	1072	640	180	60.3
34	180	0.4	45	1	0.52	1053	640	225	49.1
35	13	0.4	45	1	0.46	1034	640	270	33.1
36	540	0.34	0	1	1	828.36	817.56	0	60.89
37	513	0.34	27	1	1	828.36	817.56	0	57.33
38	486	0.34	54	1	0.9	828.36	817.56	0	65.78
39	459	0.34	81	1	0.8	828.36	817.56	0	52.44
40	400	0.33	0	3	7.84	929	654	160	68.65
41	400	0.36	2.8	3	7.84	955	673	120	57.84
42	400	0.36	5.6	3	7.84	953	671	120	58.23
43	400	0.36	8.4	3	7.84	949	669	120	64.31
44	400	0.36	11.2	3	7.84	947	667	120	50.78
45	400	0.36	14	3	7.84	943	664	120	50
46	400	0.39	2.8	3	7.84	986	695	80	60
47	400	0.39	5.6	3	7.84	983	692	80	56.47
48	400	0.39	8.4	3	7.84	980	690	80	67.05
49	400	0.39	11.2	3	7.84	977	688	80	55.29
50	400	0.39	14	3	7.84	973	686	80	57.84
51	400	0.42	5.6	3	7.84	1014	714	40	54.31
52	400	0.42	8.4	3	7.84	1010	712	40	55.49
53	400	0.42	11.2	3	7.84	1007	710	40	48.03
54	400	0.42	14	3	7.84	1004	707	40	44.5
55	400	0.46	8.4	3	7.84	1040	733	0	53.52
56	400	0.46	11.2	3	7.84	1038	731	0	47.45
57	400	0.46	14	3	7.84	1034	729	0	45.88
58	400	0.36	2.8	3	7.84	955	673	120	62.82
59	400	0.36	5.6	3	7.84	953	671	120	58.43
60	400	0.36	8.4	3	7.84	949	668	120	54.05
61	400	0.39	2.8	3	7.84	986	695	80	56.64
62	400	0.39	5.6	3	7.84	983	692	80	54.85
63	400	0.39	8.4	3	7.84	979	689	80	49.07
64	400	0.42	2.8	3	7.84	1016	706	40	42.49
65	400	0.42	5.6	3	7.84	1013	714	40	47.67
66	400	0.42	8.4	3	7.84	1010	710	40	47.87
67	400	0.46	5.6	3	7.84	1044	734	0	41.49
68	400	0.46	8.4	3	7.84	1040	732	0	40.89
69	400	0.46	11.2	3	7.84	1036	730	0	45.28
70	400	0.46	1.13	3	7.84	957	674	120	52.51
71	400	0.46	1.38	3	7.84	957	673	120	46.15
72	400	0.46	1.13	3	7.84	988	695	80	54.3

(*Continued*)

TABLE A.1 (*Continued*)
Complete Data Set Used for Model Creation

S. No	Cement	WBR	Silica	Silica Type	SP	FA	CA	Filler	CS28
73	400	0.46	1.68	3	7.84	988	695	80	44.96
74	400	0.46	1.68	3	7.84	1017	716	40	43.17
75	400	0.46	1.68	3	7.84	1048	738	0	42.18
76	400	0.46	2.24	3	7.84	1047	738	0	36.02
77	440	0.33	27.5	1	3.3	970	722	0	55.37
78	385	0.33	27.5	1	1.925	970	722	137.5	62.26
79	330	0.33	27.5	1	1.65	970	722	192.5	45.13
80	275	0.33	27.5	1	1.815	970	722	247.5	44.2
81	385	0.33	27.5	1	1.65	970	722	0	53.7
82	330	0.33	27.5	1	1.375	970	722	192.5	48.3
83	275	0.33	27.5	1	1.1	970	722	247.5	42.9
84	220	0.33	27.5	1	1.375	970	722	165	39.54
85	330	0.33	27.5	1	1.21	970	722	0	44.9
86	275	0.33	27.5	1	1.1	970	722	55	47.7
87	220	0.33	27.5	1	1.045	970	722	110	43.2
88	350	0.41	0	3	2	755	940	219	28.92
89	349.1	0.41	0.9	3	2	755	940	219	31.61
90	348.3	0.41	1.7	3	2	755	940	219	34.29
91	347.4	0.41	2.6	3	2	755	940	219	36.65
92	350	0.45	0	3	2	740	922	219	25.38
93	349.1	0.45	0.9	3	2	740	922	219	29.24
94	348.3	0.45	1.7	3	2	740	922	219	32.68
95	347.4	0.45	2.6	3	2	740	922	219	33.64
96	350	0.5	0	3	2	720	900	219	21.63
97	349.1	0.5	0.9	3	2	720	900	219	27.42
98	348.3	0.5	1.7	3	2	720	900	219	29.89
99	347.4	0.5	2.6	3	2	720	900	219	32.03

CA, coarse aggregate; FA, fine aggregate; SP, Superplasticizer; ST, silica type (1, Silica fume, 2, Nano-silica); WBR, water binder ratio.

REFERENCES

1. Okamura, H., and Ouchi, M. Self-compacting concrete. *Journal of Advanced Concrete Technology*, 2003, 1 5–15.
2. Dinakar, P., Sethy, K.P., and Sahoo, U.C. Design of self-compacting concrete with ground granulated blast furnace slag. *Materials & Design*, 2013, 43 (January) 161–169.
3. Boukendakdji, O., Kadri, E.H., and Kenai, S. Effects of granulated blast furnace slag and superplasticizer type on the fresh properties and compressive strength of self-compacting concrete. *Cement and Concrete Composites*, 2012, April 34(4) 583–590.
4. Wongkeo, W., Thongsanitgarn, P., and Chaipanich, A. Compressive strength and drying shrinkage of fly ash-bottom ash-silica fume multi-blended cement mortars. *Materials and Design*, 2012, 36 655–662.

5. Yazici, H. The effect of silica fume and high-volume Class C fly ash on mechanical properties, chloride penetration and freeze–thaw resistance of self-compacting concrete. *Construction and Building Materials*, 2008, 22 456–462.
6. Kawashima, S., Hou, P., Corr, D.J., and Shah, S.P. Modification of cement-based materials with nanoparticles. *Cement and Concrete Composites*, 2013, 36 (2013) 8–15.
7. Leon, N., Massana, J., Alonso, F., Moragues, A., and Sanchez-Espinosa, E. Effect of nano-Si_2O and nano-Al_2O_3 on cement mortars for use in agriculture and livestock production. *Biosystems Engineering*, 2014, 123 (2014) 1–11.
8. Guneyisi, E., Gesoglu, M., Al-Goody, A., and Ipek, S. Fresh and rheological behavior of nano-silica and fly ash blended self-compacting concret. *Construction and Building Materials*, 2015, 95 (2015) 29–44.
9. Jalal, M., Pouladkhan, A., Harandi, O. F., and Jafari, D. Comparative study on effects of Class F fly ash, nano silica and silica fume on properties of high performance self compacting concrete. *Construction and Building Materials*, 2015, 94 (2015) 90–104.
10. Kavitha, S., and Felix Kala, T. Evaluation of strength behaviour of self-compacting concrete using alccofine and GGBS as partial replacement of cement, *Indian Journal of Science and Technology*, 2016, 9 (22) 1–5.
11. Siddique, R., Aggarwal, P., and Aggarwal, Y. Prediction of compressive strength of self-compacting concrete containing bottom ash using artificial neural networks, *Advances in Engineering Software*, 2011, 42 (2011) 780–786.
12. Shaqadan, A. Prediction of concrete strength using support vector machines algorithm, *Materials Science Forum*, 2020, 986 9–17.
13. Han, Q., Gui, C., Xu, J., and Lacidogna, G. A generalized method to predict the compressive strength of high performance concrete by improved random forest algorithm, *Construction and Building Materials*, 2019, 226 (2019) 734–742.
14. Babajanzadeh, M., and Azizifar, V. Compressive strength prediction of self-compacting concrete incorporating silica fume using artificial intelligence methods. *Civil Engineering Journal*, 2018, July, 4(7):1542. DOI: 10.28991/cej-0309193.
15. Padmini, D., Ilamparuthi, K., and Sudhir, K.P. Ultimate bearing capacity prediction of shallow foundations on cohesionless soils using neurofuzzy models. *Computers and Geotechnics*, 2008, January 35(1) 33–46.
16. Rogers, J.L. Simulating structural analysis with neural network. *Journal of Computing in Civil Engineering*, 1994, 8(2), 252–65.
17. Chopra, P., Sharma, R.K., and Kumar, M. Predicting compressive strength of concrete for varying workability using regression models. *International Journal of Engineering and Applied Sciences*, 2014, 6(4) 10–22.
18. Smola, A.J., and Scholkopf, B. A tutorial on support vector regression. Tech. rep., NeuroCOLT2 Technical Report NC2-TR-1998-030, 1998.
19. Muller, K., Smola, A., Ratsch, G., Scholkopf, B., Kohlmorgen, J., and Vapnik, V. *Using Support Vector Machines for Time Series Prediction, Advances in Kernel Methods: Support Vector Machine*. MIT Press, USA, 1999.
20. Campbell, C., Cristianini, N. and Smola, A. J. Query learning with large margin classifiers: *Proceedings of the Seventeenth International Conference on Machine Learning*, United States, 2000, 111–118.
21. Yang H., Chan, L. and King, I. Support vector machine regression for volatile stock market prediction, IDEAL 2002(8), LNCS 24412, Springer-Verlag Berlin Heidelberg, 2002, 391–396.
22. Breiman, L. Bagging predictors. *Machine Learning*, 1996, 24(2) 123–140.
23. Breiman, L. *Random Forest-Random Features*, Technical Report 567, 1999.
24. Breiman, L. Random forests. *Machine Learning*, 2001, 45(1) 5–32.
25. Efron, B. Nonparametric estimates of standard error: The jackknife, the bootstrap and other methods. *Biometrika*, 1981, 68(3) 589– 599.

26. Wongkeo, W., Thongsanitgarn, P., Ngamjarurojana, A., and Chaipanich, A. Compressive strength and chloride resistance of self-compacting concrete containing high level fly ash and silica fume. *Materials and Design*, 2014, 64 261–269.
27. Mohan, A. and Mini, K.M. Strength and durability studies of SCC incorporating silica fume and ultra-fine GGBS. *Construction and Building Materials*, 2018, 171 919–928.
28. Durgun, M.Y., and Atahan, H.N. Strength, elastic and microstructural properties of SCCs' with colloidal nano silica addition. *Construction and Building Materials*, 2018, 158 295–307.
29. Hani, N., Nawawy, O., Ragab, K.S., Kohail, M. The effect of different water/binder ratio and nano-silica dosage on the fresh and hardened properties of self-compacting concrete. *Construction and Building Materials*, 2018, 165 504–513.
30. Choudhary, R., Gupta, R., and Nagar, R. Impact on fresh, mechanical, and microstructural properties of high strength self-compacting concrete by marble cutting slurry waste, flyash, and silica fume. *Construction and Building Materials*, 2020, 239 117888.

11 Using Soft Computing Techniques to Predict the Values of Compressive Strength of Concrete with Basalt Fiber Reinforced Concrete

Fadi Hamzeh Almohammed, Ankita Upadhya, and Ahmad Alyaseen
Shoolini University

CONTENTS

DOI: 10.1201/9781003184331-11

11.1 INTRODUCTION

In civil engineering, concrete considers an important material. And depending on various factors the property of concrete differs. Concrete strength is related to the loading, the environmental conditions, the construction methods, and the proportions of its constituents (Deepa et al., 2010). For civil engineering, concrete is an essential material for many kinds of structures. With exception of natural disasters, concrete can serve for a long period. Compressive strength must be known to simulate the behavior of the structure. However, due to its ingredients and processes, it cannot be guessed easily (Deepa et al., 2010).

Nowadays fiber reinforced polymer (FRP) composites are one of the most significant composite materials proved by civil engineering. It has begun to be employed in a variety of structural applications, including the rehabilitation and restoration of existing structures as well as new construction. FRP composites offer many merits due to good corrosion behavior, electromagnetic neutrality, and extremely high strength-to-weight ratio (Jalal, 2013). Fibers added to concrete can effectively inhibit the formation and propagation of fractures in concrete via bridging while also increasing toughness, especially under high strain-rate loading (Xu et al., 2012). Flexible (polypropylene fiber) and stiff (nylon fiber) fibers are categorized based on their elastic moduli (steel fiber, carbon fiber, and basalt fiber). Flexible fibers can work on the fracture and affect the resistance of concrete because of their high ductility, but stiff fibers add to the strength of concrete to a degree (Fu et al., 2018, John and Dharmar 2021).

Basalt fiber is one of the fibers which can be manufactured from basalt rock. And it is giving properties high modulus of elasticity, interfacial shear strength, satisfactory heat resistance, acceptable chemical resistance, and a less expensive manufacturing procedure when compared to other fibers (Mohammadyan-Yasouj and Ghaderi, 2020). Basalt fiber has been examined for its mechanical qualities and endurance in cementitious composites in the form of a single type, bundled mesh, or composite fibers (Mohammadyan-Yasouj and Ghaderi, 2020).

Basalt fiber differs greatly from other high-tech fibers in terms of properties (Militky et al., 2002). Basalt fiber is made from volcanic rock, during the manufacturing process for the basalt fiber it creates no harmful gas or waste residue and has high chemical and thermal stability. Basalt fiber has a substantially higher strength than natural fiber and synthetic fiber, making it another green material that meets environmental insurance measures (Czigány 2005).

It possesses stronger compressive and shear strength than one-dimensional linear polymer fibers, as well as good adaptability and aging resistance in hostile conditions, thanks to its three-dimensional molecules. According to the findings, concrete's hardness can be improved using basalt fiber. Due to its tensile strength and high elastic modulus, basalt fiber strengthens concrete on a microscopic level and works as a bridge at fissures (Jiang et al., 2014; Jalasutram et al., 2017, Arslan, 2016). It can also slow the spread of cracks by enhancing the energy absorption capacity of concrete and enhancing its toughness (Kizilkanat et al., 2015). When basalt fibers spread well in concrete, fibers and concrete form an outstanding combination. As a result, the strengthening impact of fiber reinforced concrete was improved, whereas fiber pull out ability was limited (Gamal and Chiadighikaobi, 2019). Furthermore, basalt fiber has a strong interfacial adhesion property with concrete, despite the fact that it is an inorganic substance. (Wang et al., 2014). Because of the physical and mechanical qualities of fibers, there is some variation in the appropriate basalt fiber content. Meanwhile, the results of the tests show some discrepancies due to the lack of a relevant toughness evaluation index.

Although basalt fiber has a low compressive strength, it is superior in conditions of tensile and flexural properties resistance. Split tensile testing is employed in the bulk of tensile properties studies. There is a paucity of research on the direct tensile test due to the test's stringent requirements (Zhou et al., 2020).

Complex plants are difficult to regulate using traditional ways because they cannot be adequately characterized by mathematical models, which is one of the challenges with traditional control systems. Soft computing, on the other hand, uses partial truth, ambiguity, and approximation to tackle challenging problems (Ibrahim, 2016, Milicevic et al., 2000). According to Dr. Zadeh, a pioneer of fuzzy logic, "Soft computing's guiding concept is to take advantage of tolerance for imprecision, uncertainty, and imprecise information to gain predictability, resilience, smaller costs, and a better relationship with reality" (Ibrahim, 2016, Maier and Dandy 2000). Because of capabilities such as nonlinear programming, optimization, intelligent control, and decision-making support, soft computing has become popular and has drawn academic attention from people from all backgrounds (Jang et al., 1997).

Traditional control system techniques are finding it increasingly challenging to regulate the increasing complexity of modern machines. Traditional methods, for example, struggle to control and stabilize a large number of nonlinear and time-variant plants with long-time delays. One of the reasons for this problem is the lack of a realistic model that depicts the plant (Ibrahim, 2016). Machine learning is proven to be a cost-effective method of controlling this sophisticated machinery. Machine learning encompasses a variety of techniques such as fuzzy logic, neural networks, and evolutionary algorithms, rather than a single method. All of these approaches are complementary to one another and can be combined to solve a specific problem (Buckley and Hayashi, 1994).

For modeling the properties of materials there are several methods, involving statistical techniques, computational modeling, and recently tools like regression analyses and artificial neural networks (ANN) (Mohammed et al., 2020a). Civil engineering professors are also interested in soft computing approaches. A neuro-fuzzy inference system and genetic programming are two examples of adaptive neuro-fuzzy inference systems (Jalal, 2013). For concrete construction, it is important to predict the compressive strength because it is given an idea about project scheduling, quality control, and time for concrete form removal. For predicting concrete strength several approaches using regression function have been proposed (Deepa et al., 2010).

Techniques like linear regression analysis, M5P modal trees, and multilayer perceptron are used in civil engineering problems in general and structural engineering in particular (Deepa et al., 2010, Baykasoğlu et al., 2004, Maier and Dandy 2000).

For the compressive strength of concrete, ANN is a suitable prediction model. ANN was also discovered to be a suitable model for predicting displacement in concrete reinforcement structures by researchers (Mohammed et al., 2020b).

Mohammed et al., 2020b found ANN to be a good model for predicting displacement in concrete reinforcing structures as well as compressive strength of concrete. ANN model successfully predicted the binding strength of FRP bars in concrete (Thakur et al., 2021).

11.2 REVIEW OF REGRESSION AND SOFT COMPUTING TECHNIQUES

11.2.1 Artificial Neural Networks

Inspired by biological cerebral activity, neural networks are a strong soft computing paradigm for linear and nonlinear approximations in a number of fields (Altunkaynak et al., 2005a; Can, 2001; Hsu et al., 1995; Kisi, 2004; Altunkaynak 2007b). The ANN principle is similar to how biological neuron cells in the brain work in that it estimates an output as a function of the input. In order to recognize patterns such as faces and sounds, this cell network can be taught and learned from earlier examples (experience). In ANN terminology, training refers to the process of estimating model parameters. The input, hidden, and output layers make up the neural network. Each layer may include numerous units that are integrated with the next layer, and each link in the system has an adjusted weight. The data is received by the input layer's nodes (dendrites), which are then processed and delivered to the hidden layer's nodes (cell body) (axon) (Sangeeta et al., 2021). The human brain contains approximately a billion neurons and various functional links that enable should be able to recognize and discriminate between various human voices. It can also tell the difference between background noises like traffic, ocean waves, and refrigerator mechanical noise, which is a troublesome assignment for most supercomputers today (Silverman and Dracup, 2000). The number of units in the hidden layer that is optimum, according to a realistic approach outlined, is the product of the number of input and output neurons divided by two-thirds. While fewer neurons may not be adequate to express complex connections between predictors and computed output, more neurons may be required, a larger number of hidden nodes may. However, the amount of time spent training must be increased, and accuracy will almost certainly suffer as a result, or the problem of an "overfitted network" may occur (Silverman and Dracup, 2000). Overfitting refers to the network's irregularities. During the training phase, the network memorizes the data, but it fails to operate with the data set even during the validation phase.

According to Equation 11.1, the total number of input parameters equals the total number of nodes in the first layer. The input values are multiplied by the weight of the link as they transit to the next layer. Every node j in the preceding layer gets signals from every node I. A weight is assigned to the incoming signal (x_i) (w_{ji}). The weighted total of all incoming signals (equation 11.1) is the effective signal (E_j) to node j. The weights (w_{ji}) are randomly assigned in the initial phase of training:

$$E_j = \sum_{i=1}^{m} X_i w_{ji} \tag{11.1}$$

The effective signal (E_j) is then transferred via a transfer function (i.e., equation 11.2) to generate the outgoing signal (y_j) of node j in the following phase. Following are non-sigmoidal transfer functions (logistic and hyperbolic tangent functions): Hard limit transfer functions include the terms linear, polynomial, rational function (polynomial ratios), and Fourier series are used interchangeably (bounded to 0 or 1). (Cosine summaries) Due to their benefit in extrapolation outside the area of the training data. In the hidden layers, sigmoidal transfer functions are used, while in the output layer, linear transfer functions are used ($y_j = E_j$) are the most often used transfer functions (Pulido-Calvo and Portela, 2007; Zealand, 1997):

$$Y_i = f\left(E_j\right) = \frac{1}{1+\exp\left(-E_j\right)} \tag{11.2}$$

The relevance of inputs in the entire estimation process is represented by the network parameters. By changing the network parameters, the fitting error (equation 11.3) between both the intended and estimated output is used as feedback to enhance the network's performance:

$$\text{Error} = \sum_{j=1}^{N}\left(Y_j - d_j\right)^2 \tag{11.3}$$

The number of output nodes is N, the calculated output is y_j, and the needed data value is d_j. This procedure is repeated until a new layer has formed (Silverman and Dracup, 2000, Koufteros (1999). In prediction, pattern recognition, and nonlinear function modeling, the feed-forward backpropagation algorithm is the most commonly used supervised technique (Zealand, 1997; Kisi, 2008). The sigmoid activation function is frequently chosen when utilizing the feed-forward backpropagation network (Demirel et al., 2009). The network is assessed using a set of data pairs (input–output) and the initial conditions are modified in each iteration phase to obtain accurate predictions. The training (calibrating) portion of the forecasting process is critical. At the output layer, the gradient for each node is determined for minimization.

$$\delta_k = d_{\sigma_k}\left(y_k - d_k\right) \tag{11.4}$$

where d_{σ_k} = the sigmoid function's derivative applied at y_k, which is specified for each kth output node. The gradient function for a concealed layer (one layer back) becomes.

$$\delta_j = d_{\sigma_j}\sum_{i=1}^{N}\delta_i w_{jk} \tag{11.5}$$

The weight value from hidden node j to output node k is w_{jk}, while the derivative of the sigmoid function is d_{rj}. The network runs once the input data is selected; the weights for each link are changed using equation (11.6) until the error is reduced to a preset error target or the requisite number of training trials is reached:

$$\Delta w_{jk} = w_{jk} - \eta \delta_k y_j \tag{11.6}$$

The rate at which each layer returns information to the network is represented by the notation. Epochs are the number of times the training data is cycled through. Prior to analysis in MATLAB® techniques, the operator can choose the number of iterations and individually tweak till a realistic result is reached in the trial-and-error phase (Demirel et al., 2009).

Unlike the SWAT model, ANN modeling does not require prior knowledge of the underlying physical processes. Furthermore, unlike traditional statistics and optimization models, there are no prerequisite conditions (e.g., normal distribution). However, ANNs have a few limitations, such being an exponential rise in the training phase as data set size increases, as a result of the network's complicated interactions (Demirel et al., 2009; Silverman and Dracup, 2000). The ANN structure is shown in Figure 11.1.

11.2.2 Random Forest

Breiman was the first to suggest the random forest (RF) algorithm (1996). It's a highly adaptable algorithm that's used successfully to a variety of engineering problems.

The RF regression method uses a set of tree predictors, with each tree being produced from the input vector using a randomly sampled vector. The RF regression uses a random selection of input variables or a mix of variables at each node to create a tree, like in this study (Singh et al., 2019; Upadhya et al., 2021).

Data from the input parameters are used to fit the output values while employing RF to solve regression problems. The data is divided into many points for each input variable, and the sum of square error for the expected and actual values at each divided point is determined. The smallest sum of square error for the node is then

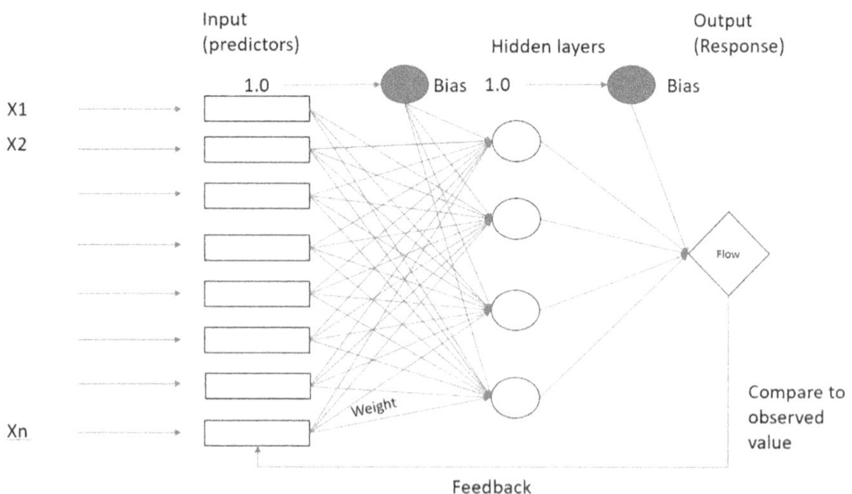

FIGURE 11.1 ANN system structure.

chosen. Furthermore, by subtracting all the input parameters' values and evaluating the variance in accuracy rate in out-of-bag samples, the variable significance may be calculated (a data set with a large number of observations not used in training, referred to as the "out-of-bag" data set) (Zhang et al., 2019).

The RF algorithm, according to (Breiman, 2001), is simple to use, has a low learning curve, and has a greater prediction accuracy. Two user-defined parameters must be utilized. The amount of user-defined input variables grows as the number of nodes develops. Models are built using the trial-and-error process. Figure 11.2 shows the structure system of random forest.

11.2.3 M5P Model

The problem is solved using the M5P decision tree technique. It was designed by Quinlan in 1992 to forecast continuous variables. In high-dimensional tasks, the M5 model's tree structure for descriptive and continuous data, and also missing values, is utilized (Ali et al., 2020).

One type of regression tree is the model tree, which has linear functions in its leaves. They are similar to nonlinear equations in small bit form. A tree with the M5P model is capable of estimating continuous mathematical features (Thakur et al., 2021; Mohammed et al., 2020; Almasi et al., 2017; Upadhya et al., 2021). For reducing the threat of overfitting, pruning is applied in this algorithm. A separation strategy is utilized to gain better information with less divergence in the cross-functional and cross-class values in each branch. Tree growth, trimming, and smoothing are the three primary processes in the M5P preparation process (Thakur et al., 2021).

A decision "tree" refers to a tree having roots, branches, and leaves to analyze the data. In general, recursion is used to find the best conclusion for a given class, and it comprises picking generating a tree for every one of the output permutations inside

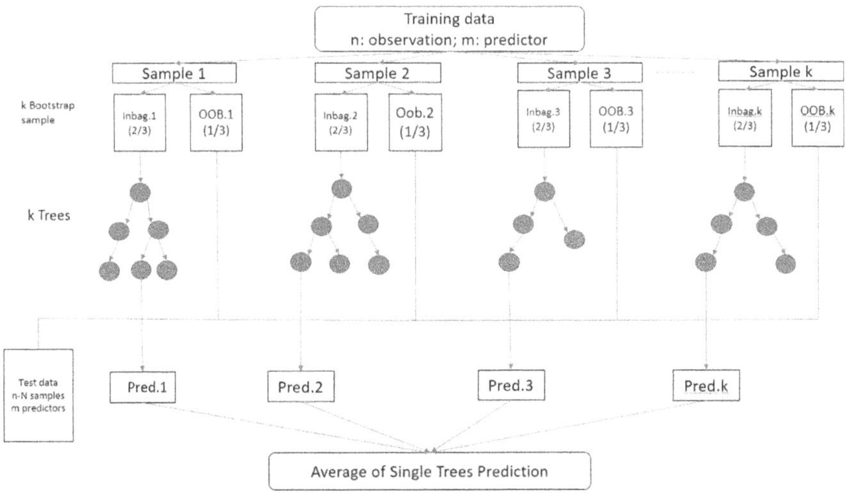

FIGURE 11.2 Random forest system structure.

the root node, and repeating the procedure for each branch, only considering the characteristics that approach the subsidiary (Ali et al., 2020).

A tree's growth can be divided into two stages. The initial phase is the tree's construction. An iterative involves dividing the data set into multiple distinct sub-bands in terms of subclasses is started based on a training sample. The forecast criterion is used to make the estimate. The tree will be pruned in the second phase. To obtain strong predictive performance, the final step removes the unrepresented branches. This operation necessitates the use of a criterion to identify which branches should be pruned. After being trimmed, the dispersion of observations utilized in the learning process is used to identify the new leaves (Blaifi et al., 2018; Ali et al., 2020).

A model tree is created using the M5P method. The aim is to create a network that connects the training instances' target values to the values of their input characteristics. The precision with which the model predicts the target values of the unknown conditions will define the model's quality (Deepa et al., 2010).

By classifying or splitting the entire data space into many subspaces, the tree algorithm allocates a linear regression function matches at a multiple regression analysis to the terminal node model to each subspace. The M5P tree method can handle very high dimensionality and deals with continuous problems rather than discrete ones. It shows the piecewise details of each linear model that was built to approximate the data set's nonlinear relationship. Information about the partition norm of the M5P model tree has been obtained based on the error calculations of each node. To calculate the error, use the confidence interval of the classifier that enters the node. To break this node, choose the attribute that maximizes the reduction of expected errors caused by the test of each attribute. The following formula is used to measure the standard deviation reduction (SDR):

$$\text{SDR} = sd(K) - \sum \frac{|K_i|}{|K|} sd(K_i) \tag{11.7}$$

where K is a collection of instances that reach the node, K is the subset of illustrations with the potential set's product, and sd is standard deviation.

11.2.4 Stochastic

Stochastic is a meta-assembly procedure that was chosen to improve the accuracy of a model based on standard methodologies. It is commonly utilized to solve complicated and nonlinear issues in a variety of engineering applications. This work used a stochastic technique to increase the accuracy of M5P models (Thakur et al., 2021).

A stochastic process is a collection of random variables whose values are derived from the data matrix described by the set of data R. A separate set of identifiers named W is used to index the gathering. The nonnegative real numbers $W=[0]$ and the natural numbers $W=0, 1, 2,\ldots$ are the most common index sets for discrete and continuous time, respectively.

As a result, the first index set yields a series of arbitrary variables $A0, A1, A2,\ldots,$ whereas the second yields a collection of arbitrary variables $A(t)$, $t0$, one for each t. For the construction of numerical models, stochastic processes are heavily used.

11.2.5 Random Tree Model

Random tree (RT)'s multi-strategy functionality allows users to generate a large number of different regression models. These models, however, are tree-based, which means that training cases would be partitioned or built for all training instances using tree-based models.

The main features of RT for tree-based models are learning trees to minimize the square error (least squares) regression tree, learning trees to minimize the absolute deviation (least absolute deviation) regression tree, and pruning the probability between different models to be used in the leaves during the prediction task based on the sequence-based regression tree. RT built a number of different tree models before deciding on one based on a set of criteria. A number of methods have been used to find the best pruning tree from a sequence of alternatives trees.

11.2.6 Performance Evaluation Indices

11.2.6.1 Correlation Coefficient (CC)

The correlation indicates how strong a linear link exists between two parameters. Given a collection of data, ($O1$, $P1$), ($O2$, $P2$),…(On, Pn), CC is represented by equation (11.8).

CC is always having a range between −1 and 1 in value, where 1 or −1 means complete correlation (in this case, all points are on a straight line). A positive correlation means a positive correlation between variables (the rise in one variable is characterized by an increase in another), while negative correlation means the negative correlation between variables (the increase in value is that one variable resembles the decrease in another variable). A small value (changeable). A correlation value close to 0 means that there is no correlation between the variables.

Since CC is calculated using a formula that standardizes the variables, changes in the scale or measurement unit have no effect on its value. As a consequence, CC is generally more useful than a graphical definition when evaluating the intensity of the correlation between two variables.

$$\text{Correlation coefficient CC} = \frac{\sum_{i=1}^{N}\left(O_i - \bar{O}\right)\left(P_i - \bar{P}\right)}{\sqrt{\sum_{i=1}^{N}\left(O_i - \bar{O}\right)^2}\sqrt{\sum_{i=1}^{N}\left(P_i - \bar{P}\right)^2}} \tag{11.8}$$

11.2.6.2 Root Mean Square Error (RMSE)

The root mean square deviation or root mean square error (RMSE) is a widely used statistical measure metric for comparing the expected and actual values of a model or estimate (sample or population value). The root mean square deviation is equal to the square average of the root.

$$\text{RMSE} = \sqrt{\frac{1}{N}\sum_{i=1}^{N}\left(P_i - O_i\right)^2} \tag{11.9}$$

11.2.6.3 Mean Absolute Error

The mean absolute error (MAE) is a statistic that measures the difference in error between pairs of observations that describe the same phenomenon. The calculation formula of MAE is represented by the equation (11.10):

$$\text{MAE} = \frac{1}{N}\sum_{i=1}^{N}\left|P_i - O_i\right| \tag{11.10}$$

11.2.6.4 Nash Sutcliffe Model Efficiency

The efficiency of Nash–Sutcliffe is calculated by dividing the ratio of the observed time series variance by the error variance of the modeled time series. In the case of a perfect model with zero expected error variance, the Nash–Sutcliffe efficiency is 1 (NSE = 1).

$$\text{NSE} = 1 - \left[\frac{\sum_{i=1}^{N}(O_i - P_i)^2}{\sum_{i=1}^{N}\left(\bar{O} - \bar{P}\right)^2}\right] \quad 0 \le \text{NSE} \le 1 \tag{11.11}$$

where O = values of the actual observations in a sample, $\bar{O}$ = mean of the values of the actual observations, P = values of the predicted observations in a sample, and $\bar{P}$ = mean of the values of the predicted observations.

11.3 MATERIALS AND METHODOLOGY

In this chapter, a collection of data for compressive strength of concrete with basalt fiber reinforced polymer (BFRP), and then applying the soft computing techniques to predict the compressive strength and trying to define which is the best technique for prediction. The data collected have nine different inputs which are cement, Fine aggregate or crushed sand, coarse aggregate, water, superplasticizer, fly ash, BFRP, length of the fiber, curing time, and compressive strength as output. By dividing the data into training and testing data set and then applying the soft computing techniques, one of these techniques will give the best values and coefficient performance. The analysis will be depending on WEKA software for the following techniques: M5P, M5P-Stochastic, ANN, RF, and RT. The next table shows the data set used (Table 11.1).

11.3.1 Data Set

Overall 100 observations had been collected from previous research papers. Out of 100, randomly separated 70 observations were chosen as training data set and 30 observations were taken for model validation and testing. Total data set consider nine input variables namely cement, fine aggregate or crushed sand, coarse aggregate, water, superplasticizer, fly ash, BFRP, length of the fiber, curing time, and compressive strength of concrete with BFRP is considered as output. The description of the feature of all overall data is shown in Table 11.2.

TABLE 11.1
The Total Data Set Collected from Lateral Review

Cement (kg/m³)	Fine Aggregate/ Crushed Sand (kg/m³)	Coarse Aggregate (kg/m³)	Water (kg/m³)	SP (kg/m³)	Fly Ash (kg/m³)	BFRP	Diameter (μm)	Length (mm)	Curing Time (Day)	Compressive Strength
514.5	942	824.4	343.5	0	0	0	0	0	28	21.3
514.5	942	824.4	343.5	0	0	0.1	16	18	28	21.9
514.5	942	824.4	343.5	0	0	0.3	16	18	28	22.4
514.5	942	824.4	343.5	0	0	0.5	16	18	28	22.95
514.5	942	824.4	343.5	0	0	1	16	18	28	23.2
514.5	942	824.4	343.5	0	0	1.5	16	18	28	22.6
350	1100	740	0.5	3.5	0	0.5	17	24	28	46.69
350	1100	740	0.5	3.5	0	1	17	24	28	45.28
350	1100	740	0.5	3.5	0	2	17	24	28	46.45
350	1100	740	0.5	3.5	0	3	17	24	28	47.17
380	720	1215.60	171	1.9	0	0	0	0	28	38.25
380	720	1215.60	171	1.9	0	0.5	16	24	28	48.5
380	720	1215.60	171	1.9	0	0.75	16	24	28	46.52
380	720	1215.60	171	1.9	0	1	16	24	28	40.28
448.84	624.35	1024.04	269.68	0	126.4	0		0	7	38.21
448.84	624.35	1024.04	269.68	0	126.4	0.05		12	7	39.63
448.84	624.35	1,024.04	269.68	0	126.4	0.1		12	7	41.27
448.84	624.35	1024.04	269.68	0	126.4	0.3		12	7	41.52
448.84	624.35	1024.04	269.68	0	126.4	0.5		12	7	40.68
448.84	624.35	1024.04	269.68	0	126.4	0.05		22	7	40.42
448.84	624.35	1024.04	269.68	0	126.4	0.1		22	7	41.23
448.84	624.35	1024.04	269.68	0	126.4	0.3		22	7	42.18
448.84	624.35	1024.04	269.68	0	126.4	0.5		22	7	41.32
448.84	624.35	1,024.04	269.68	0	126.4	0		0	28	45.12
448.84	624.35	1024.04	269.68	0	126.4	0.05		12	28	46.75

(Continued)

TABLE 11.1 (*Continued*)
The Total Data Set Collected from Lateral Review

Cement (kg/m³)	Fine Aggregate/ Crushed Sand (kg/m³)	Coarse Aggregate (kg/m³)	Water (kg/m³)	SP (kg/m³)	Fly Ash (kg/m³)	BFRP	Diameter (μm)	Length (mm)	Curing Time (Day)	Compressive Strength
448.84	624.35	1024.04	269.68	0	126.4	0.1		12	28	47.21
448.84	624.35	1,024.04	269.68	0	126.4	0.3		12	28	46.28
448.84	624.35	1024.04	269.68	0	126.4	0.5		12	28	45.02
448.84	624.35	1024.04	269.68	0	126.4	0.05		22	28	46.95
448.84	624.35	1,024.04	269.68	0	126.4	0.1		22	28	47.68
448.84	624.35	1024.04	269.68	0	126.4	0.3		22	28	46.96
448.84	624.35	1024.04	269.68	0	126.4	0.5		22	28	45.35
448.84	624.35	1,024.04	269.68	0	126.4	0		0	90	52.19
448.84	624.35	1024.04	269.68	0	126.4	0.05		12	90	53.01
448.84	624.35	1024.04	269.68	0	126.4	0.1		12	90	52.1
448.84	624.35	1,024.04	269.68	0	126.4	0.3		12	90	51.16
448.84	624.35	1024.04	269.68	0	126.4	0.5		12	90	50.45
448.84	624.35	1024.04	269.68	0	126.4	0.05		22	90	54.12
448.84	624.35	1,024.04	269.68	0	126.4	0.1		22	90	53.1
448.84	624.35	1024.04	269.68	0	126.4	0.3		22	90	51.21
448.84	624.35	1024.04	269.68	0	126.4	0.5		22	90	50.56
217	694	1416	130	2.71	54	0	0	0	28	29.9
217	694	1416	130	2.71	54	0.1	17.4	6	28	31.4
217	694	1416	130	2.71	54	0.2	17.4	6	28	33.4
217	694	1416	130	2.71	54	0.3	17.4	6	28	37.2
217	694	1416	130	2.71	54	0.4	17.4	6	28	35.2
217	694	1416	130	2.71	54	0.5	17.4	6	28	33.9
217	694	1416	130	2.71	54	0.1	17.4	12	28	30.5
217	694	1416	130	2.71	54	0.2	17.4	12	28	32.4
217	694	1416	130	2.71	54	0.3	17.4	12	28	33.5

(Continued)

TABLE 11.1 (*Continued*)
The Total Data Set Collected from Lateral Review

Cement (kg/m³)	Fine Aggregate/ Crushed Sand (kg/m³)	Coarse Aggregate (kg/m³)	Water (kg/m³)	SP (kg/m³)	Fly Ash (kg/m³)	BFRP	Diameter (μm)	Length (mm)	Curing Time (Day)	Compressive Strength
217	694	1416	130	2.71	54	0.4	17.4	12	28	26.8
217	694	1416	130	2.71	54	0.5	17.4	12	28	25.5
375	692	1207	169	3.75	0	0	0	0	28	38.4
375	692	1207	169	3.75	0	0.5	13	12.7	28	38.4
375	692	1207	169	4.87	0	1	13	12.7	28	37.4
375	692	1207	169	5.62	0	1.5	13	12.7	28	37.6
375	692	1207	169	6.75	0	2	13	12.7	28	37.3
442	678	984	156	4.16	78	0	0	0	3	31.83
442	678	984	156	4.16	78	0.5	15	18	3	32.38
442	678	984	156	4.16	78	1	15	18	3	30.32
442	678	984	156	4.16	78	1.5	15	18	3	29.53
442	678	984	156	4.16	78	2	15	18	3	27.87
442	678	984	156	4.16	78	0	0	0	7	32.46
442	678	984	156	4.16	78	0.5	15	18	7	35.86
442	678	984	156	4.16	78	1	15	18	7	34.45
442	678	984	156	4.16	78	1.5	15	18	7	30.72
442	678	984	156	4.16	78	2	15	18	7	30.88
442	678	984	156	4.16	78	0	0	0	28	35.23
442	678	984	156	4.16	78	0.5	15	18	28	38.71
442	678	984	156	4.16	78	1	15	18	28	37.37
442	678	984	156	4.16	78	1.5	15	18	28	34.83
442	678	984	156	4.16	78	2	15	18	28	32.38
442	678	984	156	4.16	78	0	0	0	60	35.36
442	678	984	156	4.16	78	0.5	15	18	60	43.75
442	678	984	156	4.16	78	1	15	18	60	39.46

(*Continued*)

TABLE 11.1 (*Continued*)
The Total Data Set Collected from Lateral Review

Cement (kg/m³)	Fine Aggregate/ Crushed Sand (kg/m³)	Coarse Aggregate (kg/m³)	Water (kg/m³)	SP (kg/m³)	Fly Ash (kg/m³)	BFRP	Diameter (μm)	Length (mm)	Curing Time (Day)	Compressive Strength
442	678	984	156	4.16	78	1.5	15	18	60	37.53
442	678	984	156	4.16	78	2	15	18	60	34.52
442	678	984	156	4.16	78	0	0	0	90	39.4
442	678	984	156	4.16	78	0.5	15	18	90	46.15
442	678	984	156	4.16	78	1	15	18	90	43.94
442	678	984	156	4.16	78	1.5	15	18	90	41.4
442	678	984	156	4.16	78	2	15	18	90	37.37
420	656	1069	210	2.4	60	0	0	0	7	41.1
420	656	1069	210	2.4	60	1	15	30	7	41.8
420	656	1069	210	2.4	60	1.5	15	30	7	40.3
420	656	1069	210	2.4	60	2	15	30	7	38.6
420	656	1069	210	2.4	60	2.5	15	30	7	30.6
420	656	1069	210	2.4	60	3	15	30	7	26.4
420	656	1069	210	2.4	60	3.5	15	30	7	34.3
420	656	1069	210	2.4	60	0	0	0	28	49.4
420	656	1069	210	2.4	60	1	15	30	28	55
420	656	1069	210	2.4	60	1.5	15	30	28	52.4
420	656	1069	210	2.4	60	2	15	30	28	48.3
420	656	1069	210	2.4	60	2.5	15	30	28	46.8
420	656	1069	210	2.4	60	3	15	30	28	44.1
420	656	1069	210	2.4	60	3.5	15	30	28	45.4
450	670	1100	180	0	0	0	0	0	28	73.89
450	670	1100	180	0	0	1	18	25	28	74.48
450	670	1100	180	0	0	2	18	25	28	77.26
450	670	1100	180	0	0	3	18	25	28	77.9

TABLE 11.2
Features of Data Set

Input Parameters[a]/ Statistics	Mean	Median	Mode	Standard Deviation	Minimum	Maximum	Confidence Level (95.0%)
Training							
Cement (kg/m^3)	409.916	442	448.84	77.106488	217	514.5	18.38539316
Fine aggregate/crushed sand (kg/m^3)	698.1186	678	624.35	110.95585	624.35	1100	26.45648819
Coarse aggregate (kg/m^3)	1060.687	1024.04	1024.04	162.11463	740	1416	38.65486966
Water (kg/m^3)	196.2534	175.5	269.68	73.211686	0.5	343.5	17.45671035
SP (kg/m^3)	2.157143	2.4	0	1.7818279	0	5.62	0.424861859
Fly ash (kg/m^3)	67.30286	78	126.4	44.93926	0	126.4	10.71538832
BFRP	0.879286	0.5	0	0.9663353	0	3.5	0.230414509
Length (mm)	16.56286	18	18	9.0614676	0	30	2.160630677
Curing time (day)	32.6	28	28	26.568151	3	90	6.334952079
Compressive strength	40.73157	39.955	32.38	11.155749	21.3	77.9	2.659994481
Testing							
Cement (kg/m^3)	414.1187	442	448.84	74.919154	217	514.5	27.97527174
Fine aggregate/crushed sand (kg/m^3)	694.305	678	624.35	108.33048	624.35	1100	40.45126708
Coarse aggregate (kg/m^3)	1058.225	1024.04	1024.04	157.01435	740	1416	58.63012243
Water (kg/m^3)	204.1873	195	269.68	73.337072	0.5	343.5	27.38451283
SP (kg/m^3)	2.091667	2.4	0	1.9515194	0	6.75	0.728709338
Fly ash (kg/m^3)	69.52	78	126.4	47.174558	0	126.4	17.61526935
BFRP	0.743333	0.5	0.5	0.7485311	0	2.5	0.279506117
Length (mm)	15.94667	18	18	8.3844986	0	30	3.130823216
Curing time (day)	32.66667	28	28	26.576284	3	90	9.923747393
Compressive strength	40.81633	39.64	#N/A	10.482538	22.4	74.48	3.914243903

[a] C = cement, CA = coarse aggregate, FA = fine aggregate, W = water, FA = flyash, BFRP = basalt fiber-reinforced polymer, CT = curing time, CS = compressive strength.

11.4 RESULTS AND DISCUSSION

11.4.1 Results of ANN Technique

ANN made it difficult to choose factors like the number of hidden neurons, the number of iterations, and the number of hidden nodes underlying weight. Because the topology of a neural network had a direct impact on its computational complexity and generalization capability, picking the proper design for a particular application is crucial. A trial and error approach where the only single hidden layer is used to train ANN is used to find the optimum network geometry. It's important to mention when enough network weights are applied, one hidden layer may approach any linear combination. Since no theory exists to forecast how many hidden units are anticipated to inexact any given function, the number of hidden layer nodes in an ANN model was determined via testing with different networking architectures. For such hidden layers and the output nodes, a backpropagation algorithm was employed.

Figures 11.3 and 11.4 show the agreement plot among observed and predict values using ANN for the training and testing stages, respectively. Overall performance of

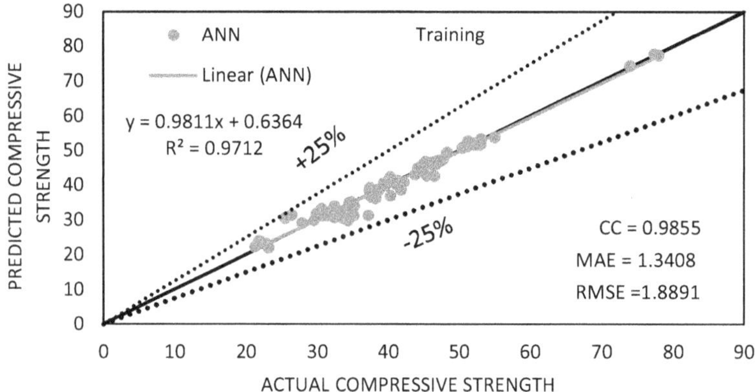

FIGURE 11.3 Actual versus predicted values using ANN using the training data set.

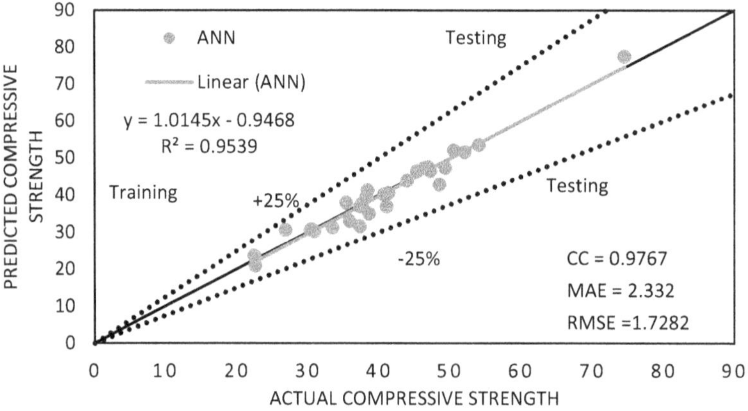

FIGURE 11.4 Actual versus predicted values using ANN using the testing data set.

ANN in predicting concrete compressive strength is satisfactory with CC (0.9855, 0.9767), RMSE (1.8891, 2.332), MAE (1.3408, 1.7282) for both training and testing stages.

11.4.2 Results of the Tree and Forest-Based Models.

In this chapter, M5P, M5P-Stochastic, RF, and RT-based models have been developed using WEKA software to predict the values of compressive strength of concrete with BFRP. Model development is a trial-and-error process. Many trials have been carried out to find the optimum value user-defined parameters of various tree and forest-based models. Figure 11.5 shows the error graphs for ANN model. Figures 11.6 and 11.7 represent the agreement plot between observed and predicted values using M5P, M5P Stochastic, RF, and for the training and testing stages, RT tree-based

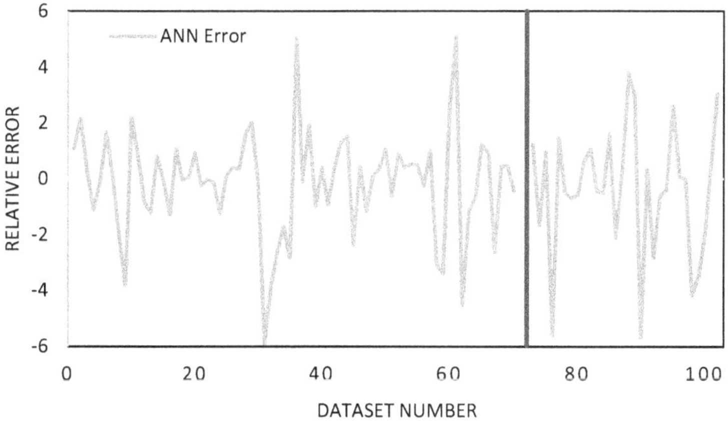

FIGURE 11.5 Error graphs for ANN model.

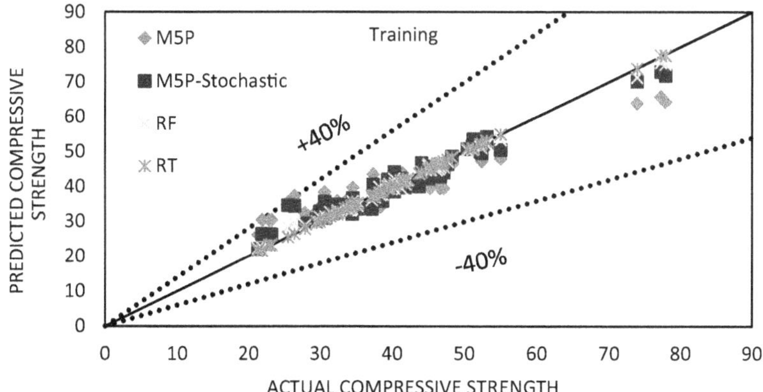

FIGURE 11.6 Actual versus predicted values using M5P, M5P-Stochastic, ANN, RF, and RT using the training data set.

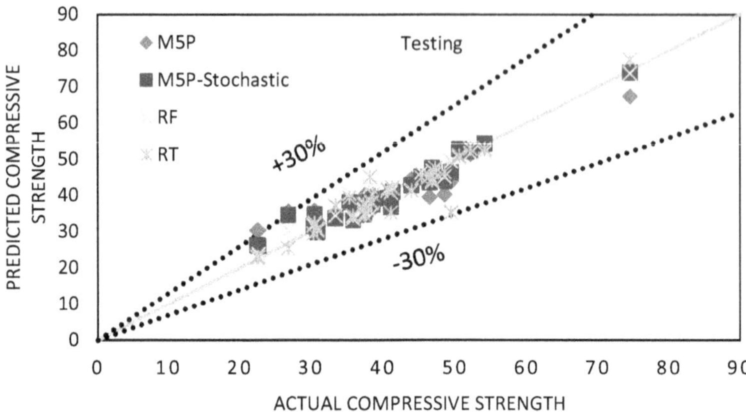

FIGURE 11.7 Actual versus predicted values using M5P, M5P-Stochastic, ANN, RF, and RT using the testing data set.

algorithms are used. Most of the predicted values using M5P, M5P-Stochastic, RF, and RT tree-based models lie within + 15% error band. Table 11.3 shows the training and testing data sets for different tree and forest-based models, as well as performance assessment index values. Table 11.3 shows that for training and testing phases, the RF-based model outperforms the M5P, M5P-Stochastic, and RT-based models, with higher CC (0.9968, 0.9813) and lower RMSE (1.0876, 2.027) and MAE values (0.8193, 1.5236). Figure 11.8 shows the shows Error Graphs for M5P, M5P-Stochastic, RF, RT model.

TABLE 11.3
Performance Evaluation Parameters M5P, M5P-Stochastic, ANN, RF, and RT

	Correlation Coefficient	Mean Absolute Error	Root Mean Square Error	Relative Absolute Error (%)	Root Relative Square Error (%)
Training					
M5P	0.9426	3.4465	4.5927	42.37	41.47
M5P-Stochastic	0.996	0.5731	1.0089	7.05	9.11
ANN	0.9855	1.3408	1.8891	16.48	17.06
RF	0.9968	0.8193	1.0876	10.07	9.82
RT	0.9999	0.0627	0.1324	0.77	1.20
Testing					
M5P	0.9353	3.1103	4.1132	39.96	39.91
M5P-Stochastic	0.9754	1.6725	2.3681	21.48	22.98
ANN	0.9767	1.7282	2.332	22.20	22.63
RF	0.9813	1.5236	2.0279	19.57	19.68
RT	0.9436	2.183	3.5274	28.04	34.22

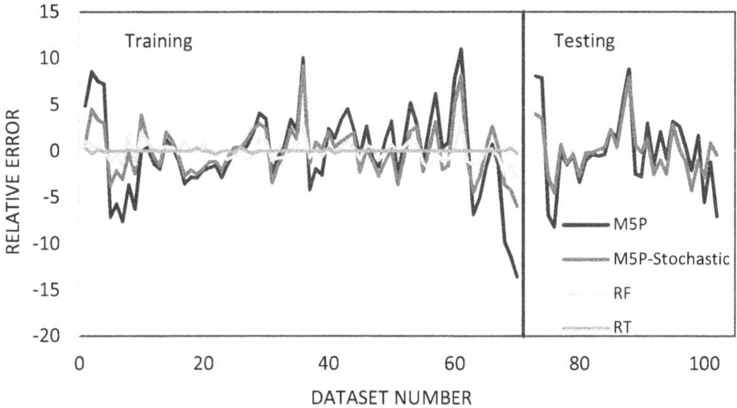

FIGURE 11.8 Error graphs for M5P, M5P-Stochastic, RF, and RT models.

11.4.3 Comparison among ANN and Soft Computing-Based Models

In this chapter predictive accuracy of M5P, M5P-Stochastic, ANN, RF, and RT-based models have been assessed and all the approaches were evaluated for the prediction of concrete compressive strength using BFRP. This, statistical performance indices are selected namely CC, RMSE, and MAE. Results of performance evaluation indices recommend that the RF model performed better than other soft computing with CC (0.9968, 0.9813), RMSE (1.0876, 2.027), and MAE (0.8193, 1.5336) for training and testing stages. Overall RF gives the best model than ANN, M5P-Stochastic, RT, and M5P.

The comparison between the error for all trees is shown in Figure 11.9.

And the comparison between the predicted values and actual values for training and testing data set for all techniques is shown in Figure 11.10.

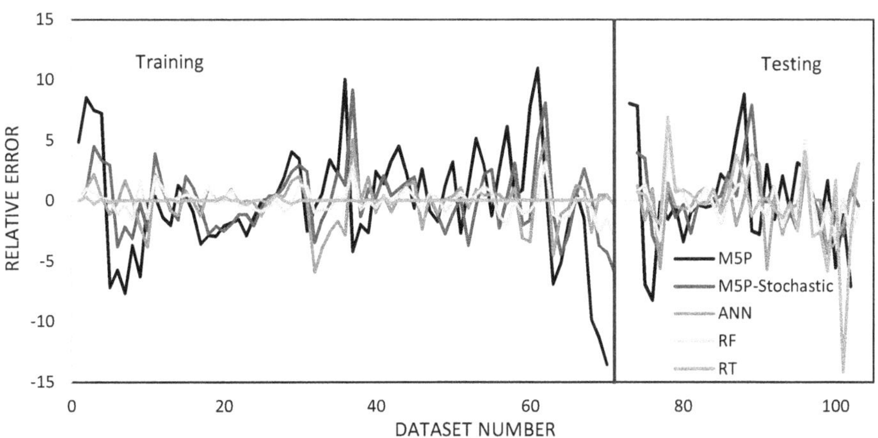

FIGURE 11.9 Comparative graph for the error between all techniques.

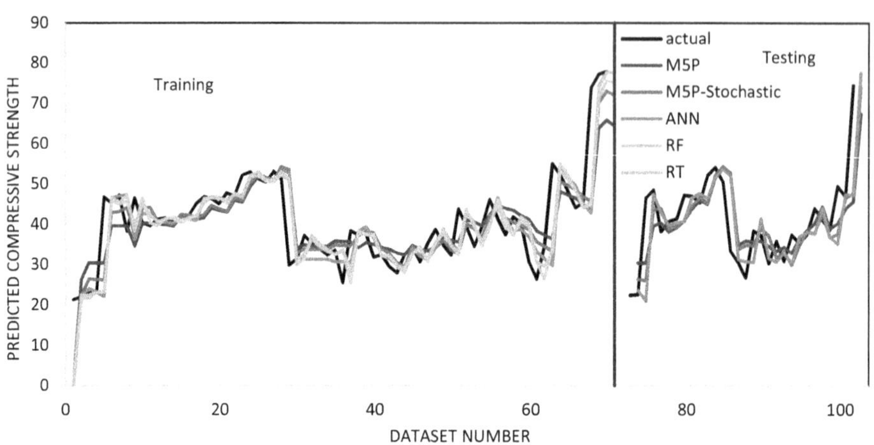

FIGURE 11.10 Comparative between the actual values and predict values for all techniques.

11.4.4 Sensitivity Analysis

To discover the main input variables in compressive strength, use the RF model for sensitivity testing. As indicated in Table 11.4, create a set of distinct test data by eliminating one input parameter at a time. The impact of each input variable on compressive strength is reported by CC, MAE, and RMSE, according to CC. Table 11.4 shows that, when compared to other input parameters in this data set, curing time plays a significant impact in forecasting concrete compressive strength.

11.5 CONCLUSION

In this chapter M5P, M5P-Stochastic, ANN, RF, and RT-based models have been developed to predict the compressive strength of concrete with BFRP. The data has been collected from different papers with different percentages of BFRP and different lengths of BFRP. The proposed RF method outperformed the other of the algorithms in a comparative study utilizing performance assessment indices (M5P, M5P-Stochastic, ANN, and RT) using given data with CC (0.9968, 0.9813), RMSE (1.0876, 2.027), MAE (0.8193, 1.5336), and lower values of CC (0.9426, 0.9353), RMSE (4.5927, 4.1132), and MAE (3.4465, 3.1103) for model development and validation period, respectively. Other major outcomes from this investigation are that ANN performs better than M5P, M5P-Stochastic, and RT models. Using this data set and the RF model, the results of the sensitivity research show that the curing time of bacterial concrete is crucial in determining its compressive strength.

This study is trying to implement the soft computing techniques in civil engineering fields to predict the compressive strength for the concrete with basalt fiber still the study for using machine learning to predict the mechanical properties for the concrete with fibers need more research and data to evaluate the accuracy and the benefit for using it.

TABLE 11.4
Sensitivity Results Using Random Forest Model

Input Variables									Output Variable	Random Forest		
Cement	Fine Aggregate	Coarse Aggregate	Water	SP	Fly ash	BFRP	Length of BFRP	Curing Time	Compressive Strength	CC	MAE	RMSE
										0.9813	1.5236	2.0279
√	x	x	x	x	x	x	x	x		0.9778	1.6123	2.1695
x	√	x	x	x	x	x	x	x		0.9797	1.7365	2.1806
x	x	√	x	x	x	x	x	x		0.9808	1.5815	2.04
x	x	x	√	x	x	x	x	x		0.977	1.7614	2.2565
x	x	x		√	x	x	x	x		0.9785	1.6865	2.1333
x	x	x	x		x	x	x	x		0.9788	1.6156	2.1306
x	x	x	x	x	x	√	x	x		0.9726	1.9251	2.3963
x	x	x	x	x	x	x	√	x		0.975	1.6477	2.3482
x	x	x	x	x	x	x	x	√		**0.8577**	**4.0839**	**5.315**

REFERENCES

Ali, M., Talha, A. and Berkouk, E.M., 2020. New M5P model tree-based control for doubly fed induction generator in wind energy conversion system. *Wind Energy*, *23*(9), pp. 1831–1845.

Almasi, S.N., Bagherpour, R., Mikaeil, R., Ozcelik, Y. and Kalhori, H., 2017. Predicting the building stone cutting rate based on rock properties and device pullback amperage in quarries using M5P model tree. *Geotechnical and Geological Engineering*, *35*(4), pp. 1311–1326.

Altunkaynak, A., 2007b. Forecasting surface water level fluctuations of Lake Van by artificial neural networks. *Water Resources Management*, *21*(2), pp. 399–408.

Altunkaynak, A., Özger, M. and Şen, Z., 2005a. Regional streamflow estimation by standard regional dependence function approach. *Journal of Hydraulic Engineering*, *131*(11), pp. 1001–1006.

Arslan, M.E., 2016. Effects of basalt and glass chopped fibers addition on fracture energy and mechanical properties of ordinary concrete: CMOD measurement. *Construction and Building Materials*, *114*, pp. 383–391.

Baykasoğlu, A., Dereli, T. and Tanış, S., 2004. Prediction of cement strength using soft computing techniques. *Cement and Concrete Research*, *34*(11), pp. 2083–2090.

Blaifi, S.A., Moulahoum, S., Benkercha, R., Taghezouit, B. and Saim, A., 2018. M5P model tree based fast fuzzy maximum power point tracker. *Solar Energy*, *163*, pp. 405–424.

Breiman, L., 1999. *Using Adaptive Bagging to Debias Regressions* (p. 16). Technical Report 547, Statistics Dept. UCB.

Breiman, L., 2001. Random forests. *Machine Learning*, *45*(1), pp. 5–32.

Buckley, J.J. and Hayashi, Y., 1994. Fuzzy neural networks: A survey. *Fuzzy Sets and Systems*, *66*(1), pp. 1–13.

Can, I., 2002. A new improved Na/K geothermometer by artificial neural networks. *Geothermics*, *31*(6), pp. 751–760.

Czigány, T., 2005. Basalt fiber reinforced hybrid polymer composites. In *Materials Science Forum* (Vol. 473, pp. 59–66). Trans Tech Publications Ltd.

Deepa, C., SathiyaKumari, K. and Sudha, V.P., 2010. Prediction of the compressive strength of high performance concrete mix using tree based modeling. *International Journal of Computer Applications*, *6*(5), pp. 18–24.

Demirel, M.C., Venancio, A. and Kahya, E., 2009. Flow forecast by SWAT model and ANN in Pracana basin, Portugal. *Advances in Engineering Software*, *40*(7), pp. 467–473.

Fu, Q., Niu, D., Zhang, J., Huang, D. and Hong, M., 2018. Impact response of concrete reinforced with hybrid basalt-polypropylene fibers. *Powder Technology*, *326*, pp. 411–424.

Gamal, T.S.F., Chiadighikaobi, P.C. Comparative analysis of reliability of non-destructive methods of strength control of concrete impregnated with vegetable oil: Basalt fiber for increasing the concrete strength. *Materials Today: Proceedings*, 2019, 19, 2479–2482. [CrossRef].

Hsu, K.L., Gupta, H.V. and Sorooshian, S., 1995. Artificial neural network modeling of the rainfall-runoff process. *Water Resources Research*, *31*(10), pp. 2517–2530.

Ibrahim, D., 2016. An overview of soft computing. *Procedia Computer Science*, *102*, pp. 34–38.

Jalal, M., 2015. Soft computing techniques for compressive strength prediction of concrete cylinders strengthened by CFRP composites. *Science and Engineering of Composite Materials*, *22*(1), pp. 97–112.

Jalasutram, S., Sahoo, D.R. and Matsagar, V., 2017. Experimental investigation of the mechanical properties of basalt fiber-reinforced concrete. *Structural Concrete*, *18*(2), pp. 292–302.

Jang, J.S.R., Sun, C.T. and Mizutani, E., 1997. Neuro-fuzzy and soft computing-a computational approach to learning and machine intelligence [Book Review]. *IEEE Transactions on Automatic Control*, *42*(10), pp. 1482–1484.

Jiang, C., Fan, K., Wu, F. and Chen, D., 2014. Experimental study on the mechanical properties and microstructure of chopped basalt fibre reinforced concrete. *Materials & Design*, *58*, pp. 187–193.

John, V.J. and Dharmar, B., 2021. Influence of basalt fibers on the mechanical behavior of concrete—A review. *Structural Concrete*, *22*(1), pp. 491–502.

Kişi, Ö., 2008. River flow forecasting and estimation using different artificial neural network techniques. *Hydrology Research*, *39*(1), pp. 27–40.

Kizilkanat, A.B., Kabay, N., Akyüncü, V., Chowdhury, S. and Akça, A.H., 2015. Mechanical properties and fracture behavior of basalt and glass fiber reinforced concrete: An experimental study. *Construction and Building Materials*, *100*, pp. 218–224.

Koufteros, X.A., 1999. Testing a model of pull production: A paradigm for manufacturing research using structural equation modeling. *Journal of operations Management*, *17*(4), pp. 467–488.

Maier, H.R. and Dandy, G.C., 2000. Neural networks for the prediction and forecasting of water resources variables: A review of modelling issues and applications. *Environmental Modelling & Software*, *15*(1), pp. 101–124.

Milicevic, M., Baranovic, M. and Zubrinic, K., 2015. Application of machine learning algorithms for the query performance prediction. *Advances in Electrical and Computer Engineering*, *15*(3), pp. 33–44.

Militky, J., Kova˘ci˘c, V. and Rubnerová, J., 2002. Influence of thermal treatment on tensile failure of basalt fibers. *Engineering Fracture Mechanics*, *69*, pp. 1025–1033.

Mohammed, A., Rafiq, S., Sihag, P., Kurda, R. and Mahmood, W., 2020a. Soft computing techniques: Systematic multiscale models to predict the compressive strength of HVFA concrete based on mix proportions and curing times. *Journal of Building Engineering*, p. 101851.

Mohammed, A., Rafiq, S., Sihag, P., Kurda, R., Mahmood, W., Ghafor, K. and Sarwar, W., 2020b. ANN, M5P-tree and nonlinear regression approaches with statistical evaluations to predict the compressive strength of cement-based mortar modified with fly ash. *Journal of Materials Research and Technology*, *9*(6), pp. 12416–12427.

Pulido-Calvo, I. and Portela, M.M., 2007. Application of neural approaches to one-step daily flow forecasting in Portuguese watersheds. *Journal of Hydrology*, *332*(1–2), pp. 1–15.

Sangeeta, Haji Seyed Asadollah, S.B., Sharafati, A., Sihag, P., Al-Ansari, N. and Chau, K.W., 2021. Machine learning model development for predicting aeration efficiency through Parshall flume. *Engineering Applications of Computational Fluid Mechanics*, *15*(1), pp. 889–901.

Silverman, D. and Dracup, J.A., 2000. Artificial neural networks and long-range precipitation prediction in California. *Journal of Applied Meteorology*, *39*(1), pp. 57–66.

Singh, B., Sihag, P., Tomar, A. and Sehgal, A., 2019. Estimation of compressive strength of high-strength concrete by random forest and M5P model tree approaches. *Journal of Materials and Engineering Structures*, *6*(4), pp. 583–592.

Thakur, M.S., Pandhiani, S.M., Kashyap, V., Upadhya, A. and Sihag, P., 2021. Predicting Bond Strength of FRP Bars in Concrete Using Soft Computing Techniques. *Arabian Journal for Science and Engineering*, *46*(5), pp. 4951–4969.

Upadhya, A., Thakur, M.S., Pandhian, S.M. and Tayal, S., 2021. Estimation of Marshall stability of asphalt concrete mix using neural network and M5P tree. In *Computational Technologies in Materials Science* (pp. 223–236). CRC Press.

Upadhya, A., Thakur, M.S., Sharma, N. and Sihag, P., 2021. Assessment of soft computing-based techniques for the prediction of marshall stability of asphalt concrete reinforced with glass fiber. *International Journal of Pavement Research and Technology*, pp. 1–20.

Wang, J., Ma, Y., Zhang, Y., and Chen, W., 2014. Experimental research and analysis on mechanical properties of chopped basalt fiber reinforced concrete. *Engineering Mechanics*, *31*, pp. 99–102.

Wang, Y., Li, Z., Zhang, C., Wang, X., Li, J., Liang, Y. and Chu, S., 2021. Bond behavior of BFRP bars with basalt-fiber reinforced cement-based composite materials. *Structural Concrete*, *22*(1), pp. 154–167.

Wu, J.S., Han, J., Annambhotla, S. and Bryant, S., 2005. Artificial neural networks for forecasting watershed runoff and stream flows. *Journal of Hydrologic Engineering*, *10*(3), pp.216–222.

Xu, Z., Hao, H. and Li, H.N., 2012. Experimental study of dynamic compressive properties of fibre reinforced concrete material with different fibres. *Materials & Design*, *33*, pp.42–55.

Zealand, C.M., Burn, D.H. and Simonovic, S.P., 1999. Short term streamflow forecasting using artificial neural networks. *Journal of Hydrology*, *214*(1–4), pp.32–48.

Zhang, J., Ma, G., Huang, Y., Aslani, F. and Nener, B., 2019. Modelling uniaxial compressive strength of lightweight self-compacting concrete using random forest regression. *Construction and Building Materials*, *210*, pp.713–719.

Zhou, L., Zheng, Y. and Taylor, S.E., 2018. Finite-element investigation of the structural behavior of basalt fiber reinforced polymer (BFRP)-reinforced self-compacting concrete (SCC) decks slabs in Thompson bridge. *Polymers*, *10*(6), p.678.

12 Soft Computing-Based Prediction of Compressive Strength of High Strength Concrete

Balraj Singh and Tanvi Singh
Panipat Institute of Engineering and Technology

CONTENTS

12.1 INTRODUCTION

The world is accepting high-strength concretes (HSCs) extensively [1,2]. HSC is concrete that meets the performance and uniformity that regular mixing, curing, and placing processes can't be attained. The consideration of HSC is different for a different country. For Europe, concrete is an HSC if it has 28th-day compressive strength (CS) of over 60 MPa and the ratio of water–binder is less than 0.40 [3]. For the USA, HSC is concrete, which gave very high strength that can't be achieved normally [4]. In India and Turkey, if the 28th-day CS is over 40 MPa, concrete is called HSC. The primary advantages of HSC are its strength and workability, which can save a good amount in construction industries [5,6]. The difference between HSC and regular concrete is mineral and chemical admixtures which ultimately decrease the

DOI: 10.1201/9781003184331-12

porosity by reducing the water content [7]. Even a shallow water content value may be achieved by introducing a very high dose of admixture in the concrete. Although, the effectiveness of admixtures depends upon the fineness, cement chemistry, and ambient temperature [8]. The admixtures have two types: chemical admixtures and mineral admixtures. As the name shows, chemical admixtures comprise chemical processes such as superplasticizers, and mineral admixtures are by-products of the industrial process such as fly ash. So, using mineral admixtures can also give economic and environmental benefits. It has increased the workability, strength, and durability of the concrete.

Using HSC boosts the working atmosphere in the construction industries, which gives way for automation in concrete construction. However, the fundamental difficulty lies in the design process because of the complex behavior of HSC. Any investigators solved this complexity and reported various standard guidelines for design mix, which ultimately enhance the use of admixtures [9–11]. It requires expert knowledge of the concrete constituent and experience because of the complexity of its design process than normal concrete. The complexity of the HSC requires an onerous mixed design process to attain the necessary characteristics. The strength of concrete is an essential part of HSC, yet variations in design specifications, admixtures, and constituents may vary from place to place [12,13]. As a result, the link between cement-to-water–binder ratio, admixtures, and aggregate sizes is unknown. These uncertainties will produce inadequacy in the strength of concrete if not appropriately managed. In this aspect, many investigators have used conventional methods by using non-linear and linear equations to predict the strength of HSC. The base of these methods are statistical analyses, but correct and precise prediction using these methods is challenging, thus need much research that conquers these complications [14]. Recently, the ideas of soft computing techniques (SCT) have overcome these complications and give the precise prediction of strength.

SCT such as, support vector machine (SVM) [15,16], Gaussian process (GP) [17,18], artificial neural network [19,20], M5P tree [21,22], adaptive neuro-fuzzy interference [23,24], random forest [25,26] and genetic engineering programming [27,28] have been broadly publicized in civil engineering [29]. Dong et al. implemented SCT in the prediction of 28th-day CS and concluded that SCTs give precise output although adaptive neuro-fuzzy interference outer perform artificial neural network in terms of performance evaluation equations [30]. Suthar [31] examined the potential of various SCT and found GP is the most suitable SCT among M5P tree, random forest, artificial neural network, and SVM. Suthar [32] investigated the GP and SVM techniques and found that the prediction of GP and SVM are closely related to each other. Park et al. used SVM as an SCT to predict concrete strength and concluded that SVM is the most effective tool. By looking at this literature, an attempt is made to predict the 28th-day CS by using three SCTs: SVM, GP, and linear regression (LR). So, this investigation aims to make an SVM and GP-based model which can predict the CS on the 28th day of concrete. Also, an equation is generated to predict the CS by LR. Furthermore, a comparison of different SCT was made using the performance evaluation equations.

12.2 SOFT COMPUTING TECHNIQUES THEORY

12.2.1 GP

The assumption for GP model is that nearby variables should give info about each other. GP are a way of stipulating a former straight above the function space. GP is on functions whereas, Gaussian distribution is on vectors. Hence, no validation is required for generalization and in GP models, because information about data and functions required was known initially.

A GP is a group of random variables, with any finite number (of which has a joint multivariate Gaussian distribution). Let $f \times g$ indicates the areas of inputs variables and outputs variables, among which x pairs were taken out (f_i, g_i) that are taken individually and uniformly distributed.

For regression function, let $g \subseteq \Re$; hence, Gaussian Process Regression (GPR) can be explained as by a mean function (12.1) and a covariance function (12.2)

$$\mu: \quad f \rightarrow \Re \tag{12.1}$$

$$k: f \times f \rightarrow \Re. \tag{12.2}$$

The most important assumption of GPR is y is given as

$$g = f(f) + \xi, \tag{12.3}$$

where $\xi \sim M(0, \sigma^2)$ and ~ represents *sampling for.*

In GPR, for every input m there is an equivalent random variable $f(f)$, and it is the value of the stochastic function f at that instance. In present work, it is presumed that the observational error ξ is independent and uniformly distributed, having a mean value of zero ($\mu(f) = 0$), a variance of σ^2 and $f(f)$ taken from GP on f specified by k. That is,

$$G = (g_1, \ldots, g_n) \sim M(0, K + \sigma^2 I), \tag{12.4}$$

where $K_{ij} = k(f_i, f_j)$, and I is the identity matrix.

Because $G/F \sim M(0, K + \sigma^2 I)$ is normal, so is the restricted dissemination of test label are given for the training and test data of $p(G/G, F, F_*)$. Then, one has $G_*/G, F, F_* \sim M(\mu, \Sigma)$, where

$$\mu = K(F_*, F)\left(K(F, F) + \sigma^2 \mathrm{I}\right)^{-1} G, \tag{12.5}$$

$$\Sigma = K(F_*, F_*) - \sigma^2 \mathrm{I} - K(F_*, F)\left(K(G, G) + \sigma^2 \mathrm{I})\right)^{-1} K(F, F_*), \tag{12.6}$$

If there are n training data and n_* test data, then $K(F, F_*)$ represents the $n \times n_*$ matrix of covariance, which is assessed for every pairs of training and test data sets, and this is same for the other values of $K\left(F,F\right)$, $K\left(F_*,F\right)$ and $K\left(F_*,F_*\right)$; here F and G are the vector chosen from the training data. labels g_i, whereas F_* is the vector of the test data.

12.2.2 SVM

SVM uses a set of input variables, which is referred as the training data set, and calculates appropriate weights to acquire the ruling function. SVM was first established by Vapnik [36]. Specific amount input variables (which is lesser than of total variables) are carefully chosen in SVM, and these are termed as support vectors. The input vectors which are chosen are utilized to calculate the parameters of SVM with least error. After training process have been completed transfer function obtained are as follows:

$$s = \sum_{b=1}^{a} \left\{ m_i \times n\left(C_i, D\right) \right\} + k(1) \tag{12.7}$$

in which s = output, a = number of support vectors, C_i & m_i = the ith support vector and its weight, D = the input sample vector for which an appropriate solution s is sought, and $n(C_i, D)$ = the kernel function.

While training of input variable using SVM main focus is to find the optimum values for k, m, b, C_i, and a by varying the parameters and observing the error obtained and the combination of different parameter which gives the least error are finalized.

12.3 DATA REPRESENTATION SUPERPLASTICIZER AND DESCRIPTION

Total 357 observations were used to model the CS of HSCs. The data was collected from the previously published article [34]. Although, the purpose was to use the database to predict the optimized quantities rather than going for hit and trial in experimental work. Out of the total database, 70% of data was used to train the model, whereas 30% was for testing. Random choosing criteria were adopted for splitting the entire database [35]. Also, the total variables were divided into two parts: input and output variables. The input variables are cement, coarse aggregate, fine aggregate, water, and superplasticizer, and the output variable is 28th-day CS. The descriptive details of the training and testing database are shown in Table 12.1. The parameters of the descriptive analysis are mean, standard error, median, mode, standard deviation, sample variance, kurtosis, skewness, range, minimum, maximum, and 95% confidence level. It provides a way to understand the complex database in a sensible way. Figure 12.1 provides the pairs of input and output variables. The significant advantage of giving the pair of variables is to know the situation of the outliers, which ultimately affects the prediction of SCT. Also, it provides the interrelationship within the variables. At last, a correlation plot was provided in Figure 12.2, which gave the correlation between the input and output variables.

TABLE 12.1
Descriptive Analysis of Parameters

Parameters[a]	C	CA	FA	W	SP	CS
		Training Database				
Mean	390.426	865.489	799.607	172.793	2.159	51.989
Standard error	6.039	6.886	7.685	0.896	0.159	0.638
Median	378	898	801	170	1.25	49.25
Mode	360	950	628	170	1	41.5
Standard deviation	93.169	106.234	118.565	13.833	2.463	9.844
Sample variance	8680.636	11285.66	14057.8	191.354	6.070	96.905
Kurtosis	−0.328	6.945	1.167	4.692	3.618	1.229
Skewness	0.147	−0.677	0.216	1.118	1.932	1.078
Range	440	986	793	111.9	12	51.8
Minimum	160	500	342	132	0	39.5
Maximum	600	1486	1135	243.9	12	91.3
Confidence level (95.0%)	11.897	13.565	15.140	1.766	0.314	1.257
		Testing Database				
Mean	372.195	849.985	819.407	175.116	2.719	52.041
Standard error	8.421	8.545	9.366	1.697	0.281	0.988
Median	360	880	820	170	1.25	49
Mode	360	845	916	170	1	45
Standard deviation	91.868	93.221	102.181	18.513	3.076	10.788
Sample variance	8439.815	8690.277	10441.09	342.736	9.462	116.389
Kurtosis	−0.434	3.527	0.849	19.906	1.703	1.976
Skewness	0.135	−1.668	0.275	3.400	1.536	1.434
Range	407	489	583	162.08	12	50.99
Minimum	160	500	552	140	0	40
Maximum	567	989	1135	302.08	12	90.99
Confidence level (95.0%)	16.677	16.922	18.549	3.360	0.558	1.958

[a] C = Cement, CA = Coarse aggregate, FA = Fine aggregate, W = Water, SP = Superplasticizer, CS = Compressive Strength.

12.4 RESULTS AND DISCUSSION

WEKA3.8.5 executed SVM and GP with three different kernels and LR. The first step to model the CS is to set up the primary parameters. The primary parameters are the parameter that decides the accuracy of the SCT. The direct setup is set by the trial & error process, and optimum parameters know the most accurate set of parameters. The values of optimum parameters for this investigation are tabulated in Table 12.2.

Four performance evaluation equations were used to check the potential of the SCT. These equations are correlation coefficient (CC), mean absolute error (MAE),

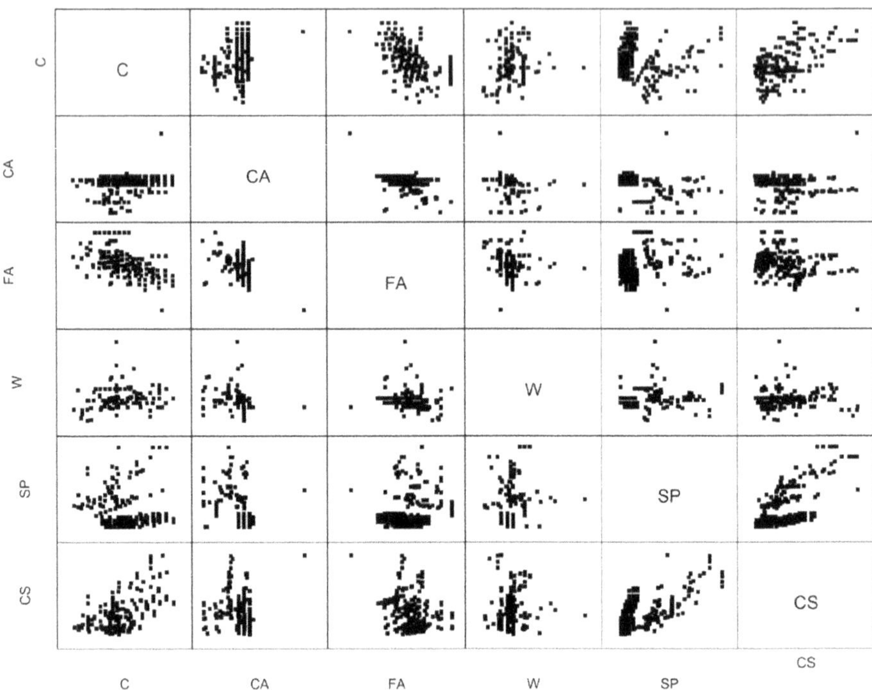

FIGURE 12.1 Pairs of variables.

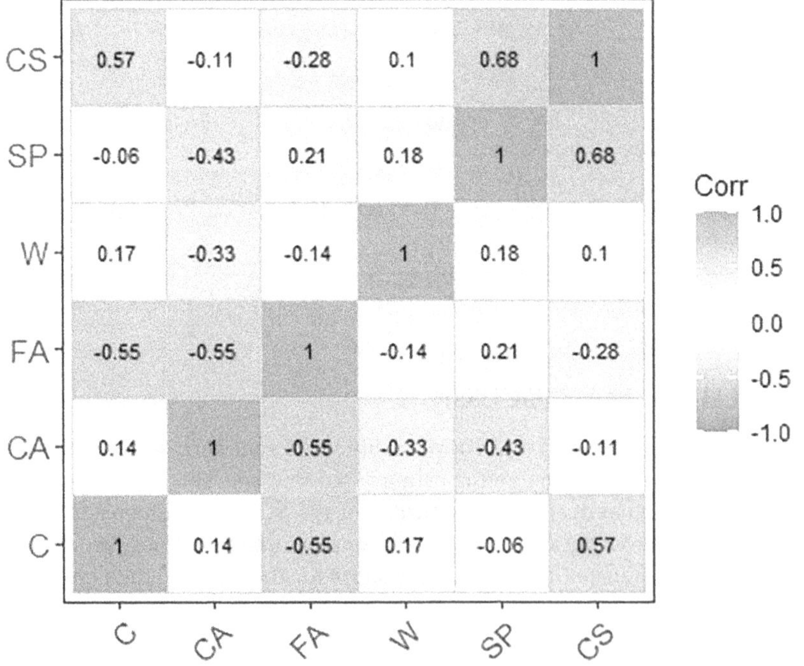

FIGURE 12.2 Correlation plot of variables.

TABLE 12.2
Optimum Parameters for SCT

SCT	Optimum Parameters
GP with poly kernel	$C=1; E=3$
GP with PUK kernel	Gaussian noise $=0.1$, $\omega=0.5$, $\sigma=2.0$
GP with RBF kernel	Gaussian noise $=0.1$, $\gamma=3.0$
SVM with poly kernel	$L=0.1$, $E=3$
SVM with PUK kernel	$C=10$, $\omega=0.5$, $\sigma=2.0$
SVM with RBF kernel	$C=10$, $\gamma=3.0$

root mean square error (RMSE), and relative absolute error (RAE). The unit of the RMSE is the same as the output variable, which is MPa; CC and MAE are unitless equations while values of RAE are in percentage. Table 12.3 provides the details of the performance evaluation equations for different SCT and SCT with the kernels: LR, GP with poly, puk, and rbf (GP_POLY, GP_PUK, and GP_RBF, respectively) and support vector regression with poly, puk, and rbf (SVM_POLY, SVM_PUK, and SVM_RBF, respectively).

12.4.1 Prediction of CS by GP

In the prediction of CS by GP, three different kernel functions were used. The details of the optimum parameters and the performance evaluation equations were discussed in Tables 12.2 and 12.3, respectively. Figure 12.3 described the scatter plot of actual and predicted CS using GP, showing that GP_PUK gave the best results among GP_POLY and GP_RBF kernels. The same trends were also be suggested by the performance evaluation equations (Table 12.3). According to Table 12.3, GP_PUK kernels were having the values of CC, MAE, RMSE and RAE (0.957, 2.526, 3.427 MPa

TABLE 12.3
Results of the Performance Evaluation Equations for Different SCT

SCTs	GP_ POLY	GP_ PUK	GP_ RBF	SVM_ POLY	SVM_ PUK	SVM_ RBF	LR
			Training Data Set				
Correlation coefficient	0.935	0.966	0.904	0.926	0.990	0.913	0.928
Mean absolute error	2.952	2.071	6.920	2.961	0.808	4.200	3.015
Root mean square error	3.567	2.727	8.505	3.753	1.371	5.314	3.654
Relative absolute error (%)	36.640	25.703	85.878	36.748	10.028	52.119	37.414
			Testing Data Set				
Correlation coefficient	0.926	0.957	0.925	0.924	0.960	0.933	0.936
Mean absolute error	3.136	2.526	7.108	3.066	2.044	4.155	2.979
Root mean square error	4.058	3.427	9.206	4.220	3.037	5.362	3.824
Relative absolute error (%)	37.809	30.456	85.691	36.969	24.645	50.087	35.914

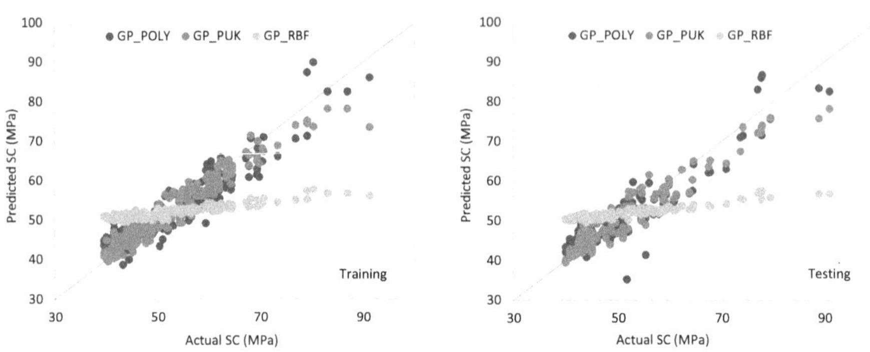

FIGURE 12.3 Scatter plot of CS using GP.

and 30.456%) which were more accurate than GP_POLY (0.926, 3.136, 4.058 and 37.809) and GP_RBF (0.925, 7.108, 9.206 MPa and 85.691%. The overall trends of GP showed that GP with PUK kernel is more precise, followed by GP_POLY and GP_RBF kernels. Thus, GP_PUK is the best kernel that will be used for further comparison in the prediction of CS.

12.4.2 Prediction of CS by SVM

In SVM, poly, puk, and rbf kernels were also used to predict the CS. Figure 12.4 provided the scatter plot of GP in the prediction of CS. The trends from Figure 12.4 suggested that SVM_PUK outer performs in predicting CS than SVM_POLY and SVM_RBF. Table 12.3 also showed the same tendency of results. SVM_PUK gave the most precise result with CC, MAE, RMSE, and RAE values 0.960, 2.044, 3.037 MPa, and 24.645%. In between the poly and rbf kernels, the poly kernel gave better results than rbf (values of CC, MAE, RMSE, and RAE for poly and rbf = 0.924, 3.066, 4.220 MPa and 36.969% and 0.933, 4.155, 5.362 MPa and 50.087%, respectively). The order% of performance in SVM was also the same as GP. SVM_PUK was most efficient, followed by SVM_POLY and SVM_RBF. Hence, SVM_PUK was the most accurate kernel, which will be used for further comparison.

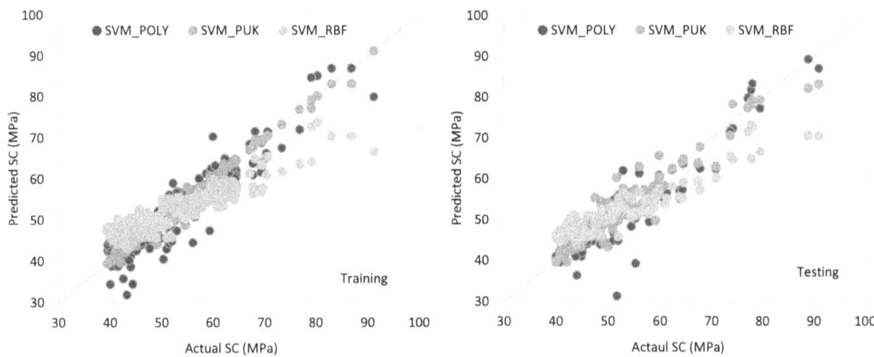

FIGURE 12.4 Scatter plot of CS using SVM.

12.4.3 Prediction of CS by LR

In the prediction of CS by LR, it developed a general equation. The LR equation is:

$$CS = 0.0648 * C - 0.0125 * FA - 0.1334 * W + 2.92 * SP + 53.4338 \quad (12.8)$$

The results of the performance evaluation equation for the LR equation were also tabulated in Table 12.3. The CC, MAE, RMSE, and RAE values were 0.936, 2.979, 3.824 MPa, and 35.914%.

12.4.4 Comparison of Results

A comparison of the best-selected kernels from GP (GP_PUK) and SVM (SVM_PUK) was also made with the LR. Figure 12.5 represented the performance of the best-selected model along with LR. The trends from Figure 12.5 suggested that SVM_PUK is the best SCT in predicting CS among GP_PUK and LR. The values obtained from the performance evaluation equations (Table 12.3) also showed that SVM_PUK is the most precise and accurate SCT in the prediction of CS. The values of CC, MAE, RMSE and RAE was 0.960, 2.044, 3.037 MPa and 24.645% for SVM_PUK which is much ideal than GP_PUK (0.957, 2.526, 3.427 MPa and

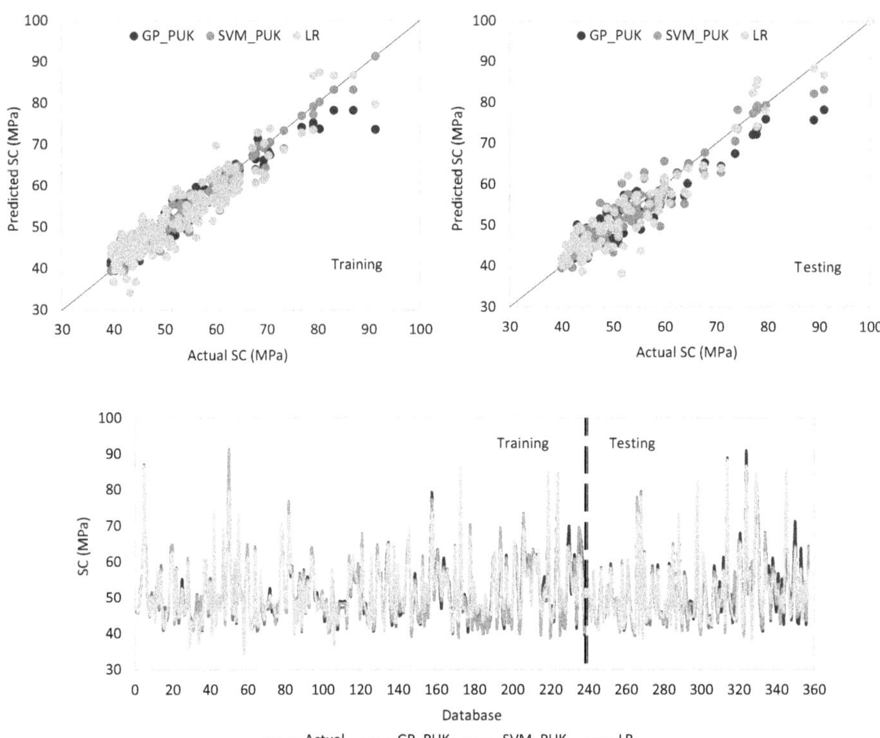

FIGURE 12.5 Performance of best-selected kernel function.

30.456%) and LR (0.936, 2.979, 3.824 MPa and 35.914%). Hence, Table 12.3 and Figure 12.5 represented that SVM_PUK is the most capable SCT, predicting the CS of HSC. Although all the SCT can predict the CS, SVM_PUK is the most precise SCT. The puk kernel is more efficient than poly and rbf with this database than three different kernel functions.

12.5 CONCLUSION

The SCT provides uncompromising and precise results between experimental and modeled data, which helps the researchers to focus pre-design phase than experimentation. The SCT's use in the prediction of CS can make the procedure smooth, reduce human efforts, and save time for experimentation. Three SCT, SVM, and GP with three kernels (poly, puk, and rbf) and LR were implemented to predict the CS of HSC. The potential of these SCT was checked using performance evaluation equations and performance plots, which suggested that SVM _PUK showed the best agreement between the actual and predicted values of CS. The performance orders of different SCT were SVM_PUK > GP_PUK > LR > SVM_POLY > GP_POLY > SVM_RBF > GP_RBF. The puk kernel was the most efficient, followed by poly and rbf kernels. Also, an equation was generated using LR by which anyone can predict the values of CS by putting the values of input variables. Thus, SVM_PUK is the most accurate technique and can predict the CS of HSC when the experimentation data is not available.

REFERENCES

1. Shah, S.P. (1993). Recent trends in the science and technology of concrete, concrete technology, new trends, industrial applications. In: *Proceedings of the International RILEM Workshop*, London, E & FN Spon; 1–18.
2. Bache, H.H. (1981). Densified cement/ultrafine particle based materials. In: *The Second International Conference on Superplasticizers in Concrete*, Ottawa.
3. Öztaş, A., Pala, M., Özbay, E., Kanca, E., Caglar, N., & Bhatti, M.A. (2006). Predicting the compressive strength and slump of high strength concrete using neural network. *Construction and Building Materials*, 20(9), 769–775.
4. Camoes, A., Aguiar, B., & Jalali, S. (2003). Durability of low cost high performance fly ash concrete. In: *International Ash Symposium*, KY; paper no. 43.
5. Mowlaei, R., Lin, J., de Souza, F.B., Fouladi, A., Korayem, A.H., Shamsaei, E., & Duan, W. (2021). The effects of graphene oxide-silica nanohybrids on the workability, hydration, and mechanical properties of Portland cement paste. *Construction and Building Materials*, 266, 121016.
6. Sobolev, K. (2004). The development of a new method for the proportioning of high performance concrete mixtures. *Cement and Concrete Research*, 26(7), 901–907.
7. Mittal, A., & Basu, D.C. (1999). Development of HPC for PC Dome of NPP. *Indian Concrete Journal*, 73(3), 671–679.
8. Yeh, I.C. (1998). Modeling of strength of HPC using ANN. *Cement and Concrete Research*, 28(12), 1797–1808.
9. Persson, B. (2001). A comparison between mechanical properties of self-compacting concrete and the corresponding properties of normal concrete. *Cement and Concrete Research*, 31(2), 193–198.

10. Sahmaran, M., Yurtseven, A., & OzgurYaman, I. (2005). Workability of hybrid fiber reinforced self-compacting concrete. *Building and Environment*, 40(12), 1672–1677.
11. Grdic, Z.J., Toplicic-Curcic, G.A., Despotovic, I.M., & Ristic, N.S. (2010). Properties of self-compacting concrete prepared with coarse recycled concrete aggregate. *Construction and Building Materials*, 24(7), 1129–1133.
12. Akbar, A., Farooq, F., Shafique, M., Aslam, F., Alyousef, R., & Alabduljabbar, H. (2021). Sugarcane bagasse ash-based engineered geopolymer mortar incorporating propylene fibers. *Journal of Building Engineering*, 33, 101492.
13. Khan, M.A., Memon, S.A., Farooq, F., Javed, M.F., Aslam, F., & Alyousef, R. (2021). Compressive strength of fly-ash-based geopolymer concrete by gene expression programming and random forest. *Advances in Civil Engineering*, 1(9) 2021.
14. Ali Khan, M., Zafar, A., Akbar, A., Javed, M.F., & Mosavi, A. (2021). Application of Gene Expression Programming (GEP) for the prediction of compressive strength of geopolymer concrete. *Materials*, 14(5), 1106.
15. Singh, B., Sihag, P., Pandhiani, S. M., Debnath, S., & Gautam, S. (2021). Estimation of permeability of soil using easy measured soil parameters: assessing the artificial intelligence-based models. *ISH Journal of Hydraulic Engineering*, 27(supl), 38–48.
16. Vand, A.S., Sihag, P., Singh, B., & Zand, M. (2018). Comparative evaluation of infiltration models. *KSCE Journal of Civil Engineering*, 22(10), 4173–4184.
17. Sihag, P., Kumar, M., & Singh, B. (2021). Assessment of infiltration models developed using soft computing techniques. *Geology, Ecology, and Landscapes*, 5(4), 241–251.
18. Sihag, P., Singh, B., Gautam, S., & Debnath, S. (2018). Evaluation of the impact of fly ash on infiltration characteristics using different soft computing techniques. *Applied Water Science*, 8(6), 1–10.
19. Singh, B. (2020). Prediction of the sodium absorption ratio using data-driven models: A case study in Iran. *Geology, Ecology, and Landscapes*, 4(1), 1–10.
20. Sihag, P., Singh, B., SepahVand, A., & Mehdipour, V. (2020). Modeling the infiltration process with soft computing techniques. *ISH Journal of Hydraulic Engineering*, 26(2), 138–152.
21. Singh, B., Sihag, P., & Deswal, S. (2019). Modelling of the impact of water quality on the infiltration rate of the soil. *Applied Water Science*, 9(1), 15.
22. Arora, S., Singh, B., & Bhardwaj, B. (2019). Strength performance of recycled aggregate concretes containing mineral admixtures and their performance prediction through various modeling techniques. *Journal of Building Engineering*, 24, 100741.
23. Singh, B., Sihag, P., Singh, K., & Kumar, S. (2021). Estimation of trapping efficiency of a vortex tube silt ejector. *International Journal of River Basin Management*, 19(3), 261–269.
24. Sihag, P., Esmaeilbeiki, F., Singh, B., Ebtehaj, I., & Bonakdari, H. (2019). Modeling unsaturated hydraulic conductivity by hybrid soft computing techniques. *Soft Computing*, 23(23), 12897–12910.
25. Singh, B., Sihag, P., Singh, V. P., Sepahvand, A., & Singh, K. (2021). Soft computing technique-based prediction of water quality index. *Water Supply*, 21(8), 4015–4029.
26. Singh, B., Sihag, P., & Singh, K. (2017). Modelling of impact of water quality on infiltration rate of soil by random forest regression. *Modeling Earth Systems and Environment*, 3(3), 999–1004.
27. Singh, B., Sihag, P., Tomar, A., & Sehgal, A. (2019). Estimation of compressive strength of high-strength concrete by random forest and M5P model tree approaches. *Journal of Materials and Engineering Structures*, 6(4), 583–592.
28. Azimi, H., Bonakdari, H., & Ebtehaj, I. (2021). Gene expression programming-based approach for predicting the roller length of a hydraulic jump on a rough bed. *ISH Journal of Hydraulic Engineering*, 27(supl), 77–87.

29. Azimi, H., Bonakdari, H., & Ebtehaj, I. (2017). A highly efficient gene expression programming model for predicting the discharge coefficient in a side weir along a trapezoidal canal. *Irrigation and Drainage*, 66(4), 655–666.
30. Sepahvand, A., Singh, B., Ghobadi, M., & Sihag, P. (2021). Estimation of infiltration rate using data-driven models. *Arabian Journal of Geosciences*, 14(1), 1–11.
31. Ly, H.B., Nguyen, T.A., & Tran, V.Q. (2021). Development of deep neural network model to predict the compressive strength of rubber concrete. *Construction and Building Materials*, 301, 124081.
32. Suthar, M. (2020). Applying several machine learning approaches for prediction of unconfined compressive strength of stabilized pond ashes. *Neural Computing and Applications*, *32*(13), 9019–9028.
33. Vapnik, V.N. (1995). *The Nature of Statistical Learning Theory*. Springer, New York.
34. Aslam, F., Farooq, F., Amin, M. N., Khan, K., Waheed, A., Akbar, A., ... & Alabdulijabbar, H. (2020). Applications of gene expression programming for estimating compressive strength of high-strength concrete. *Advances in Civil Engineering*, 26.
35. Pandhiani, S.M., Sihag, P., Shabri, A.B., Singh, B., & Pham, Q.B. (2020). Time-series prediction of streamflows of Malaysian rivers using data-driven techniques. *Journal of Irrigation and Drainage Engineering*, 146(7), 04020013.
36. Vapnik, V. (1998). *Statistical Learning Theory* New York: Wiley, 1, 2.

13 Forecasting Compressive Strength of Concrete Containing Nano-Silica Using Particle Swarm Optimization Algorithm and Genetic Algorithm

Deepika Garg
G.D Goenka University

Sakshi Gupta
ASET, Amity University

CONTENTS

DOI: 10.1201/9781003184331-13

13.1 INTRODUCTION

Concrete is one of the primogenital and most useful materials in the building industry, combining paste and aggregates. Concrete manufacturing is a complicated method that requires many processing parameters depending on the concrete's quality control. It is connected to different characteristics, such as workability, compressive strength (CS), shear strength, and durability. Such criteria are useful in generating a single strength quantity of CS. For civil construction, nanotechnology gives additional opportunities to improve properties of the materials. Attracting civil engineers to embrace nanotechnology could allow them to bring in a ground-breaking solution to today's tangled construction problems.

Nano-silica is a profoundly beneficial pozzolanic element and comprises typically of extremely fine glassy particles relatively 1000 times less than the typical cement particles. It has demonstrated to be a prominent admixture for the cement to enhance the CS and durability and diminishing permeability (Aitcin et al., 1981). Far-reaching of the concrete materials in primary designing and structural engineering since many years has led to a wide range of enhancement of the design and overall performance of concrete by utilizing various optimization problems. This could enhance the concrete application in various practices, reducing the amount of hazardous material and improving construction use.

Throughout the most recent decades, diverse data mining techniques such as the fuzzy logic, artificial neural network (ANN) and genetic algorithm (GA) have gotten far and wide. These days, such techniques are utilized by a number of researchers for varied kinds of engineering applications. To overcome and deal with uncertainty over the years, the hypothesis of probability was set up and effectively applied to numerous areas of engineering and technology (Akkurt et al., 2004; Demir, 2005; Unal et al., 2007; Topcu and Sarıdemir, 2008, Peyman et al., 2021). Soft computing techniques do not necessitate the internal system knowledge provided that a condensed solution for multi-variable problems is obtained. The quality of the soft computing techniques' prediction depends on various parameters such as error, the methodology used, and problems in estimation with respect to the forecasting method (Mehrabi et al., 2021). Artificial intelligence methods are anticipated to alleviate various estimation problems. Swarm intelligence models are computational models enthused by animal social behaviour. Various computational intelligence models, such as optimization techniques and particle swarm optimization (PSO), have been proposed in the literature. PSO, artificial bee colony, bacterial foraging, cat swarm optimization, artificial immune system and glow worm swarm optimization. Chen et al., (2018) utilized the ANN-PSO algorithm in predicting the shear strength of reinforced concrete walls in their study. ANN-PSO, ANN-GA, ANN-ICA, and ANN-ABC were evaluated and compared in a study conducted by Toghroli et al., (2020). The study results revealed the greater competence of the ANN-PSO model over the other models that were studied.

Evolutionary algorithms have drawn many researchers' attention as an essential tool for solving numerous optimization problems. GA is ubiquitous in different fields, mostly in light of its purpose, execution, and the capacity to tackle multifaceted difficulties typically found in engineering frameworks. GA has a high execution cost and normally requires a larger number of iterations which is one of the limitations of this technique. PSO is a comparatively new heuristic algorithm that focuses on the behaviour of living organisms' swarming characteristics. PSO is fundamentally the same as the GA. This implies that PSO and the GA vary from a set of focuses to another arrangement of focuses during an iteration with a noticeable enhancement using specific probabilistic and deterministic rules from the previous values.

PSO is a knowledge developmental strategy which imitates the social conduct of bird flocking. In 1997, PSO algorithm was proposed by Kennedy and Eberhart (1997) and since then many researchers have used the technique in various computational problems. The applications of PSO are varied ranging from the critical thinking of multi-objective integer programming, optimization, clustering to classification, combinatorial optimization, and min–max problems, or various kinds of engineering applications. PSO algorithm has an extremely quick convergence rate in contrast with the other evolutionary algorithms (Khan and Ahai, 2012). Consequently, it has been effectively utilized to find the solutions for diverse engineering applications (Bao et al., 2013; Mohandes, 2012; Hasanipanah et al., 2016; Mohamad et al., 2018).

The current study was carried out to establish a relationship between different input parameters and an output parameter, i.e., 28-days CS, using GA and PSO techniques. The objective was to study GA and PSO's application for the prediction of the CS of concrete containing nano-silica as a partial replacement of cement. This work is used to study these two evolutionary algorithms' effectiveness in the concrete strength at 28th day. The purpose is to provide a methodology for predicting the CS of concrete where the partial replacement of cement is done with nano-silica which uses the features of PSO and GA, and it is presented as an improved approach.

13.2 PROBLEM FORMULATION

13.2.1 Database

The database was possessed from obtainable literature on concrete containing nano-silica as a partial replacement of cement, as summarized in Table 13.1. The different literatures were studied and similar behaving results were assessed carefully and then the list was prepared. The accomplishment of the models' rests on the exhaustiveness and completeness of the data. Thus, huge data variations were possessed, and out of 57 data collected only 32 datasets were used in the present study for the analysis. The missing value data were removed to get the best dataset for the analysis. For example, in some cases the superplasticizer and the diameter of the nano-particles were missing and therefore that particular data had to be removed. The basic parameters considered are the amount of cement, coarse aggregate, fine aggregate, nano-silica, water-to-binder ratio, diameter of nano-silica particles, and dosage of superplasticizer. The elimination of one or the other property of concrete in a few studies and the uncertainty of mix proportions along with the testing techniques was accountable for demarcating the conditions for identifying the required data. The magnitude of the training data is of prime importance as the accomplishment of the model rely on the same. The predicted results for 28-day CS were equated with the values attained by experimentation as per the dataset given in Table 13.1.

TABLE 13.1
Dataset for the GA and PSO Modelling Input Parameters[a]

Cement Content (kg/m³) (x1)	Coarse Aggregates (kg/m³) (x3)	Fine Aggregates (kg/m³) (x2)	SP (kg/m³) (x5)	NS (kg/m³) (x6)	W/B Ratio (x4)	D (nm) (x7)	Output 28-Day CS (MPa) (y)	Researcher (Year)
396.6	722	826	7	16.5	0.37	15	75.2	Beigi et al., (2013)
380	722	826	7	33	0.35	15	86.1	
363.5	722	826	7	49.6	0.33	15	85.4	
318.4	1040	840	2.71	1.6	0.5	15	36.8	Zhang and Hui (2011)
316.8	1040	840	4.75	3.2	0.5	15	40.2	
356.4	1260	650	5.4	3.6	0.42	10	66.36	Heidari (2013)
349.2	1260	650	7.2	10.8	0.42	10	61.16	
447.75	1148	492	0	2.25	0.4	80	39.2	Gupta (2013)
445.5	1148	492	0	4.5	0.4	80	40.3	
443.25	1148	492	0	6.75	0.4	80	41.2	
441	1148	492	0	9	0.4	80	38.1	
447.75	1148	492	0	2.25	0.4	15	42.7	
445.5	1148	492	0	4.5	0.4	15	43.6	
443.25	1148	492	0	6.75	0.4	15	42.9	
441	1148	492	0	9	0.4	15	39.7	
394	915	811	1.68	12	0.45	15	53.8	
388	915	811	2.32	24	0.45	15	56.5	
382	915	811	3	36	0.45	15	60	Nili et al., (2010)
247.5	0	625	4.5	7.5	0.5	40	54.3	Jo et al., (2007)
240.6	0	626	5.8	14.4	0.5	40	61.9	
241.8	0	627	7	23.2	0.5	40	68.2	
227.7	0	628	7.5	27.3	0.5	40	68.8	
370	1088	647	13.5	13.9	0.49	15	44	
568.36	0	1757.8	8.85	17.5	0.5	15	32.9	Li et al., (2004)
556.64	0	1757.8	14.58	29.3	0.5	15	33.8	
527.34	0	1757.8	29.3	58.59	0.5	15	36.4	
556.64	0	1757.8	10.28	11.71	0.5	15	35.4	
480	1140	647	10	20	0.28	10	75.8	Li (2004)
390	1175	783	1.78	23.4	0.4	35	70	Said et al., (2012)
390	1162	774	3.56	46.8	0.4	35	76	
390	1154	769	1.27	23.4	0.4	35	60	
390	1143	762	2.54	46.8	0.4	35	66	

Note: In this investigation, all forms of SP were assumed to be the same.

[a] C = Cement, FA = Fine Aggregate, SP = Superplasticizer, NS = Nanisilica, W/B ratio = Water binder ratio, CS = Compressive Strength.

TABLE 13.2
Bounds of Input Variables

Input Parameters	Abbreviation	Parameter Ranges	
		Min.	Max.
Cement (kg/m^3)	Cement	227	569
Fine aggregate (kg/m^3)	FA	492	1758
Nano-silica (kg/m^3)	NS	1.60	59
Diameter of nano-silica (nm)	D	10	80
Coarse aggregate (kg/m^3)	CA	0	1260
Superplasticizer (kg/m^3)	SP	0	30
Water-to-binder ratio	W/B ratio	0.28	0.50

The variables used in the present study are cement, fine aggregates, coarse aggregates, water-to-binder ratio, superplasticizer, nano-silica and diameter of nano-silica particles.

The ranges of different input parameters used in the present study are given in Table 13.2.

13.2.2 Function Approximation

13.2.2.1 Function Approximation: Algorithm

Linear regression represents the association among a dependent variable y (i.e., response variable) and one or more independent variables x_1, x_2..., x_n (i.e., predictor variables).

$$y = \alpha_0 + \alpha_1 x + \alpha_2 x \ldots + \in (1) \tag{13.1}$$

where α_0 is the y-intercept, α_i (i)other than $i = 0$ are regression coefficient, and $\in$ is the error term.

The algorithm used in the function approximation is shown below:

1. To start with, take a set of n observed values of x and y given by (p_1,q_1), (p_2,q_2),..., (x_n,y_n).
2. The observed values arrange into a system of linear equations, using simple linear regression method. Then, denote the equations obtained in matrix form as

$$\begin{bmatrix} p_1 \\ p_2 \\ \vdots \\ p_n \end{bmatrix} = \begin{bmatrix} 1 & q_1 \\ 1 & q_2 \\ \vdots & \vdots \\ 1 & q_n \end{bmatrix} \begin{bmatrix} \alpha_1 \\ \alpha_2 \end{bmatrix} \tag{13.2}$$

where

$$p = \begin{bmatrix} p_1 \\ p_2 \\ \vdots \\ p_n \end{bmatrix}, \quad Q = \begin{bmatrix} 1 & q_1 \\ 1 & q_2 \\ \vdots & \vdots \\ 1 & q_n \end{bmatrix}, \quad A = \begin{bmatrix} \alpha_1 \\ \alpha_2 \end{bmatrix} \tag{13.3}$$

$$\text{The relation is} \quad p = QA \tag{13.4}$$

$$A = p - 1Q$$

3. To authenticate the model, the maximum of the absolute value of the deviation of the data is to be found out by using the following formula:

$$P = Q * A$$

4. Validation is the last step which is done by using the relation: MaxErr = max (abs ($P - p$))

13.2.2.2 Function Approximation: Model

By applying the function approximation on the data mentioned in Table 13.1. Equation connecting the y (output variable) and input variable ($x_1, x_2, x_3, x_4, x_5, x_6, x_7$) is shown as below:

$$y = a_0 + a_1x_1 + a_2x_2 + a_3x_3 + a_4x_4 + a_5x_5 + a_6x_6 + a_7x_7 \tag{13.5}$$

By solving above equation from above algorithm values of $a_0, a_1, a_2, a_3, a_4, a_5, a_6, a_7$

$$a_0 = 221.364265226088$$

$$a_1 = 0.175546705711193$$

$$a_2 = 0.00411624974491635$$

$$a_3 = 0.00411624974491635$$

$$a_4 = 260.894997850725$$

$$a_5 = 0.207507335152316$$

$$a_6 = 0.232271784522169$$

$$a_7 = -0.0731846372091189$$

$$\text{MaxErr} = 10.9937*10^{-1} \tag{13.6}$$

This value is considerably lesser than somewhat any of the data values, representing that the present model precisely trails the dataset and shall be taken further for optimizing the values using GA and PSO techniques.

13.2.2.3 Function Optimization: Model

- Maximize y given in equation (13.2), for $x_i^{min} \le x_i \le x_i^{max}$ where $i = 1,2,3...7$.
- For the input variable, the bound is shown in Table 13.2.

13.3 METHODOLOGY

13.3.1 Genetic Algorithm (GA)

Data mining has a goal to extract knowledge from large databases. GA belongs to the bigger group of evolutionary algorithms that creates solutions for optimization problems thereby utilizing methods enthused by natural development, like inheritance, mutation, selection, and crossover. With GA as an adaptive heuristic search algorithm introduced on natural selection and genetic evolutionary ideas, it becomes easier in data mining. The elementary idea of GA is intended to stimulate developments in natural arrangement vital for evolution, principally the ones that trail the principles that were first laid down by Charles Darwin stating about the survival of the fittest. GA is a technique of refinement of the computer programs and solutions to optimize the problems with stimulated evolution, procedures based on natural selection, crossover, and mutation are recurrently applied to a population representing probable solutions. With the passage of time, the numbers of above-average individuals are generated, until an excellent or the best fit solution to the problem is generated. Nevertheless, GA has shortcomings, such as lower local convergence speed and inkling to premature convergence and related problems. It finds application in numerous fields like computational science, engineering, bioinformatics, economics, physics, chemistry, manufacturing, mathematics, and many others.

There are various steps involved in GA algorithm which have been depicted in Figure 13.1. The steps can be described as follows:

- Outline a preliminary population arbitrarily or heuristically.
- Estimate the fitness value for each associate exclusively for the population.
- Allocate the chosen probability for each associate to compare it to its fitness value.
- Articulate the subsequent generation from the present generation by choosing the anticipated entities to harvest the offsprings.
- Reiteration all the steps of the process stated above to find the optimized and the most suitable solution (best fit).

It must be noted that GA describes a group of particles which are known as the population, and each associate is known as a chromosome. These chromosomes are then assessed using the cost function known as the fitness function, which is the given problem's objective function.

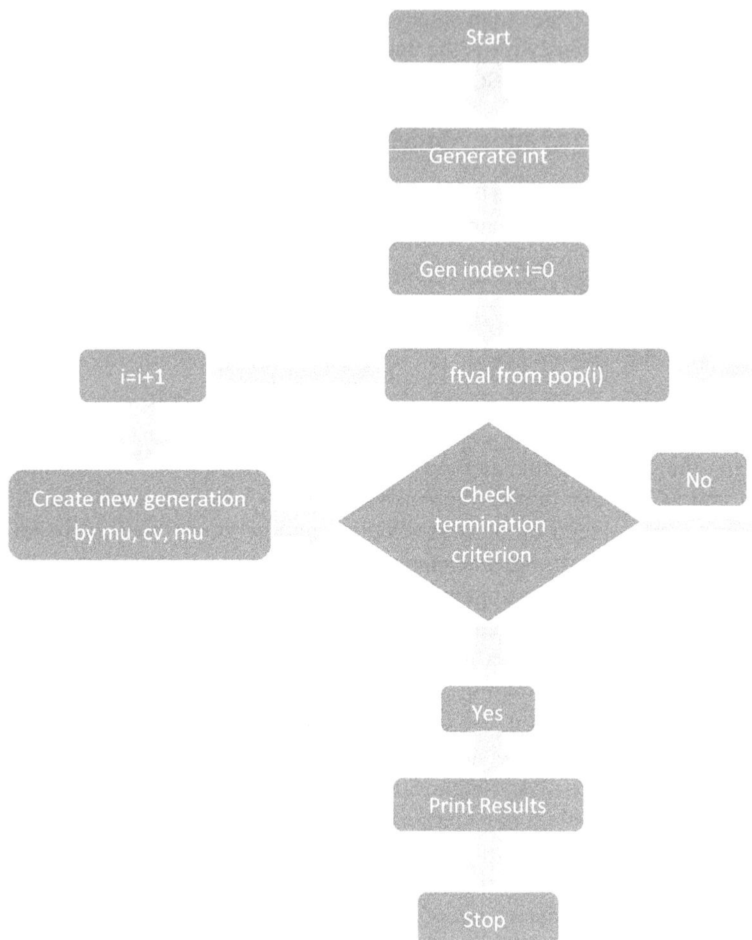

FIGURE 13.1 Flowchart depicting the action of GA technique.

Number of Abbreviations used:

Pop(i)	Population on iteration i
Int	Initial
gen_index	Generation Index
new_pop	New Population
et	Etism
cv	Crossover operator
mu	Mutation operator
ftval	Fitness value

The pseudo codes for GA technique are described below:

```
1   current population = Pop ()
2   current_ population.generate_initial()
3   gen_index = 1
4   while True:
    new_pop= deepcopy(current_population)
    parents = new_population.select_parents()
    new_pop.cv(parents)
    new_pop.mu()
    new_pop.et()
5   if new_pop.get_best_fitness() < current_population.get_best_fitness():
    current_population = new_pop
    gen_index += 1
    else print result
```

13.3.2 Particle Swarm Optimization

A cost function that ought to be maximized or minimized is explicitly demarcated in PSO algorithm. Once the cost function is defined for the problem, a swarm of particles is generated and dispersed in the dimensional space of the same. The pseudocode for the PSO algorithm is explained below. PSO necessitates substantial implementation time to acquire the solutions for extensive engineering applications, even though it has demonstrated its potential on numerous facets for explaining diverse optimization problems.

However, PSO is similar to GA with respect to population initialization with arbitrary solutions and probing for global optima in consecutive generations. In case of PSO algorithm, there is no crossover and mutation, but the particles transverse the problem space trailing the present optimum particles. The fundamental idea is that for each instant of time, the velocity of every particle, i.e., the potential solution, fluctuates between its personal best (pbest) and global best (gbest) positions. Statistically, the initialization of the swarm of particles is arbitrary over the search space and travel via the D-dimensional space to hunt for better and novel solutions.

Let x_k^i and v_k^i respectively be the location and velocity of ith particle hunting space at kth iteration, then its velocity and location of this particle at (k+1)th iteration are rationalized using the following equations:

$$v_{k+1}^i = w \cdot v_k^i + c_1 \cdot r_1 \left(p_k^i - x_k^i\right) + c_2 \cdot r_{2.} \left(p_k^g - x_k^i\right) \tag{13.7}$$

Here, $w \cdot v_k^i$ is the inertia, $c_1 \cdot r_1 \left(p_k^i - x_k^i\right)$ is personal influence and $c_2 \cdot r_2 \left(p_k^g - x_k^i\right)$ is social influence.

$$x_{k+1}^i = x_{k+}^i + v_{k+1}^i \tag{13.8}$$

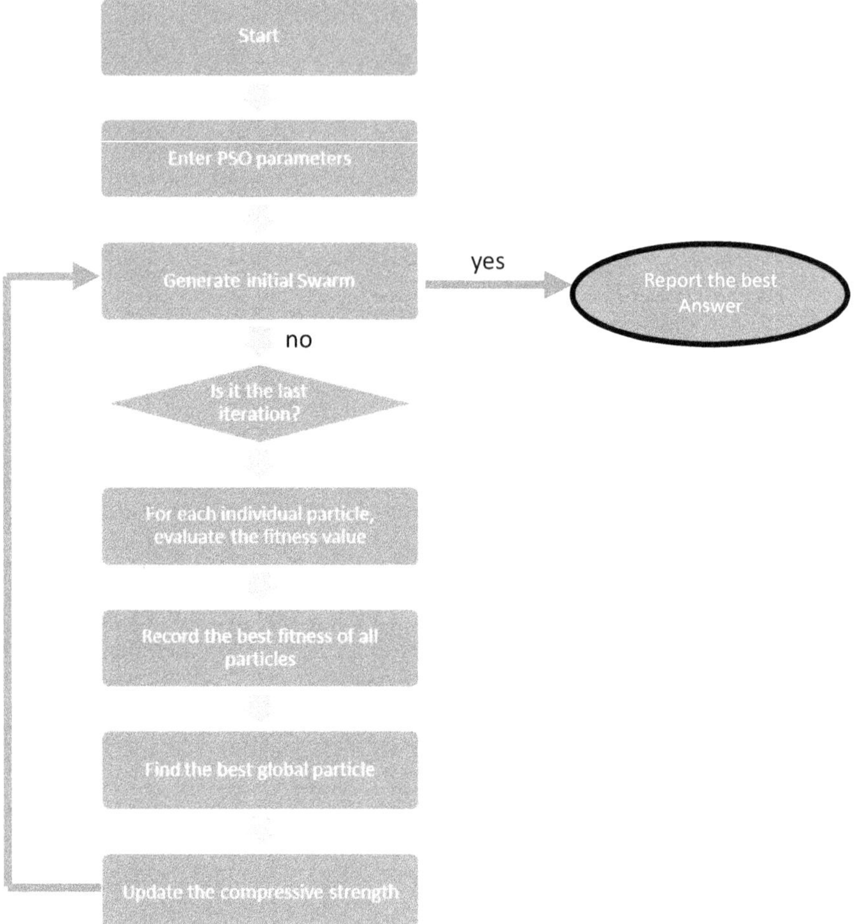

FIGURE 13.2 Flowchart of the PSO algorithm.

where r_1 and r_2 denote arbitrary numbers between 0 and 1, c_1 and c_2 are constants, p_k^i denotes the best ever position of *i*th particle, and p_k^g resembles the global best position in the swarm up to *k*th iteration. The flowchart of the PSO algorithm has been explained in Figure 13.2 which is self-explanatory.

Number of Abbreviation used:

$vel(t)$	Particle's velocity at time t
$pos(t)$	Particle's position at time t
Iw	Inertia weight
$c_{1,}$ c_2	accelerating factor
rand	uniformly distributed random number between 0 and 1
pos_{pbest}	particle's best position
pos_{gbest}	global best position

The pseudo codes for PSO algorithm are depicted by the following steps:

1	Objective function: $f(x)$, $x = (x_1, x_2, \ldots x_D)$
2	set $k = 1$ and Initialize pos(t), *vel*(t)
3	gen_index = 1
4	for $i = 1$ to *number of particles* do
	Calculate the fitness function f
	Update pos_{pbest} and pos_{pbest}
	Update *vel*(t)equation 13.7
	Update the *pos*(t) using equation 13.8
5	if new_pop.get_best_fitness() < current_population.get_best_fitness():
	current_population = new_pop
	gen_index += 1
	else print result

13.4 RESULTS

13.4.1 GA Technique

Flow chart representing the action of GA for parameter optimization in this study (Figure 13.2) can be described as below:

In this technique, firstly a MATLAB® program was formulated for 28-day CS-based optimization using GA and the optimum value of the CS corresponding to explicit value of parameters was considered. The optimum value of CS of concrete is found to be 84.1225 MPa on applying GA technique. The best possible combination values of cement are shown in Table 13.3. Figure 13.3 depicts the best fit value of the variables using GA, while Figure 13.4 shows the variation of objective function with iteration.

13.4.2 Optimization Using PSO Technique

Flowchart depicting the action of PSO technique was given in Figure 13.2. The steps were followed as per the algorithm stated below:

- In the search space, set the particles with approximate random velocities and positions.
- For the swarm particles, calculate the equivalent value of fitness function.

TABLE 13.3
Best Fit Values of the Variables Using GA

Cement (*x*1)	FA (*x*2)	CA (*x*3)	W/B (*x*4)	SP (*x*5)	NS (*x*6)	*D* (*x*7)
500	499.9999992	1	0.28	1	58.58999985	10

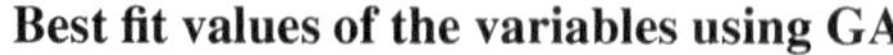

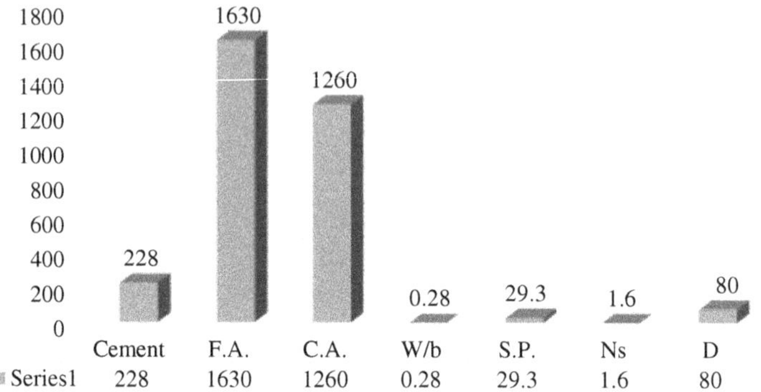

FIGURE 13.3 Best fit value of the variables obtained on using GA.

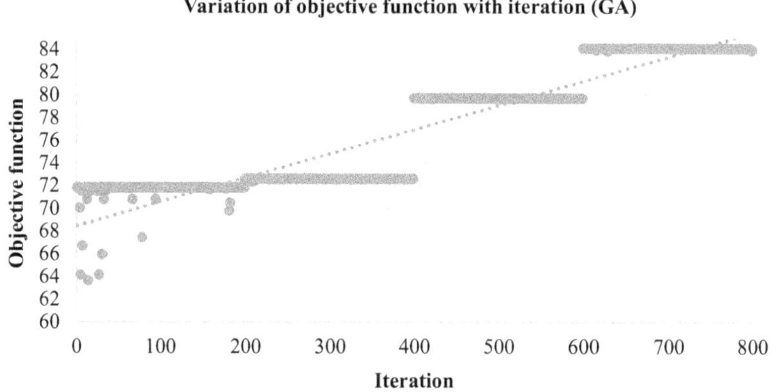

FIGURE 13.4 Variation of objective function with iteration using GA.

- Compare the fitness value evaluated with the present the value of particle's pbest. If present value comes out to be better than pbest, then set it as a new pbest value and set the pbest location to the present location in n-dimensional space.
- After this, compare the fitness value with the preceding overall best. If present value is better than gbest, then reset the gbest to the present particle's array index and value.
- To conclude, allocate these values to the matching position and velocity of the swarm particle.

The range of input parameters were same as that used in GA technique as per Table 13.2. The best fit values obtained for the variables using PSO technique is represented in Table 13.4.

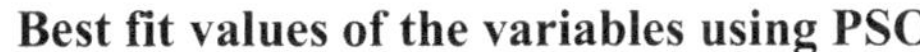

TABLE 13.4
Best Fit Values of the Variables Using PSO

Cement ($x1$)	FA ($x2$)	CA ($x3$)	W/B ($x4$)	SP ($x5$)	NS ($x6$)	D ($x7$)
228	1630	1260	0.2800	29.30000	1.6000	80

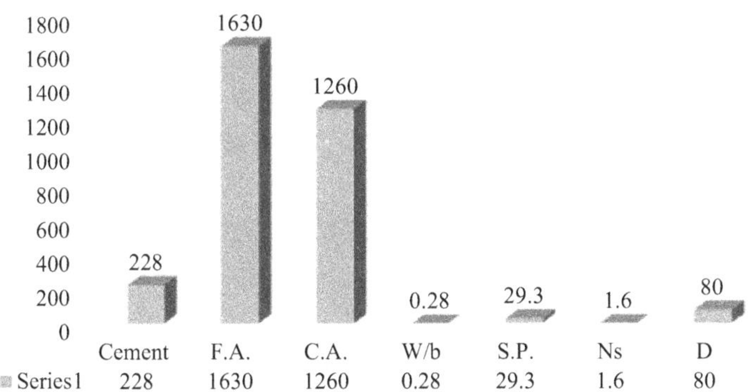

FIGURE 13.5 Best fit value of the variables obtained on using PSO.

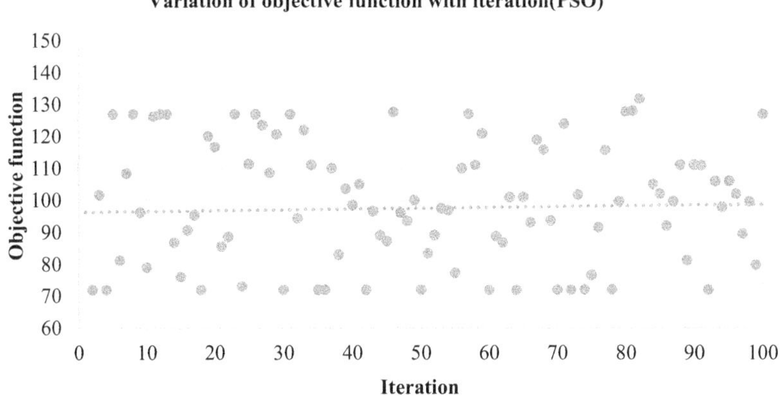

FIGURE 13.6 Variation of objective function with iteration using PSO.

On carrying out the PSO technique, the best fit CS is found to be 127.1286 MPa. Figures 13.5 and 13.6 represents the best fit values of the variables using PSO technique and the variation of objective function with iteration, respectively.

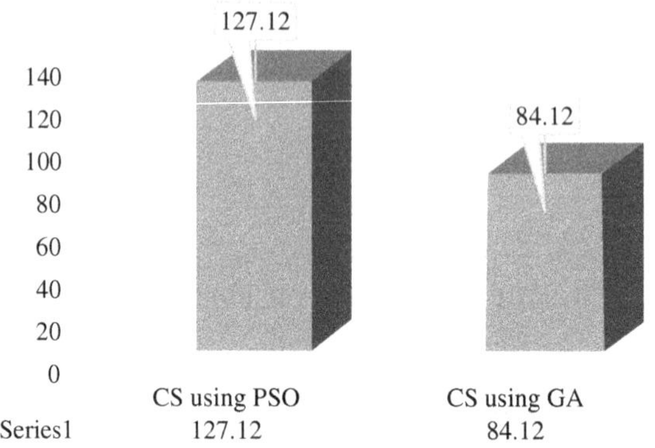

FIGURE 13.7 Comparative values of CS.

13.4.3 Comparison of the Results from GA and PSO Techniques

On comparing the results obtained in the present study, it is found that PSO technique performed better as compared to GA. The best fit CS is found to be 127.12 MPa while from GA technique it was found to be 84.12 MPa (Figure 13.7). Thus, from the present study we can assess the difference between the two techniques. GA is distinct in nature that means that it converts the parameters into binary 0s and 1s, and consequently, effortlessly manage the discrete problems. PSO is continuous and therefore it must be improved to manage discrete problems. Contrasting to GA, the different parameters in PSO technique can utilize any value grounded on their present position in the particle space and the equivalent velocity vector. GAs does not manage complication in a competent manner as the quantity of elements experiencing mutation is huge leading to a substantial surge in the search space. So, in this case PSO is regarded as the best alternative as it necessitates less quantity of variables and similarly lesser number of iterations. GA typically converges to a local optimum or arbitrary points rather than the global optimum of the problem, while PSO technique attempts to find the global optima.

13.5 CONCLUSIONS

The GA and PSO models for 28-day CS have been established in the present study. The models were trained with input and output experimental data which was procured from various literature. In this study, there are seven inputs and one output (28-day CS). As a result, CS values of concrete can be forecasted in GA models without endeavouring any experiments in a relatively lesser time interval with very less error rates. The GA results were compared with the random optimization where the prediction of function was made using PSO technique for 28-day CS of concrete. It is found that the optimized value for 28-day CS for concrete containing nano-silica using GA based optimization is 84.12 MPa as compared to PSO technique, which was 127.12 MPa. The nano-silica particles plug the voids and act as a nucleus to firmly bond with C–S–H gel particles, which helps in making the binding paste matrix denser. This results in a rise in long-term strength and durability of concrete. Thus, the utilization of nanoparticle materials in concrete can enhance several properties which are directly associated with the durability of cementitious materials. Also, it is likely to decrease the amounts of cement in the composite gaining the optimized cost benefits and material saving aspects.

Though both GA and PSO form a significant portion of evolutionary algorithms, they undergo a few shortcomings limiting their practice to only a few problems/applications. A combination of both GA and PSO can be utilized for enhancing the inclusive performance and also overcome the problems associated with them individually. Amalgamation of GA and PSO algorithms will produce a compound algorithm having more applied value by combining the advantages of both PSO and GA techniques. Therefore, a hybrid algorithm of GA and PSO can be studied for further research in the area.

13.6 CONFLICT OF INTEREST

There are no conflicts of interest with any of the author.

REFERENCES

Aitcin, P. C. Hershey, P. A., and Pinsonneault. Effect of the addition of condensed silica fume on the CS of mortars and concrete, *Am Ceramic Soc*, 1981, 22, pp. 286–290.

Akkurt, S., Tayfur, G., and Can, S. Fuzzy logic model for prediction of cement compressive strength, *Cement Concr Res*, 2004, 34(8), pp.1429–1433.

Bao, Y., Xiong, T., Hu, Z. PSO-MISMO modeling strategy for multistep-ahead time series prediction. *IEEE Trans Cybern*, 2013, 44(5), pp. 655–668.

Beigi, M. H., Berenjian, J., Omran, O. L., Nik, A. S. and Nikbin, I. M. An experimental survey on combined effects of fibers and nanosilica on the mechanical, rheological, and durability properties of self-compacting concrete. *Mater Design*, 2013, 50, pp. 1019–1029.

Chen, X. L., Fu, J. P., Yao, J. L., Gan, J. F. Prediction of shear strength for squat R.C. walls using a hybrid ANN–PSO model. *Eng Comput*, 2018, 34(2), pp. 367–383.

Demir, F. A new way of prediction elastic modulus of normal and high strength concrete-fuzzy logic, *Cement Concr Res*, 2005, 35(8), pp. 1531–1538.

Gupta, S. Using Artificial Neural Network to Predict the Compressive Strength of Concrete containing Nano-silica. *Civil Eng Architec*, 2013, 1(3), pp. 96–102, Doi: 10.13189/cea.2013.010306.

Hasanipanah, M., Noorian-Bidgoli, M., Armaghani, D. J., Khamesi, H. Feasibility of PSO-ANN model for predicting surface settlement caused by tunneling. *Eng Comput*, 2016, 32(4), pp. 705–715.

Heidari, A., Younesi, H., Mehraban, Z., Heikkinen, H. Selective adsorption of Pb (II), Cd (II), and Ni (II) ions from aqueous solution using chitosan–MAA nanoparticles. *Int. J. Biol. Macromol*, 2013, 61, pp. 251–263.

Jo, B. W., Kim, C. H., Tae, G., Park, J. B. Characteristics of cement mortar with nano- SiO_2 particles. *Construct Build Mater*, 2007, 21(6), pp. 1351–1355.

Kennedy, J. and Eberhart, R. C. A discrete binary version of the particle swarm algorithm. In 1997 IEEE International conference on sys-tems, man, and cybernetics. *Comput Cybern Simul*, 1997, 5, pp. 4104–4108.

Khan, K., Ahai, A. A comparison of B.A., GA, PSO, B.P. and L.M. for training feed forward neural networks in e-learning context. *Int J Intell Syst Appl*, 2012, 4(7), p. 23.

Li, G. Properties of high-volume fly ash concrete incorporating nano-SiO_2. *Cement Concr Res*, 2004, 34(6), pp. 1043–1049.

Li, H., Gang, H., Jie, X., Yuan, J., Ou, J. Microstructure of cement mortar with nanoparticles. *Compos Part B: Eng*, 2004, 35(2), pp. 185–189.

Mohamad, E. T., Armaghani, D.J., Momeni, E., Yazdavar, A. H., Ebrahimi, M. Rock strength estimation: A PSO-based BP approach. *Neural Comput Appl*, 2018, 30(5), pp. 1635–1646.

Mohandes, M. A. Modeling global solar radiation using Particle Swarm Optimization (PSO). *Sol Energy*, 2012, 86(11), pp. 3137–3145.

Mehrabi, N., Morstatter, F., Saxena, N., Lerman, K., Galstyan, A. A survey on bias and fairness in machine learning. *ACM Computing Surveys (CSUR)*, 2021, 54(6), pp. 1–35.

Nili, M., Ehsani A., and Shabani, K. Influence of Nano-$SiO_2$and Micro-silica on Concrete Performance. *Second International Conference on Sustainable Construction Materials and Technologies*, Italy, 2010.

Peyman, M., Soheil, H., Shervin, R., Soheil, J., Mohsen, A., Bidgoli, Seismic response prediction of FRC rectangular columns using intelligent fuzzy-based hybrid metaheuristic techniques, *J Ambient Intell Humanized Comput*, 2021, Doi: 10.1007/s12652-020-02776-4.

Said A. M., Zeidan M. S., Bassuoni M. T., Tian Y. Properties of concrete incorporating nanosilica. *Construct Build Mater*, 2012, 36, pp. 834–844.

Toghroli, A., Mehrabi, P., Shariati, M., Trung, N. T., Jahandari, S., and Rasekh, H. Evaluating the use of recycled concrete aggregate and pozzolanic additives in fiber-reinforced pervious concrete with industrial and recycled fibers. *Constr Build Mater*, 2020, 252, p. 118997.

Topcu, I. B., and Sarıdemir, M. Prediction of mechanical properties of recycled aggregate concretes containing silica fume using artificial neural networks and fuzzy logic, *Comput Mater Sci*, 2008, 42(1), pp. 74–82.

Unal, O., Demir, F., and Uygunoglu, T. Fuzzy logic approach to predict stress–strain curves of steel fiber-reinforced concretes in compression, *Build Environ*, 2007, 42(10), pp. 3589–3595.

Zhang, M.H., and Hui L. Pore structure and chloride permeability of concrete containing nanoparticles for pavement. *Construct Build Mater*, 2011, 25, pp. 608–616.

14 Prediction of Ultrasonic Pulse Velocity of Concrete

Tanvi Singh and Balraj Singh
Panipat Institute of Engineering and Technology

Sunita Bansal
Manav Rachna University

Khyati Saggu
Panipat Institute of Engineering and Technology

CONTENTS

DOI: 10.1201/9781003184331-14

14.1 INTRODUCTION

The use of non-destructive tests (NDT) is increasing day by day in the civil engineering testing system. It has been becoming an essential tool for quality checks of any building. These tests are the center of attraction for the researchers. The main reason for adopting these tests is their method, which does not disturb the performance as well as the appearance of the building [1]. Also, these tests are performed at the same place which makes a nonstop monitoring of the structure and a possible variation of concrete is determined for a certain period. After monitoring the structure, the prediction of life span can be done which can recover and treat economically [2,3]. Ultrasonic pulse velocity test (UPV) is one of the famous tests of NDT by which the compressive strength of the concrete can be measured [4]. UPV is one of the most widely used non-destructive methods to inspect the mechanical behavior of the concrete [5–10]. The performance of concrete, defects, elastic modulus, counting strength, depth of surface cracks, porosity, and damages caused by fire and chemical attacks can be calculated using UPV. It is altered by various traits: water-cement ratio, type of cement and its content, the temperature of the measuring medium, curing method, concrete age, aggregate size and type, and measuring distance length.

Nowadays, the use of the modeling approach has gained a trend in civil engineers for the analysis of experimental data. The traditional modeling methods are created using empirical formulas derivative from experimental data. Various models have been suggested depending on varying experimental conditions, to forecasts the compressive and shear strength of beams and columns. Studies have shown that these models have data specific nature. Designing any structural component may involve an iterative practice, in which the implicit model behavior meets with the experimental behavior, which causes the cost-efficient computational technique [11]. Atıcı [12] used multiple regression analysis and artificial neural networks for predicting the compressive strength of concrete containing mineral-admixture and data collection was done using rebound hammer and UPV. Alexandridis et al. [13] employed artificial neural networks for predicting the compressive strength of cement-based materials data was collected using the info veiled in weak electrical signals observed in samples imposed by mechanical stress. Similarly, various researchers used soft computing techniques on civil engineering problems, including the behavior of concrete materials [14–18] and results were promising. Soft computing techniques are helpful to conclude significant outcomes from extensive data. For data analysis, model trees are one of the most effective methods. Trees function is employed when numeric prediction is like conventional decision trees also known as regression trees. Among all the soft computing techniques, M5 rule (M5PR), M5P tree (M5PT), and random forest (Ran-For) gained popularity nowadays. Keeping this in mind present study aims at using the M5PR, M5PT, and Ran-For models (to compare the performance of the rule, tree, and regression model on present data) to develop a relationship between various tests performed and to compare the compatibility of different modeling approach with present data WEKA workbench was implemented to evaluate the data.

14.2 DETAILS OF MODELING APPROACHES USED

Three modeling techniques were used in the present study: M5PT, M5PR, and Ran-For. Then, if then rule approach, model tree, and regression methods are used and their performances were compared.

14.2.1 M5P Tree (M5PT)

Quinlan [19] introduced the concept of "model tree" in 1992. M5PT is the amalgamation of the decision tree and linear regression functions at the leaves. Young and Witten [20] have developed M5PT which give better results on data set also they made some enhancement on M5PT algorithm to reduce the size of the tree. Working on an M5PT tree model includes building a model tree using, decision-tree algorithm. Afterward Pruning of the tree is done, and this was introduced by Breiman [21] and Quinlan [19]. Breiman and Quinlanre commended a new decision tree algorithm known as classifying and regression tree which was used for classification and solution of regression problems both it can use both numerical and nominal data types. Dividing criteria must be well defined to build an M5PT tree. It is based on the standard deviation of the attribute values. The root of the tree comprises an attribute that diminishes estimated error. The formula of standard deviation reduction is given as:

$$\mathrm{SDR} = sd(V) - \sum_{i} \frac{Vi}{V} \times sd(V_i) \tag{14.1}$$

where V is set of attributes values, V_i is attribute value that taken from divided node according to the selected attribute, V is average value of the sets of V attribute, and $sd\ (V)$ is standard deviation of V.

14.2.2 M5 Rule (M5PR)

M5PR can predict the nominal and numeric values, and it is a rule-based learning method. M5PR sets are created from model trees and these predict numeric and nominal values. The working principle of the rule algorithm is that it repeats building model tree building and then selecting the best rule at each cycle. M5PR creates the rules from M5PT using the partial and regression tree algorithm offered by Frank and Witten [22]. The principle of M5PR is as follows:

1. An M5PR learner is deployed for the entire training data.
2. The tree is pruned, similar to M5PT.
3. Leaf with minimum error is turned into the rule.
4. Above steps carry on till the entire instances are built-in the rules. An instance can be included by different rules at the same time.
5. In contrast to the partial and regression tree, which applies the same approach for categorical prediction, M5PR forms full trees in place of partially explored trees.

14.2.3 Random Forest Regression (Ran-For)

Ran-For regression working principle comprises a fusion of tree predictors and every single tree is produced from the input vector using a random vector sampled only. Ran-For regression comprises the grouping of variables at every single node to produce a tree or by haphazardly selected input variable adopted in the present study. To create a training data set, bagging, which haphazardly pulls *T* training samples with replacement, where *T* is the size of the actual training set (Breiman, [21]), or a haphazardly selected portion of the training set which was employed for the building of every singletree for a random feature combination. Approximately one-third of the data are left behind at each grown tree in case of bagging (bootstrap) actual training set is 67% of the original training set and the remaining data left are called out-of-bag (out of the bootstrap sampling). Choosing the selection measure and a pruning method is essential for designing a tree predictor. For variable selection, the Gini Index (Brieman et al., [23]) has been used in this study. Gini Index measures the adulteration of the variable compared to the output.

Two user-defined parameters are essential for random forest regression:

1. The number of input variables (*m*) employed at each node to produce a tree.
2. The number of trees to be grown (*k*).

14.3 EXPERIMENTS PERFORMED AND DATA SET

The old hostel of Panipat institute of Engineering and Technology (PIET) (Samalkha) was found to have some cracks and distress signs which ultimately require structural rehabilitation and repayment. We have undertaken the assessment work for the complete building and the tests performed are as follows:

1. Half-cell potential test
2. Ultrasonic pulse velocity test
3. Carbonation test
4. Core test
5. Chloride test

In total, 38 locations were identified and all five tests were carried out at each location details of each test location are given in Table 14.1.

14.3.1 Ultrasonic Pulse Velocity Test (UPV)

UPV is the most modern test of concrete which is based on wave propagation. It transmitted the ultrasonic pulse into the concrete and noted the wave transit time. The velocity in km/s was calculated by dividing the path length by wave transit time. The velocity can be associated with the quality of concrete (elaborated in Table 14.2). With the help of the UPV test, the concrete condition can be evaluated to its relative quantity, homogeneity, presence of voids, and integrity between or within the members. In the present study, UPV readings were recorded at different time intervals at each location.

TABLE 14.1
Test Location Details

S. No.	Structural Location
1	BasementP1, 11(K)-12(K)
2	BasementP1, 9(A)-10(A)
3	BasementP1, Column 16(A)
4	BasementP1, Column 10(G)
5	BasementP1, Column 10(K)
6	BasementP1, Column 19(E)
7	BasementP1, Column 18(E)
8	BasementP1, Column 16(D)
9	BasementP1, Column 15(E)
10	BasementP1, Column 13(C)
11	South Block Entry (D) wall
12	South Block Slab 37(A)-38(A)
13	South Block Beam 37(A)-38(A)
14	South Block Column 38(A),
15	South Block Entry Column 37(B)
16	South Block Entry Column 31(A)
17	BasementP2, Column-30(D)
18	BasementP2, Column-29(D)
19	BasementP2, Column-30(C)
20	BasementP2, Column 22(A)
21	BasementP2, Column 20(O)
22	BasementP2, Slab 21(I)-21(J)
23	BasementP2, Slab 21(I)-2(I)
24	BasementP2, Slab 21(J) From Wall
25	BasementP2, Slab 20(J)-21(J)
26	BasementP2, Beam 21(A)-22(A)
27	BasementP2, Slab 21(A)-22(A)
28	BasementP2, Slab 22(A)-23(A)
29	BasementP2, Beam 22(A)-wall
30	BasementP2, Slab 21(A)-22(A)
31	BasementP2, P(20)-wall
32	BasementP2, Beam C(29)-Wall
33	BasementP2, Slab 29(E)-29(D)
34	BasementP2, Slab 28(E)-28(D)
35	BasementP2, Beam
36	BasementP2, Slab 29(A)-29(B)
37	BasementP1, Beam 10(I)-11(I)
38	BasementP1, Slab 10(I)-11(I)

TABLE 14.2
Relation of Grading of Concrete with Ultrasonic Pulse Velocity

Velocity Generated by Ultrasonic Pulse Velocity (km/s)	Grading of Concrete
Over 4.5	Excellent
Between 3.5 and 4.5	Good
Between 3.0 and 3.5	Medium
Less than 3.0	Poor or doubtful

14.3.2 Core Tests

The core samples of 75 mm were taken from the structure. It was tested for compressive strength in compression testing machine. The machine used for core sampling was a portable diamond core cutting machine. With the help of this test, the compressive strength of concrete (*in situ*) is calculated.

14.3.3 Half-Cell Potential Test (HCP)

The HCP test is one of the most used tests to find out the presence of corrosion. It is based on the electrochemical method. As corrosion takes place, there is a flow of electrons and ions. The electrons move through the bar to the cathode where they combine with the water and oxygen in the concrete and is the reduction reaction occurs. Thus, a potential difference exists between rebar and concrete electrolyte at electrode across the electrical, called the electrode potential. From the measured potential possible anodic sites can be differentiated from cathodic sites based on the empirically determined risk of corrosion criteria given in Table 14.3.

14.3.4 Carbonation Test

The natural alkalinity in concrete provides a protective layer against corrosion. Carbon dioxide in the air reacts with the Calcium salts, resulting in reducing the pH of concrete to below 9. The depth of carbonation is determined using phenolphthalein as an indicator. As a pH indicator, for pH above 9, Phenolphthalein changes color from colorless to pink.

TABLE 14.3
Risk of Corrosion with HCP Values

HCP	Risk for Corrosion (%)
Values more than −200	10
Values between −200 and −350	50
Values less than −350	90

14.3.5 Chloride Test

In the chloride test, a test on the selected concrete sample is done for checking the free water-soluble chloride content on the building site. Using this method, free chloride content is calculated by the titration method by using the Silver Nitrate.

14.3.6 Data Set and Analysis

The goal of this research is to model UPV utilising (V) in Km/s using M5PR, M5PT, and Ran- Approaches to modeling. For which compressive strength value of concrete, half-cell potential, chloride, carbonation, and transit time was used as input parameter whereas UPV as an output parameter. One hundred and forty-nine data samples were present, out of which approximately one-third of randomly selected data which is equal to 50 data samples was taken as test data remaining was considered as training data set. The summary of each attribute is given in Table 14.4.

To attain the finest value of user-defined parameters for different modeling approaches, an enormous amount of hits and trials were carried out. Table 14.5 gives the optimal value of user-defined parameters with M5PR, M5PT, and Ran-For regression techniques. To compare the performance of all three techniques, correlation coefficient (CC) and root mean square error (RMSE) values were obtained and compared in Table 14.6.

14.4 RESULTS AND DISCUSSION

14.4.1 Results from M5PR

The M5PR was implemented by WEKA3.8.5. Figure 14.1 indicates actual UPV and predicted UPV using the M5PR rules algorithm. It can be inferred from the plot that predicted values show minor variation from actual values. Total Nine rules were obtained for the present data set (Table 14.7). Table 14.6 indicate value of CC = 0.8492 and RMSE = 0.4197 which is satisfactory.

TABLE 14.4
Summary of Each Attribute[a]

Attribute	Minimum Value	Maximum Value	Mean	Std. Dev.
Compressive value of concrete (CCS)	−277	47.6	18.741	26.099
Half-cell potential (H)	−589.8	−76	−224.21	97.049
Chloride (CL)	0.053	1.628	0.547	0.519
Carbonation (CA)	9	32	22.25	7.063
Travel time (T)	59.9	302	114.737	43.904
UPV	0.99	5.01	3.712	0.762

[a] CCS = Compressive value of concrete, H = Half-cell potential, CL = Chloride, CA = Carbonation, T = Travel time, UPV = Ultrasonic pulse velocity.

TABLE 14.5
User-Defined Parameters

Modeling Approach	User-Defined Parameter
M5PR	$M = 4.0$,
M5PT	$M = 4$
Ran-For	$k = 100$, $m = 1$

TABLE 14.6
Values of CC and RMSE for Modeling Approach

	Modeling Approach		
Performance Evaluation Parameter	M5PR	M5PT	Ran-For
Correlation coefficient	0.8492	0.8246	0.9541
Root mean squared error	0.4197	0.4694	0.2565

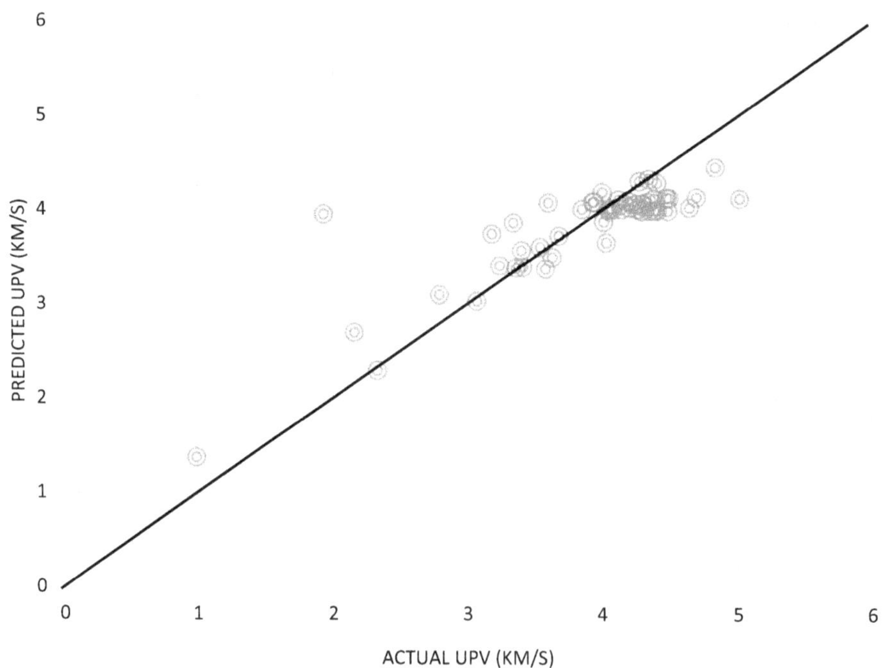

FIGURE 14.1 Plot between actual UPV and predicted UPV from M5 rules.

14.4.2 M5PT

The M5PT was implemented by WEKA3.8.5. Figure 14.2 indicates actual UPV and predicted UPV using the M5PT function. It can be observed from the plot that predicted values are following actual values. M5 tree gives 18 equations depending on different input data sets. (Table 14.7). Table 14.6 indicate value of CC = 0.8246 and RMSE = 0.4197, which is slightly lesser than M5 rule.

14.4.3 Result from Ran-Forest

The Ran-For was also implemented by WEKA3.8.5. Figure 14.3 shows the plot between the actual UPV and predicted UPV using Ran-For regression. It can be concluded from the plot that predicted values are following actual values. Table 14.6 indicates the value of CC = 0.9541 and RMSE = 0.2565, which indicates that Ran-For gives the best performance with present data.

14.4.4 Comparison of Results

A comparison of the three soft computing models was done in Figure 14.4, which shows a comparison of actual and predicted values of testing data. It suggests that the Ran-For model is the best model as compared to M5PR and M5PT. Also, the comparison is done with the box plot (Figure 14.5). The box plot is the plot that is used by modern data scientists and researchers for the comparison of soft computing tools. The box of Actual values is similar to the Ran-For (shown in Figure 14.5)

TABLE 14.7
Rules Set Obtained from M5Rules Algorithm

Rule: 1	
IF	
	Travel Time (T) > 82.5
	chloride (CL) ≤ 0.667
	half-cell potential (H) > −250.25
	half-cell potential (H) > −208
	Compressive value of concrete (CCS) > 12.25
THEN	
UPV =	
	−0.0151 * Compressive value of concrete (CCS)
	+ 0.0012 * half-cell potential (H)
	−0.2032 * chloride (CL)
	+ 0.0091 * carbonation (CA)
	−0.001 * Travel Time (T)
	+ 4.4749 [19/21.428%]

(Continued)

TABLE 14.7 (*Continued*)
Rules Set Obtained from M5Rules Algorithm

Rule: 2

IF

Travel Time (T) > 82.5

Travel Time (T) ≤ 100

Travel Time (T) ≤ 89.7

THEN

UPV =

0.0018 * Compressive value of concrete (CCS)

+ 0.0008 * half-cell potential (H)

−0.3333 * chloride (CL)

−0.0235 * Travel Time (T)

+ 5.9101 [18/0.46%]

Rule: 3

IF

Travel Time (T) > 86.25

chloride (CL) > 0.425

Travel Time (T) ≤ 114.65

THEN

UPV =

−0.0004 * Compressive value of concrete (CCS)

+ 0.0008 * half-cell potential (H)

−0.4078 * chloride (CL)

−0.013 * Travel Time (T)

+ 4.9889 [20/1.163%]

Rule: 4

IF

Travel Time (T) ≤ 116.2

Compressive value of concrete (CCS) > 18

Travel Time (T) > 90.8

carbonation (CA) ≤ 22.5

THEN

UPV =

0.001 * Compressive value of concrete (CCS)

+ 0.0016 * half-cell potential (H)

(*Continued*)

TABLE 14.7 (*Continued*)
Rules Set Obtained from M5Rules Algorithm

	−0.0787 * chloride (CL)
	+ 0.0044 * carbonation (CA)
	−0.0057 * Travel Time (T)
	+ 5.0659 [9/0.128%]
Rule: 5	
IF	
	Travel Time (T) > 104.5
	chloride (CL) ≤ 0.43
	half-cell potential (H) > −223.25
THEN	
UPV =	
	−0.0457 * Compressive value of concrete (CCS)
	−0.0479 * half-cell potential (H)
	−0.4701 * chloride (CL)
	−0.0066 * Travel Time (T)
	−4.1136 [15/1.816%]
Rule: 6	
IF	
	Travel Time (T) ≤ 116.2
	chloride (CL) > 0.153
	chloride (CL) > 0.184
	Travel Time (T) > 69.15
	Travel Time (T) ≤ 78.45
THEN	
UPV =	
	0.0042 * Compressive value of concrete (CCS)
	+ 0.0005 * half-cell potential (H)
	+ 0.0997 * chloride (CL)
	−0.0152 * Travel Time (T)
	+ 5.2332 [15/0.602%]
Rule: 7	
IF	
	Travel Time (T) ≤ 116.2
	chloride (CL) > 0.132

(*Continued*)

TABLE 14.7 (*Continued*)
Rules Set Obtained from M5Rules Algorithm

THEN	
UPV =	
	0.0231 * Compressive value of concrete (CCS)
	+ 0.0013 * half-cell potential (H)
	+ 0.4289 * chloride (CL)
	−0.0028 * Travel Time (T)
	+ 4.0144 [25/28.29%]
Rule: 8	
IF	
	chloride (CL) ≤ 0.43
	half-cell potential (H) > −310.4
	half-cell potential (H) ≤ −250.25
	Compressive value of concrete (CCS) ≤ 28.5
THEN	
UPV =	
	0.0018 * Compressive value of concrete (CCS)
	+ 0.0018 * half-cell potential (H)
	−0.3088 * chloride (CL)
	−0.0025 * Travel Time (T)
	+ 4.1174 [7/24.021%]
Rule: 9	
UPV =	
	−0.0112 * Travel Time (T)
	+ 4.0433 [20/52.538%]

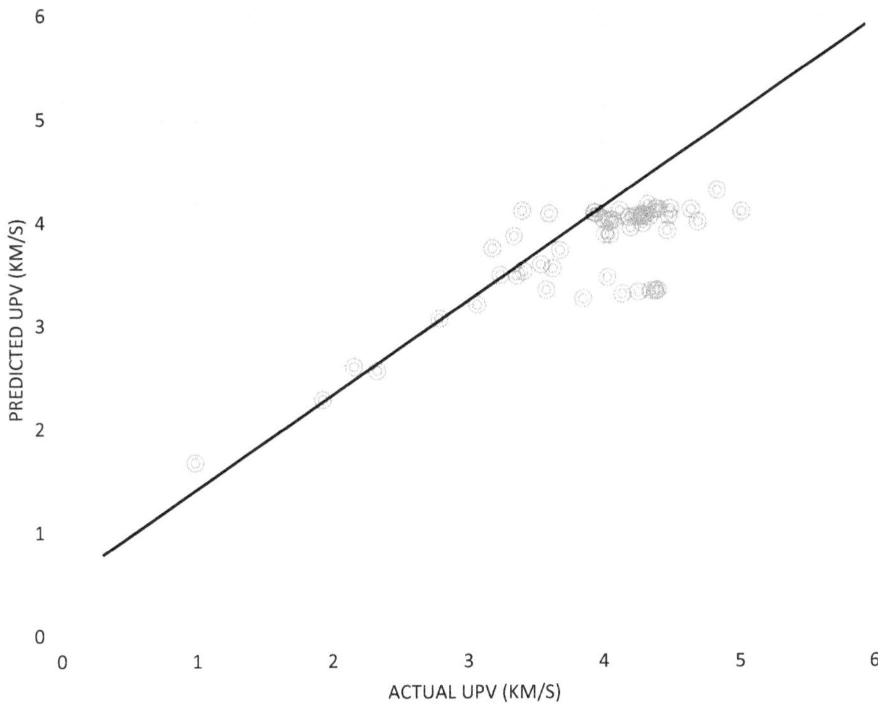

FIGURE 14.2 Plot between actual UPV and predicted UPV from M5PT.

TABLE 14.8
Tree Obtained from M5PT

Travel Time (T) ≤ 82.5 :
| Travel Time (T) ≤ 69.15: LM1 (8/1.247%)
| Travel Time (T) > 69.15 :
| | Travel Time (T) ≤ 79.6 :
| | | Travel Time (T) ≤ 75.4: LM2 (9/0.588%)
| | | Travel Time (T) > 75.4: LM3 (10/0.322%)
| | Travel Time (T) > 79.6 :
| | | half-cell potential (H) ≤ −101.35: LM4 (4/2.871%)
| | | half-cell potential (H) > −101.35: LM5 (2/5.269%)
Travel Time (T) > 82.5 :
| chloride (CL) ≤ 0.667 :
| | half-cell potential (H) ≤ −250.25 :
| | | Travel Time (T) ≤ 97.25: LM6 (6/0.539%)
| | | Travel Time (T) > 97.25 :
| | | | Compressive value of concrete (CCS) ≤ 28.5 :

(Continued)

TABLE 14.8 (*Continued*)
Tree Obtained from M5PT

| | | | | Compressive value of concrete (CCS) ≤ 22: LM7 (5/1.666%)
| | | | | Compressive value of concrete (CCS) > 22: LM8 (10/0.477%)
| | | | Compressive value of concrete (CCS) > 28.5: LM9 (4/4.165%)
| | half-cell potential (H) > −250.25 :
| | | half-cell potential (H) ≤ −208 :
| | | | carbonation (CA) ≤ 25.5 :
| | | | | Travel Time (T) ≤ 107.2: LM10 (13/0.289%)
| | | | | Travel Time (T) > 107.2: LM11 (7/13.37%)
| | | | carbonation (CA) > 25.5: LM12 (10/8.443%)
| | | half-cell potential (H) > −208 :
| | | | Compressive value of concrete (CCS) ≤ 12.25: LM13 (7/0.812%)
| | | | Compressive value of concrete (CCS) > 12.25: LM14 (19/21.428%)
| chloride (CL) > 0.667 :
| | Travel Time (T) ≤ 114.65 :
| | | Travel Time (T) ≤ 95.25: LM15 (12/0.786%)
| | | Travel Time (T) > 95.25: LM16 (13/0.482%)
| | Travel Time (T) > 114.65 :
| | | Travel Time (T) ≤ 221.55: LM17 (7/1.488%)
| | | Travel Time (T) > 221.55: LM18 (2/3.293%)

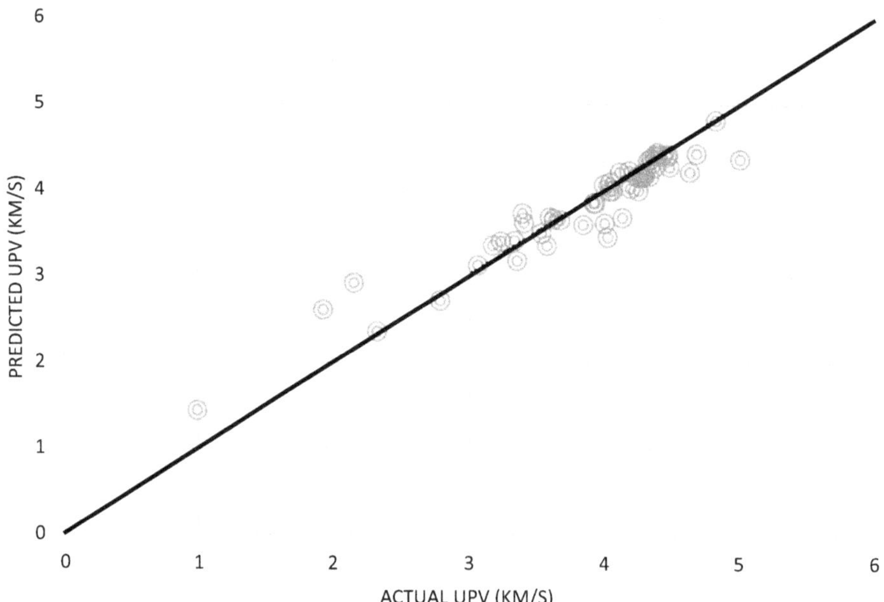

FIGURE 14.3 Plot between the actual UPV and predicted UPV from Ran-For.

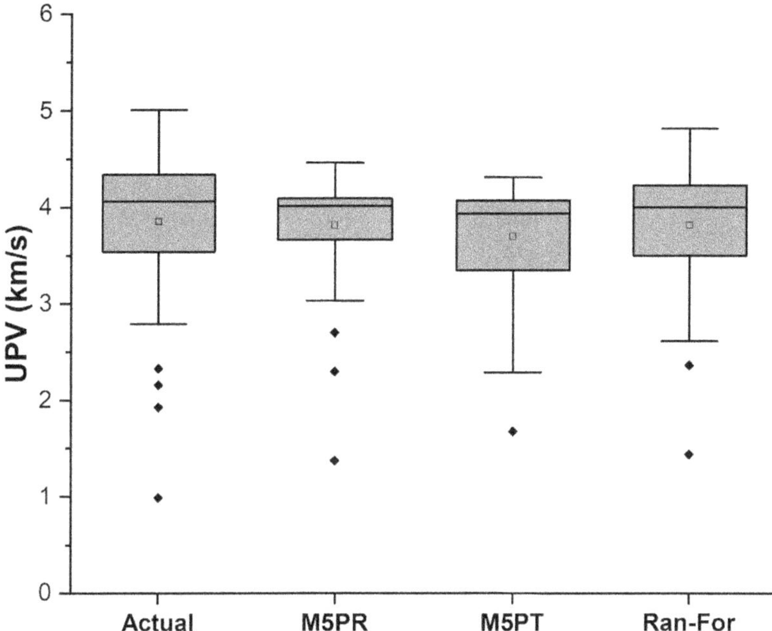

FIGURE 14.4 Box-plot of the applied soft computing models.

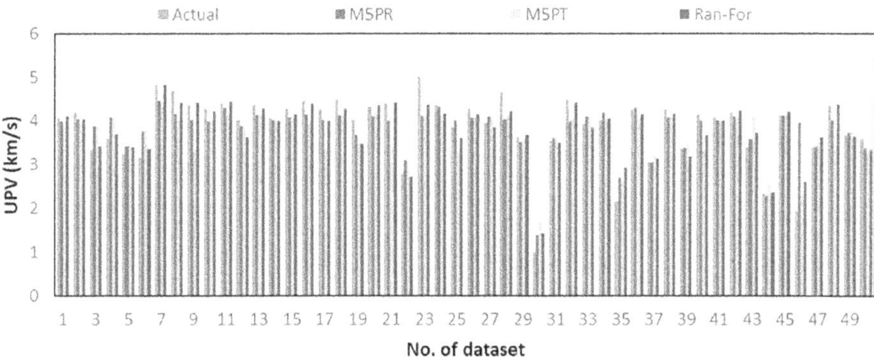

FIGURE 14.5 Variation in predicted values of UPV approach in comparison to actual values of UPV.

which shows that Ran-For is the best efficient model. Hence, Table 14.6, Figures 14.4 and 14.5 interpret that Ran-For is the most accurate model than M5PR and M5PT, which can predict the UPV precisely.

14.5 CONCLUSION

To predict the UPV of the concrete, the experimentation tests were performed on the old hostel of PIET, and tests carried out were UPV, core test, half-cell potential test, carbonation test, chloride test. Three modeling approaches were used: M5 Rule, M5 tree, and RF regression.

The results suggested that all the soft computing models can predict the UPV. But Ran-For were the models which can predict the UPV accurately. The values of CC and RMSE are 0.9541 and 0.2565 respectively for the Ran-For which was much ideal than the values of M5PR (0.8492 and 0.4197) and M5PT (0.8246 and 0.4694). While the performance of M5PR and M5PT is approximately the same. Hence, Ran-For was the model which can be used in the prediction of UPV.

Although the experimentation of the UPV is not a simple task, the values of the UPV can easily be predicted using the soft computing models, which is also the most reliable method. So, the soft computing models can be used in the prediction of UPV of concrete.

REFERENCES

1. Jamle, S., Delmiya, N., & Singh, R. (2020). Efficient use of UPV meter: A non destructive test of concrete by fragmentation analysis. *Journal of Xi'an University of Architecture & Technology*, ISSN, 1006-7930.
2. Davis, A. G., Ansari, F., Gaynor, R. D., Lozen, K. M., Rowe, T. J., Caratin, H., ... & Sansalone, M. J. (1998). Nondestructive test methods for evaluation of concrete in structures. *American Concrete Institute, ACI*, 228, 4.
3. Chrisp, T. M., Waldron, P., & Wood, J. G. M. (1993). Development of a non-destructive test to quantify damage in deteriorated concrete. *Magazine of Concrete Research*, 45(165), 247–256.
4. Malhotra, V. M., & Carino, N. J. (2003). *Handbook on Nondestructive Testing of Concrete*. Boca Raton, FL: CRC Press.
5. Demirboğa, R., Türkmen, İ., & Karakoc, M. B. (2004). Relationship between ultrasonic velocity and compressive strength for high-volume mineral-admixtured concrete. *Cement and Concrete Research*, 34(12), 2329–2336.
6. Pfister, V., Tundo, A., & Luprano, V. A. (2014). Evaluation of concrete strength by means of ultrasonic waves: A method for the selection of coring position. *Construction and Building Materials*, 61, 278–284.
7. Bogas, J. A., Gomes, M. G., & Gomes, A. (2013). Compressive strength evaluation of structural lightweight concrete by non-destructive ultrasonic pulse velocity method. *Ultrasonics*, 53(5), 962–972.
8. Nik, A. S., & Omran, O. L. (2013). Estimation of compressive strength of self-compacted concrete with fibers consisting nano-SiO_2 using ultrasonic pulse velocity. *Construction and Building Materials*, 44, 654–662.
9. Jain, A., Kathuria, A., Kumar, A., Verma, Y., & Murari, K. (2013). Combined use of non-destructive tests for assessment of strength of concrete in structure. *Procedia Engineering*, 54, 241–251.
10. Yılmaz, T., Ercikdi, B., Karaman, K., & Külekçi, G. (2014). Assessment of strength properties of cemented paste backfill by ultrasonic pulse velocity test. *Ultrasonics*, 54(5), 1386–1394.
11. Kreimeyer, M. F. (2010). *A structural measurement system for engineering design processes* (Doctoral dissertation, Technische Universität München).

12. Atici, U. (2011). Prediction of the strength of mineral admixture concrete using multivariable regression analysis and an artificial neural network. *Expert Systems with Applications*, 38(8), 9609–9618.
13. Alexandridis, A., Triantis, D., Stavrakas, I., & Stergiopoulos, C. (2012). A neural network approach for compressive strength prediction in cement-based materials through the study of pressure-stimulated electrical signals. *Construction and Building Materials*, 30, 294–300.
14. Fairbairn, E. M. R., Ebecken, N. F. F., Paz, C. N. M., & Ulm, F. J. (2000). Determination of probabilistic parameters of concrete: Solving the inverse problem by using artificial neural networks. *Computers & Structures*, 78(1–3), 497–503.
15. Fairbairn, E. M., Silvoso, M. M., Toledo Filho, R. D., Alves, J. L., & Ebecken, N. F. (2004). Optimization of mass concrete construction using genetic algorithms. *Computers & Structures*, 82(2–3), 281–299.
16. Hadi, M. N. (2003). Neural networks applications in concrete structures. *Computers & Structures*, 81(6), 373–381.
17. Akkurt, S., Ozdemir, S., Tayfur, G., & Akyol, B. (2003). The use of GA–ANNs in the modelling of compressive strength of cement mortar. *Cement and Concrete Research*, 33(7), 973–979.
18. Evsukoff, A. G., Fairbairn, E. M., Faria, É. F., Silvoso, M. M., & Toledo Filho, R. D. (2006). Modeling adiabatic temperature rise during concrete hydration: A data mining approach. *Computers & Structures*, 84(31–32), 2351–2362.
19. Quinlan, J. R. (1986). Simplifying decision trees. *Proceeding of Workshop on Knowledge Acquisition for Knowledge-based Systems*, Banff, Canada.
20. Wang, Y., & Witten, I. H. (1997). April. Inducing model trees for continuous classes. *Proceedings of the Ninth European Conference on Machine Learning*, pp. 128–137.
21. Breiman, L. (1996). Bagging predictors. *Machine Learning*, 24(421), 123–140. Doi: 10.1007/BF00058655.
22. Frank, E., & Witten, I. H. (1998). Generating accurate rule sets without global optimization. *Proceedings of the Fifteenth International Conference on Machine Learning*, Morgan Kaufmann, pp. 144–151.
23. Breiman, L, Friedman, J. H., Olshen, R. A., & Stone, C. J. (1983). *Classification and Regression Trees, Wadsworth, Belmont, CA, Since 1993 This Book Has Been Published by* Chapman & Hall, New York.

15 Evaluation of ANN and Tree-Based Techniques for Predicting the Compressive Strength of Granite Powder Reinforced Concrete

Bhupender Kumar and Navsal Kumar
Shoolini University

Ahmed Elbeltagi
Mansoura University

Fadi Hamzeh Almohammed
Shoolini University

CONTENTS

DOI: 10.1201/9781003184331-15

15.1 INTRODUCTION

One of the most broadly used construction materials on the planet is concrete. It is employed in a variety of applications, including buildings, highway pavements, bridges, walkways, housing construction, dams and more. The proportions of the various components in the mix are the key to solid and long-lasting concrete. More voids mean low strength and resilience, while greater cement paste means less durability and more shrinkage. The gradation of fine and coarse particles, as well as the ratio of fine to coarse aggregates, can have an impact on strength and porosity. The design mix should also provide the preferred workability of concrete in order to stop segregation and enable placing. To bear the imposed loads, concrete should have sufficient flexural and compressive strength. It should also be sturdy to extend its design life and save maintenance costs (Ghannam et al., 2016). Fine aggregate is an important part of the concrete mix. The most commonly used fine aggregate is river sand. The worldwide depletion of natural river sand is particularly enormous due to the extensive use of concrete. Natural river sand, in particular, is in high demand due to infrastructure development in industrialized countries. In many parts of India, lack of adequate quantities of ordinary river sand for manufacturing concrete is limiting the growth of the construction sector.

Over time, the granite waste produced by the industry has accrued. Only little amounts have been used, and the remainder has been discarded in an unethical manner, causing an environmental hazard. With the massive growth in the amount of garbage that needs to be disposed of, a severe shortage of dumping sites, and a steep rise in shipping and dumping expenses that harm the environment thus ecological growth is impossible. The garbage which is growing substantially has its disposal a critical issue. Therefore provocative efforts are required to create a new building material out of granite scrap, which is an industrial waste, that can be utilized to substitute fine aggregate in concrete. It is possible to achieve the goal of reducing construction costs as well as the difficulty of disposing of granite, which contributes to the region's environmental problems. Substituting different materials might cause performance attributes to vary, which may or may not be satisfactory for high-performance concrete. The use of granite powder instead of sand has been found to improve the mechanical characteristics of concrete, and the values of plastic and drying shrinkage were negligible in granite powder concrete compared to regular concrete (Felixkala and Partheeban, 2010).

The construction and environmental industries are concerned about the growing volume of solid, liquid and gaseous industrial waste. Tamil Nadu, India's southernmost state, holds 45% of the country's total granite reserves. Granite blocks are provided to local processing companies after being manufactured from quarries using various cutting procedures. The stones are then industrially prepared before being used as decorative stones, such as sawing and polishing. The fine granite particles and water mingled together during this industrial operation results in the generation of granite colloidal waste (Vijayalakshmi et al., 2012). When the stone slurry is disposed of in landfills, the amount of water in the trash is greatly reduced, and the waste is reduced to a dry mud consisting of extremely fine powder that humans and animals may easily inhale. Additionally, it is a non-biodegradable, unwanted material that pollutes and destroys the environment. According to published data, the granite industry creates 20%–25% of total production waste at various stages of production, resulting in millions of tons of colloidal waste each year, and disposing of these fine wastes is one of the world's most pressing environmental issues today. With increasing constraints on landfills in the surrounding area, the cost of disposal is rising, and enterprises are being compelled to find new ways to reuse trash. Although granite waste has been repurposed, the quantity of waste repurposed in this manner is still insignificant. As a result, its application in other industries has become critical (Vijayalakshmi and Sekar, 2013).

The use of biological admixtures, such as superplasticizers, diminishes the water content of the hydrated cement paste, lowering the porosity. Mineral admixtures, such as blast furnace slag, silica fume and fly ash are used as cement replacement materials. When they come into contact with free lime, they become pozzolanic and take on the same cementing properties as conventional Portland cement. The strength and durability of concrete can be improved by using these materials alone or in conjunction with cement and a sufficient dose of superplasticizer. To achieve excellent performance, admixtures can be used in cement concrete as a limited replacement for cement, combined with superplasticizer as a water reducer (Arivumangai et al., 2014). With the privatization in the early 1990s, the granite sector has risen dramatically in recent decades as the world's most thriving building business. As a result, the amount of waste generated by excavating and handling has increased. The Indian granite stone industry now produces over 18 million tons of dense granite waste, of which 12 million tons are prohibited at industrial sites, 5.2 million tons are in the form of carvings or undersize materials, and 0.4 million tons are treated and refined. The granite garbage that has been disposed of in an unethical manner has caused environmental problems (Arivumangai et al., 2016). During the manufacture of granite tiles, tons of granite dust/slurry is produced. When placed on land, these wastes soil productivity due to diminished porosity, water absorption and water percolation, among other things. This industrial waste has several reuses and recycling options, both in the experimental stage and in real-world applications. The granite stone industry's economy will be boosted by reducing waste output by producing value-added items from granite stone trash. Granite powder in high-performance concrete could turn a waste product into a valuable resource while also assisting in environmental protection (Srinivasa et al., 2015).

There are several mix design processes that rely on empirical relationships derived from previous experience. The theoretical computation of mix design is complicated and time-consuming due to various assumptions and formulae. A number of trial mixes must be created in order to get the appropriate mix proportion for strength and workability. As a result, achieving an acceptable mix proportion takes a long time (Das et al., 2015).

Artificial neural network (ANN) is a model of bio-neural networks and is a sophisticated machine learning data processing progression. This learning approach tries to replicate the human brain's information acquisition and inference processes (Montavon et al., 2020). ANN has been broadly employed to solve nonlinear regression problems (Adeloye and Rustum, 2012). The most typical ANN training approach is the backpropagation neural network, which is commonly used for regression analysis and other practical applications (Mai et al., 2021; Kumar et al., 2020).

In terms of the arrangement, precision and training time, the random forest (RF) classification is on par with the support vector machine (SVM). The amount of user-defined factors essential for RF classifiers is smaller than that of SVMs, making them easier to define (Pal, 2005). The attribute filter's effectiveness on the algorithms (random tree and REP tree) is clear. After adding the attribute filter to algorithms (random tree and REP tree), true positive rate (TP), precision and recall increased in both methods, whereas false positive rate (FP) declined in the J48 algorithm (Hamoud et al., 2018). Different optimization algorithms, such as the genetic algorithm, the salp swarm algorithm and the grasshopper optimization algorithm, are each hybridized with an ANN to expect the concrete compressive strength comprising reused aggregate, and the ANNs' efficiency is tested using the M5P tree model. When compared to other models, the findings of this investigation suggest that the modified ANN with salp swarm algorithm is more efficient. The numerical indicators of hybrid ANNs with salp swarm algorithm, grasshopper optimization algorithm and genetic algorithm on the other hand, are extremely close to each other (Kandiri et al., 2021).

Several researchers have used various soft computing algorithms to evaluate this design factor for various varieties of concretes such as ANN, genetic programming, SVM, multi expression programming, adaptive network-based fuzzy inference system, bio-geographical-based programming, artificial bee colony programming, classification trees and so on (Golafshani and Behnood, 2018). With the performance of all soft computing approaches in mind, the compressive strength of granite powder reinforced concrete was explored using ANN, RF, RT and REP tree in this work.

15.2 SOFT COMPUTING TECHNIQUES

15.2.1 Artificial Neural Network (ANN)

A computational model encouraged by the assembly and functioning of the human mind is known as an ANN. ANNs, like the human brain, can learn from previous data. By recapping the learning method for a number of repetitions, the network learns the previous data. Once an ANN has been properly trained on past data, it can accurately anticipate outcomes from unknown inputs (Das et al., 2015; Kumar et al., 2021a). The ANN is a nonlinear arithmetic methodology for solving issues that are not accessible using conventional methods (Adeloye et al., 2012). Despite their success in finding complicated nonlinear correlations and modelling a variety of applications, their performance is mostly dependent on the superiority of the database used to train them (Kumar et al., 2021b). Additionally, in order to develop a successful ANN model, the linkage must be able to correctly predict the output when accomplished with a set of inputs that lie within the training data's practical range (Sonebi et al., 2016).

ANN has a layered design with a parallel structure that is interrelated. The input and output layers of an ANN are referred to as input and output layers, respectively, while the layers in between are referred to as hidden layers. Each layer has one or more neurons that are coupled via weighted connections to the nodes in the following layer (Mokhtar et al., 2021). In addition, there is a node for the bias in the input and hidden layer. The input layer nodes are non-computing nodes that solely accept information from the outside world. The number of nodes in the input and output layers of an ANN assembly is equivalent to the number of effective factors and the number of method outputs, respectively, regardless of the bias nodes (Elbeltagi et al., 2020; Kumar et al., 2021c). The hidden and output layers nodes are computing nodes that perform two main calculation processes (Golafshani and Behnood, 2018).

Cement, fine aggregate, coarse aggregate, water, granite powder and curing days are regarded as input data for all ANN structures, whereas compressive strength is treated as output data. Following a little investigation, it was determined to employ one hidden layer with ten neurons in the simulations. Figure 15.1 depicts the system employed in the ANN prototype for the first simulation. The testing and training data are normalized separately, that is, over their own ranges rather than the entire range of all inputs. This data preprocessing is complete to scale all of the data to a range of zero to one. After that, the input and output nodes are normalized to the zero and one ranges, respectively (Topçu et al., 2008).

15.2.2 Random Forest (RF)

The RF classifier uses a number of tree classifiers to categorize an input vector. Each tree uses a random vector sampled separately from the input vector to construct a unit vote for the most popular class (Pal, 2005). The RF approach is developed using a technique known as bagging (short for bootstrap aggregating), in which trees are constructed independently using a bootstrap sample of the whole data set. RF is a classifier that combines L tree-structured base classifiers $h(P,n)$, $N = 1,2,3,\ldots L$, where P is the input data and n is a group of identical and reliant distributed random vectors.

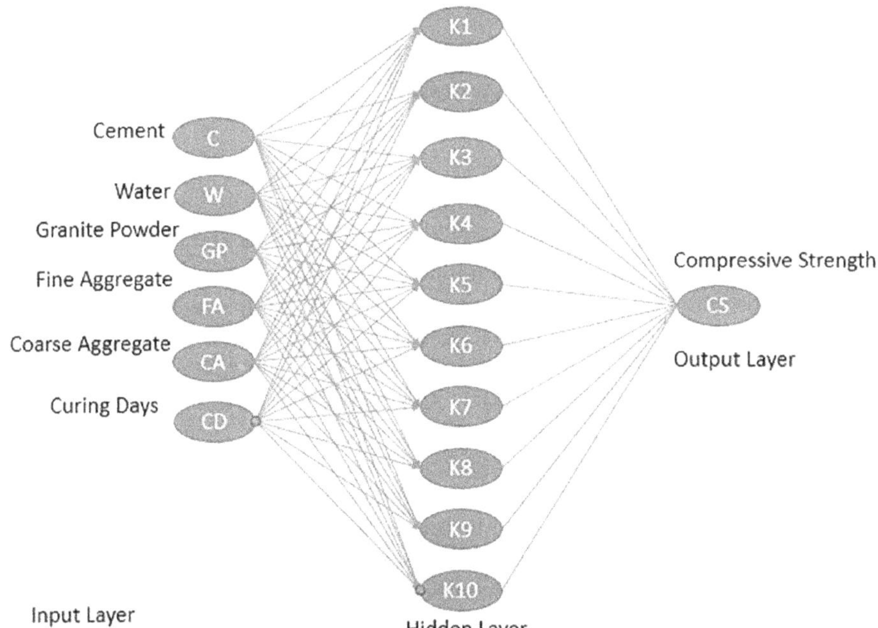

FIGURE 15.1 ANN model for simulation.

Each decision tree is created by picking data at random from the given data. A decision tree is created in RF by randomly picking a feature subset and/or a training data subset for every decision tree (Shaqadan, 2016).

Because of its advantages over other algorithms, RF is widely used. It can control data with a huge number of features and assess the value of the characteristics, frequently with great precision in classification (or regression) and quick learning practice. Each tree in the RF chooses only a minor collection of characteristics during construction; this process allows the RF to calculate a data set with a huge number of features in a sensible amount of time. The number of attributes used to form trees in the forest can be set by default; for classification problems, the best default is p, and for regression problems, it's $p/3$ (p is the number of all properties of the original data set). The forest's tree count should be high enough to ensure that all attributes are used many times. Typically, 500 trees are used for classification and 1000 trees are used for regression. The sub data sets have around two-thirds of the trials that do not intersect for tree production due to the usage of the bootstrap method of random return sampling. Around a third of the remaining samples are classed as out-of-bag since they do not participate in tree construction. Overall, the RF model is chosen because of its many advantages, including forecast accuracy, fast replication speed, robustness to noise, over fitting, and ease of parallelization, which makes it particularly valuable for Monte Carlo simulations, error approximation and variable importance determination (Mai et al., 2021). The RF algorithm is relatively unaffected by training set characteristics and can attain high prediction precision. It calls for the use of two user-defined factors: the number of cultivated trees and the number of input characteristics. A trial-and-error method is considered to construct models. The RF-based prototype in this experiment was developed using the WEKA 3.9 programme (Sihag et al., 2021, Upadhya et al., 2021).

15.2.3 Random Tree (RT)

The RT operator works similarly to the decision tree operator, with one exception: each split has only an arbitrary collection of characteristics. The RT is responsible for decision tree learning from both nominal and calculated input. The decision tree is a simple to understand efficient classifier model. A variable subset ratio is used to describe the subset size. The advantage of expressing the data as a tree is that it is easier to interpret. The goal is to create a classifier model that can identify label values based on the example set's various input attributes. Each tree's internal node corresponds to one of the input features. The number of possible standards of the individual input property is equal to the edge count of an internal node. The value of the label provided to the values of the input attributes indicated by a path from the root to leaf is indicated by every leaf node. Following that, the leaf nodes that are not involved in the decision tree's discriminative influence are pruned out. It is performed in order to improve the categorization process of a previously unknown data set. Pre-pruning is a type of pruning that occurs concurrently with the tree-generation process. When the tree-generation method is finished, post-pruning is done (Veeramanikandan and Jeyakarthic, 2019).

15.2.4 Reduced Error Pruning (REP) Tree

REP tree is a fast decision tree learning method that generates a decision tree based on information gain or alteration reduction. It is a decision tree splitter and pruner that uses information gain as the splitting criterion to split and prune a decision/regression tree (Srinivasan et al., 2014). This is a quick decision tree learner algorithm. It builds a decision tree using information alteration and prunes it using reduced-error pruning. Numeric attribute values are only sorted once by the algorithm (Goyal et al., 2013). REP tree creates a large number of trees and uses reversion tree logic to change iterations. The algorithm then chooses the best tree from all of the produced trees. The technique creates a regression decision tree based on variance and result data. Furthermore, this approach prunes the trees utilizing back-fitting and reduced-error pruning techniques (Hamoud et al., 2018).

15.2.5 Performance Assessment Parameters

The accuracy of the ANN, RF, RT and REP tree-based models for predicting the compressive strength of granite powder reinforced concrete was evaluated through three performance indicators: coefficient of correlation (CC), root mean square error (RMSE) and mean absolute error (MAE). The range of CC is −1 to 1 and it is zero to infinity for MAE and RMSE. These performance evaluation indicators can be calculated utilizing equations (15.1)–(15.3).

$$\text{CC} = \frac{\sum_{i=1}^{N}\left(R_i - \bar{R}\right)\left(T_i - \bar{T}\right)}{\sqrt{\sum_{i=1}^{N}\left(R_i - \bar{R}\right)^2 \sum_{i=1}^{N}\left(T_i - \bar{T}\right)^2}} \tag{15.1}$$

$$\text{RMSE} = \sqrt{\frac{1}{N}\sum_{i=1}^{N}\left(R_i - T_i\right)^2} \tag{15.2}$$

$$\text{MAE} = \frac{1}{N}\sum_{i=1}^{N}\left|R_i - T_i\right| \tag{15.3}$$

where T_i and R_i are the observed and the predicted sediment transport rate, $\bar{T}$ is the average observed sediment transport rate and n is the number of observations.

15.2.6 Data Set

In Table 15.1, a total of 108 observations of granite powder reinforced concrete compressive strength were used for model building and validation. The entire data set was separated into two groups, randomly. The bigger group (75 observations) was used as the training data set for model building, while the remaining group (33 observations) was used to validate the model. Cement, fine aggregate, coarse aggregate, water,

TABLE 15.1
Range of Data Set

Sr. no	Author	Input and Output Variables[a] C	FA	CA	W	GP	CD	CS (N/mm²)
1	Azunna (2019)	393.33–410.00	406.66–813.33	1220	165–226.66	0–406.66	3–28	14.89–48.8
2	Vijayalakshmi et al., (2012)	465	451–603	1086	186	0–159.71	7–90	22.75–42.07
3	Felixkala and Partheeban (2010)	343–480	0–533	1086	192	0–533	1–90	6–48
4	Upadhyaya (2018)	390	409–681.6	1214	195	0–259.2	7–28	13.6–39.56

[a] C = Cement, FA = Fine Aggregate, CA = coarse Aggregate, W = Water, GP = Granite Powder, CD = Curing Days, CS = Compressive Strength.

TABLE 15.2
Characteristics of the Data Utilized for Model Growth and Validation

Range	C	FA	CA	W	GP	CD	CS (N/mm²)	Data Set
Mean	398.79	460.43	1152.40	198.12	177.97	26.81	28.22	Training
	393.64	435.48	1160.90	199.41	218.22	29.78	29.03	Testing
Median	393.33	525.00	1086.00	192.00	129.60	14.00	26.40	Training
	393.33	482.00	1214.00	192.00	133.00	28.00	26.40	Testing
Standard Deviation	46.45	226.34	70.15	19.04	163.16	27.09	10.65	Training
	41.27	214.86	70.04	19.88	180.21	27.78	10.54	Testing
Minimum	343.00	0	1086.00	165.00	0	1.00	6.00	Training
	343.00	0	1086.00	165.00	0	1.00	6.40	Testing
Maximum	480.00	813.33	1250.00	226.66	533.00	90.00	48.90	Training
	465.00	681.60	1250.00	226.66	533.00	90.00	48.00	Testing
Confidence Level(95.0%)	480.00	813.33	1250.00	226.66	533.00	90.00	48.90	Training
	14.63	76.18	24.83	7.05	63.90	9.85	3.74	Testing

granite powder and curing days were used as input variables, and compressive strength was used as the target variable for model development and validation. Table 15.2 shows the characteristics of the data utilized for model growth and validation.

15.3 RESULT AND DISCUSSION

In this investigation potential of ANN, RF, RT and REP tree-based soft computing models are assessed for the prediction of compressive strength of granite powder concrete. The performance of these soft computing-based models is evaluated using the three performance indicators CC, MAE and RMSE.

15.3.1 Results of ANN-Based Models

The construction of an ANN-based model is a trial-and-error practice. The training data set is considered to construct the model, while the testing data set is used to validate it. Weka 3.9 software is used for the analysis. Agreement plots between observed and predicted compressive strength of granite powder concrete for training and testing stages are shown in Figure 15.2. Results in terms of the goodness fit from Table 15.3 indices for the different shapes of membership function (MF)-based ANN models with CC, MAE and RMSE values are 0.9891, 1.1925 and 1.5781 for training stage and 0.9572, 2.1025 and 3.0914 for testing stage, respectively.

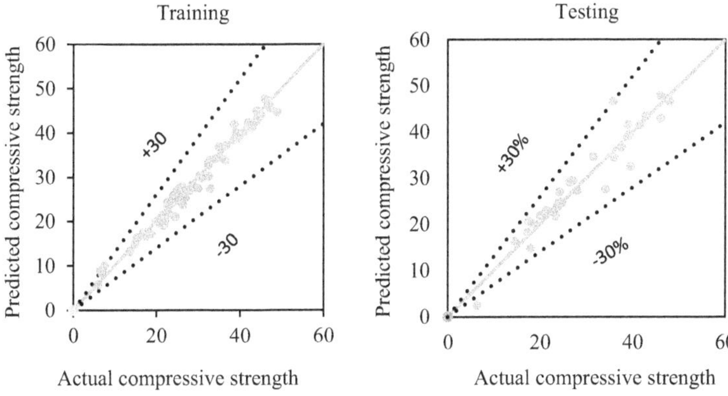

FIGURE 15.2 Observed versus estimated compressive strength of granite powder concrete using ANN-based models for training and testing stages.

TABLE 15.3
Performance Indices for ANN, RT, RF and REP Tree-Based Models

Approaches	CC	MAE	RMSE
	Training Data Set		
ANN	0.9891	1.1925	1.5781
RF	0.9954	0.8975	1.2403
RT	0.9999	0.0384	0.1089
REP tree	0.9598	2.2235	2.9693
	Testing Data Set		
ANN	0.9572	2.1025	3.0914
RF	0.9631	2.0923	2.9093
RT	0.9591	2.2158	2.9983
REP tree	0.9214	3.5663	4.0971

15.3.2 Results of RF-Based Models

For the training and testing stages, Figure 15.3 shows the agreement plot between observed and predicted compressive strength of granite powder concrete using an RF-based model. The RF-based model's predicted values are quite close to the observed ones. The performance of the RF technique is suitable for the prediction of compressive strength of granite powder concrete, as shown in Table 15.3, with CC, MAE and RMSE values of 0.9954, 0.8975 and 1.2403 for the training stage and 0.9631, 2.0923 and 2.9093 for the testing stage, respectively.

15.3.3 Results of RT-Based Models

The construction of RT-based models is a trial-and-error practice. This is a model that is based on trees. Figure 15.4 shows agreement graphs between real and anticipated compressive strength values of granite powder concrete utilizing RT with training and testing data. Table 15.4

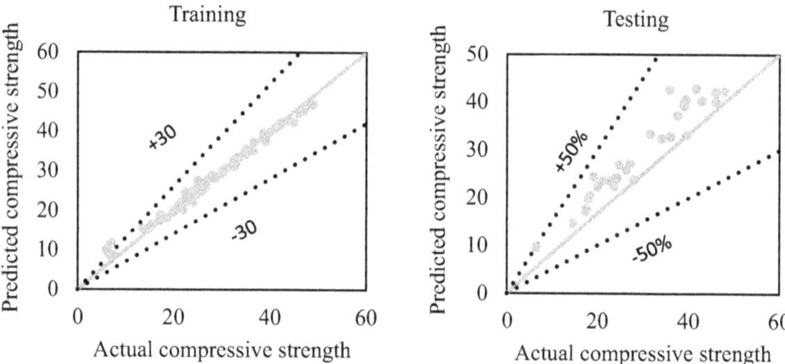

FIGURE 15.3 Observed versus estimated compressive strength of granite powder concrete using RF-based models for training and testing stages.

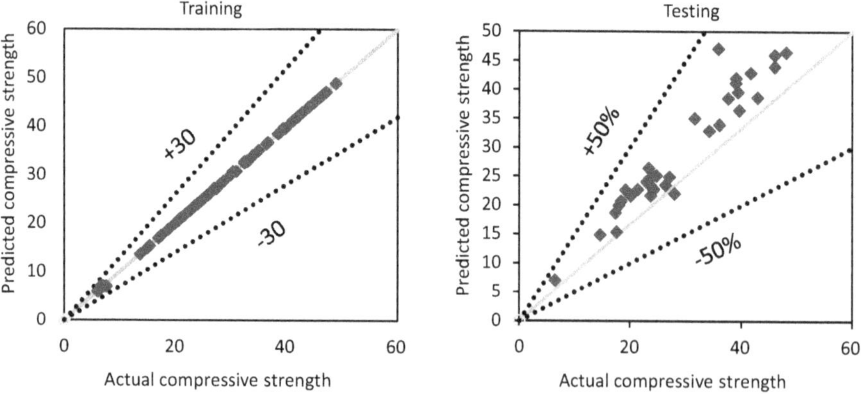

FIGURE 15.4 Observed versus estimated compressive strength of granite powder concrete using RT-based models for training and testing stages.

shows the results of the RT-based model's performance evaluation parameters, with CC, MAE and RMSE values of 0.9999, 0.0384 and 0.1089 for the training stage and 0.9591, 2.2158 and 2.9983 for the testing stage, respectively.

15.3.4 Results of REP Tree-Based Models

Figure 15.5 depicts the agreement plot between observed and estimated compressive strength of granite powder concrete for training and testing stages. Table 15.3 shows the results of the REP tree-based model's performance evaluation parameters, with CC, MAE and RMSE values of 0.9598, 2.2235 and 2.9693 for the training stage and 0.9214, 3.5663 and 4.0971 for the testing stage, respectively.

15.4 ASSESSMENT OR COMPARISON AMONG SOFT COMPUTING-BASED APPLIED MODELS

When RF-based models are compared to other regression and soft computing models (Table 15.3), it is clear that RF-based models perform better. Figure 15.6 depicts the agreement performance of regression and soft computing-based models for forecasting the compressive strength of granite powder concrete during the testing stage. Table 15.3 shows that the RF-based modelling approach performs well for the prediction of compressive strength of granite powder reinforced concrete, with CC, MAE and RMSE values of 0.9954, 0.8975 and 1.2403 for the training stage, and 0.9631, 2.0923 and 2.9093 for the testing stage, respectively.

15.5 SENSITIVITY ANALYSIS

The most significant input variable in compressive strength of granite powder concrete was determined by sensitivity analysis. The analysis was done for the most efficient model (i.e. RF). After deleting one input variable at a time, different training

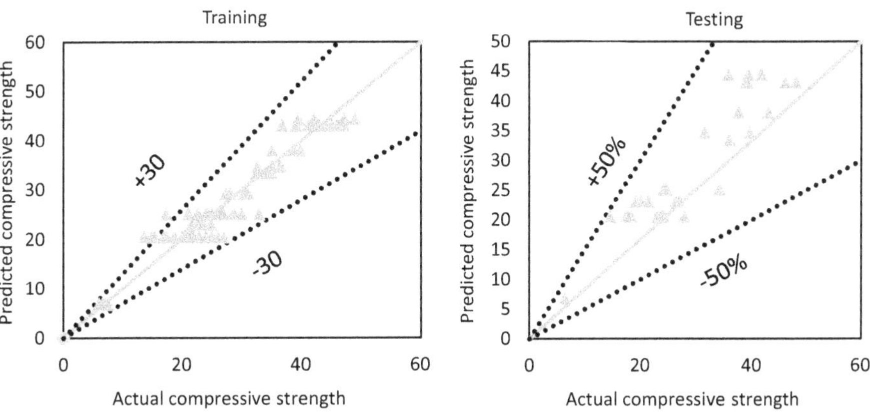

FIGURE 15.5 Observed versus estimated compressive strength of granite powder concrete using REP tree-based models for training and testing stages.

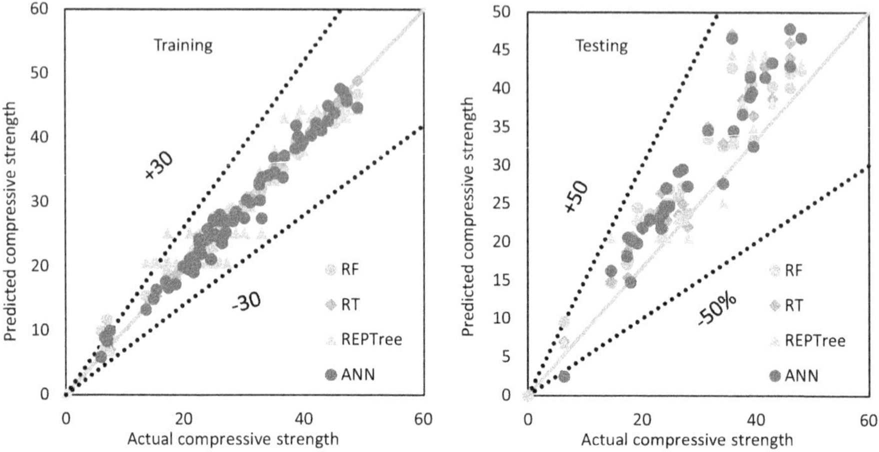

FIGURE 15.6 Comparison of actual versus predicted compressive strength of granite powder concrete using ANN, RF, RT and REP tree-based models for training and testing stages.

data sets were employed. The results were presented in terms of the CC, RMSE and MAE. Table 15.3 suggests that the most significant input variable for determining the compressive strength of granite powder reinforced concrete is curing days.

15.6 CONCLUSION

The compressive strength of the granite powder reinforced concrete was explored in this chapter using several soft computing approaches such as RF, RT, ANN and REP tree. The following conclusions are drawn from the results of the study:

- RF-based modelling approach performs better than other approaches in predicting the compressive strength, with the CC, MAE and RMSE values of 0.9954, 0.8975 and 1.2403 for the training stage, and 0.9631, 2.0923 and 2.9093 for the testing stage.
- REP tree-based model has the lowest CC values of 0.9598 and 0.9214 for the training and testing stages, respectively, among all the approaches considered in this chapter.
- Sensitivity analysis suggests that the curing days is the most significant input variable for determining the compressive strength of granite powder reinforced concrete.

TABLE 15.4
Sensitivity Analysis of ANN and Tree-Based Techniques

Input Variables						Output Variable	RF		
Cement (C)	Fine Aggregate (FA)	Coarse Aggregate (CA)	Water (W)	Granite Powder (GP)	Curing Days (CD)	Compressive Strength N/mm^2	CC	MAE	RMSE
✓	✓	✓	✓	✓	✓	✓	**0.9631**	**2.0923**	**2.9093**
×	✓	✓	✓	✓	✓	✓	0.9641	1.8962	2.7696
✓	×	✓	✓	✓	✓	✓	0.9588	2.2362	2.9787
✓	✓	×	✓	✓	✓	✓	0.9633	2.0352	2.818
✓	✓	✓	×	✓	✓	✓	0.9632	1.9742	2.8311
✓	✓	✓	✓	×	✓	✓	0.9629	1.8964	2.819
✓	✓	✓	✓	✓	×	✓	**0.2047**	**8.8139**	**12.027**

REFERENCES

Adeloye, A. J., & Rustum, R., (2012). Self-organising map rainfall-runoff multivariate modelling for runoff reconstruction in inadequately gauged basins. *Hydrology Research, 43*(5), pp. 603–617.

Adeloye, A. J., Rustum, R., & Kariyama, I. D., (2012). Neural computing modeling of the reference crop evapotranspiration. *Environmental Modelling & Software, 29*(1), pp. 61–73.

Arivumangai, A., & Felixkala, T., (2014). Strength and durability properties of granite powder concrete. *Journal of Civil Engineering Research, 4*(2A), pp. 1–6.

Arivumangai, A., & Felixkala, T., (2016). Experimental investigation on fire resistance of granite powder concrete. *International Journal of Applied Engineering Research, 10*(27), p. 2015.

Azunna, S. U., (2019). Compressive strength of concrete with palm kernel shell as a partial replacement for coarse aggregate. *SN Applied Sciences, 1*(4), pp. 1–10.

Das, S., Pal, P. and Singh, R.M., (2015). Prediction of concrete mix proportion using ANN technique. *International Research Journal of Engineering and Technology, 2*(5), pp. 820–825.

Elbeltagi, A., Deng, J., Wang, K., & Hong, Y., (2020). Crop Water footprint estimation and modeling using an artificial neural network approach in the Nile Delta, Egypt. *Agricultural Water Management, 235*, p. 106080.

Felixkala, T. and Partheeban, P., (2010). Granite powder concrete. *Indian Journal of Science and Technology, 3*(3), pp. 311–317.

Ghannam, S., Najm, H. and Vasconez, R., (2016). Experimental study of concrete made with granite and iron powders as partial replacement of sand. *Sustainable Materials and Technologies, 9*, pp. 1–9.

Golafshani, E.M. and Behnood, A., (2018). Application of soft computing methods for predicting the elastic modulus of recycled aggregate concrete. *Journal of Cleaner Production, 176*, pp. 1163–1176.

Goyal, M.K., Ojha, C.S.P., Singh, R.D., Swamee, P.K. and Nema, R.K., (2013). Application of ANN, fuzzy logic and decision tree algorithms for the development of reservoir operating rules. *Water Resources Management, 27*(3), pp. 911–925.

Hamoud, A., Hashim, A.S. and Awadh, W.A., (2018). Predicting student performance in higher education institutions using decision tree analysis. *International Journal of Interactive Multimedia and Artificial Intelligence, 5*, pp. 26–31.

Kumar, M., Kumari, A., Kumar, D., Al-Ansari, N., Ali, R., Kumar, R., … Kuriqi, A., (2021a). The superiority of data-driven techniques for estimation of daily pan evaporation. *Atmosphere*, 12(6), p. 701.

Kumar, N., Adeloye, A. J., Shankar, V., & Rustum, R., (2020). Neural computing modelling of the crop water stress index. *Agricultural Water Management*, 239, p. 106259.

Kumar, N., Rustum, R., Shankar, V., & Adeloye, A. J., (2021b). Self-organizing map estimator for the crop water stress index. *Computers and Electronics in Agriculture, 187*, p. 106232.

Kumar, N., Shankar, V., Rustum, R., & Adeloye, A. J., (2021c) Evaluating the performance of self-organizing maps to estimate well-watered canopy temperature for calculating crop water stress index in Indian mustard (Brassica Juncea). *ASCE Journal of Irrigation and Drainage Engineering, 147*(2), p. 04020040.

Kandiri, A., Sartipi, F., & Kioumarsi, M., (2021). Predicting compressive strength of concrete containing recycled aggregate using modified ANN with different optimization algorithms. *Applied Sciences, 11*(2), p. 485.

Mai, H. V. T., Nguyen, T. A., Ly, H. B., & Tran, V. Q., (2021). Investigation of ANN model containing one hidden layer for predicting compressive strength of concrete with blast-furnace slag and fly ash. *Advances in Materials Science and Engineering, 2021*, pp. 1–17.

Mai, H. V. T., Nguyen, T. A., Ly, H. B., & Tran, V. Q. (2021). Prediction compressive strength of concrete containing GGBFS using random forest model. *Advances in Civil Engineering*, *2021*, pp. 1–12.

Mokhtar, A., Jalali, M., He, H., Al-Ansari, N., Elbeltagi, A., Alsafadi, K., ... Rodrigo-Comino, J., (2021). Estimation of SPEI meteorological drought using machine learning algorithms. *IEEE Access*, *9*, pp. 65503–65523.

Montavon, G. (2020). Introduction to neural networks. In: Schütt, K., Chmiela, S., von Lilienfeld, O., Tkatchenko, A., Tsuda, K., Müller, KR. (eds.) *Machine Learning Meets Quantum Physics*. Lecture Notes in Physics, 968, pp. 37–62. Springer, Cham.

Pal, M., (2005). Random forest classifier for remote sensing classification. *International Journal of Remote Sensing*, *26*(1), pp. 217–222.

Shaqadan, A., (2016). Prediction of concrete mix strength using random forest model. *International Journal of Applied Engineering Research*, *11*(22), pp. 11024–11029.

Sihag, P., Dursun, O. F., Sammen, S. S., Malik, A., & Chauhan, A. (2021). Prediction of aeration efficiency of parshall and modified venturi flumes: application of soft computing versus regression models. *Water Supply*, *21*(8), pp. 4068–4085.

Sonebi, M., Grünewald, S., Cevik, A., & Walraven, J., (2016). Modelling fresh properties of self-compacting concrete using neural network technique. *Computers and Concrete*, *18*(4), pp. 903–920.

Srinivasa, C. H., & Venkatesh, (2015, January). Optimization of granite powder used as partial replacement to cement in the design of ready mix concrete of M20 Grade using IS10262:2009" *International Journal of Engineering Research & Technology (IJERT)*, *4*, (01), pp. 104–111.

Topçu, İ.B., Boğa, A.R., & Hocaoğlu, F.O., (2009). Modeling corrosion currents of reinforced concrete using ANN. *Automation in Construction*, *18*(2), pp. 145–152.

Upadhyaya, S., Nanda, B., & Panigrahi, R. (2019). Experimental analysis on partial replacement of fine aggregate by granite dust in concrete. In: Das, B. B., and Neithalath, N. (eds.) *Sustainable construction and building materials*, pp. 335–344. Springer, Singapore.

Upadhya, A., Thakur, M. S., Sharma, N., & Sihag, P. (2021). Assessment of soft computing-based techniques for the prediction of marshall stability of asphalt concrete reinforced with glass fiber. *International Journal of Pavement Research and Technology*, 1–20.

Vijayalakshmi, M. & Sekar, A.S.S., (2013). Strength and durability properties of concrete made with granite industry waste. *Construction and Building Materials*, *46*, pp. 1–7.

Vijayalakshmi, M., Sekar, A. S. S., Sivabharathy, M., & Ganesh Prabhu, G. (2012). Utilization of granite powder waste in concrete production. In *Defect and Diffusion Forum* (Vol. 330, pp. 49–61). Trans Tech Publications Ltd., Switzerland.

16 Predicting Recycled Aggregates Compressive Strength in High-Performance Concrete Using Artificial Neural Networks

Ahmad Alyaseen, Arunava Poddar, Fadi Almohammed, and Salwan Tajjour
Shoolini University

Karam Hammadeh
Vel Tech Rangarajan Dr. Sagunthala R&D Institute of Science and Technology

Hussain Alahmad
KTH Royal Institute of Technology

CONTENTS

DOI: 10.1201/9781003184331-16

16.1 INTRODUCTION

For almost a century, concrete has been the most in-demand building material in the civil engineering industry (Zamora-Castro et al., 2021; Jain et al., 2021). It is responsible for the increasing demand for cement, which pollutes the environment via CO_2 emissions (Fernando et al., 2021; Etim et al., 2021). As a result, an alternate approach is required to reduce the cement usage in concrete manufacturing. The waste utilization idea is often utilized in concrete. As a result, many factors should be addressed when producing concrete from waste, such as the percentage of water–cement ratios, aggregates, admixture superplasticizer, cement, waste, and so on, as determined by the concrete needs for a specific building (Yazdanbakhsh et al., 2018; Gencel et al., 2020). It is difficult to take many trials to obtain the required grade, explaining the lengthy time and high price. A difficult task, such as mix proportioning, maybe more straightforward by using an artificial neural network (ANN) (Getahun et al., 2018; Shahmansouri et al., 2021), so ANN may be used to improve concrete manufacture. Different inputs and hidden layers are fixed for each parameter. On the other hand, ANN's mix-proportioning may save time and money by eliminating concrete mix trials and labor expenses (Kapadia and Jariwala, 2021).

In 1986, Japan developed high-performance concrete with excellent segregation and deformability (Okamura and Ouchi, 2003) highly complicated in nature (Ghezal and Khayat, 2002; Yeh, 1998). ANNs can solve complex problems utilizing networked computer technologies (Yeh, 1998). They are also used in civil engineering (Basma, 1999). Other concrete components, besides the water–cement ratio, affect concrete strength (Ouchi, 2000). The ANN model predicted concrete compressive strength (Plevris and Asteris, 2014; Akbari et al., 2021). Adaptative neuro fuzzy inference system (ANFIS) (Nazari and Riahi, 2012) and fuzzy logic are currently being used in civil engineering (Sarıdemir, 2009). Concrete free expansion strain, cross-sectional area, and cure time may all be studied using ANN (Wang et al., 2015). From 13 to 15, ANN was used to estimate a few industrial wastes (Mukherjee and Biswas, 1997; Karakoç et al., 2011; Ahmadi-Nedushan, 2012).

In this book chapter, an ANN is used to evaluate the strength of a high-performance concrete mix that includes concrete waste. In order to conduct correlation research to assess how well ANNs perform in terms of correlation coefficient (R). In addition to the standard methods for predicting the compressive strength of concrete that comprises both natural and recycled aggregates, ANN models are evaluated using numerous methodologies, including mean square error (MSE), k-fold cross-validation, and sensitivity analysis.

16.2 ARTIFICIAL NEURAL NETWORKS (ANN) (YADOLLAHI 2016)

ANNs are computing techniques that understand how neurons in the brain work and their design mimic that of biological neural cells. Instead, these models are simplified representations of genuine neurons that neglect biological, chemical, and physical processes common to all neurons. While some parts of information are not specifically represented in the model, such as combining or pattern recognition, the system can nevertheless show how an actual neuron would behave if subjected to these inputs (Flood and Kartam, 1994; Thakur et al., 2021; Upadhya et al., 2021a, b, Badagha and Modhera, 2017). Like 'learn by doing' replaces 'programming,' the adaptation of these networks benefits significantly from this flexible approach (Bandyopadhyay and Chattopadhyay, 2007). When sufficient data is available, a neural network will always produce a solution based on regularity (Bandyopadhyay and Chattopadhyay, 2007; Upadhya et al., 2021a, b). A critical characteristic of ANN is the parallel computing design, which makes ANN systems faster in specialized hardware implementations (Bandyopadhyay and Chattopadhyay, 2007). Neural networks provide several distinct advantages over traditional digital computing methods in that they allow for massively parallel processing, use of distributed storage, resistance to error, and exceptional robustness after training.

Another thing to keep in mind is that neural networks learn through examples (Schalkoff, 1997). An example that has been seen before will make it easier to "acquire" information about the issue (Schalkoff, 1997). A properly trained network may help solve problems that no one knows how to solve, or no training is available (Schalkoff, 1997).

16.3 DESIGN OF NEURAL NETWORKS MODELS

The neural networks are trained using input data for forecasting concrete compressive strength. The study's data pattern correlates to nine parameters. The feed-forward back-propagation is among many designs and paradigms. It is easy and effective for more advanced human tasks, such as categorization, decision-making, and prediction. A simple and popular algorithm for complicated and multi-layered networks is widely used and effective. With the training of a multi-layered feed-forward back-propagation neural network, it is possible to forecast the strength. This technique generalizes Levenberg–Marquardt's method to multiple layer networks and is based on the concept of neural networks.

When it comes to back-propagation, there are three main components: the input layer, the output layer, and potentially another layer or layers concealed from observation. There are often just one or two layers of a network that are concealed from observation. With the use of weighted connections, Figure 16.1 shows the link between the visible and hidden input and output layers.

The system must identify which input provided the most incorrect output and how that error is corrected. The network's input layer is given training inputs to address this issue, and the network's output layer is used to find the most favorable result. Until the input layer is reached, the procedure continues for the preceding layer. As the weights are refined, the same data set is processed multiple times during the network training. The illustration shown in Figure 16.2 depicts the order in which neural networks learn.

Defining a performance function provides a numerical explanation for the network. A critical algorithm is the choosing of an error function. Also, these errors, also known as parameters, are frequently used for training feed-forward neural network and applied in the research.

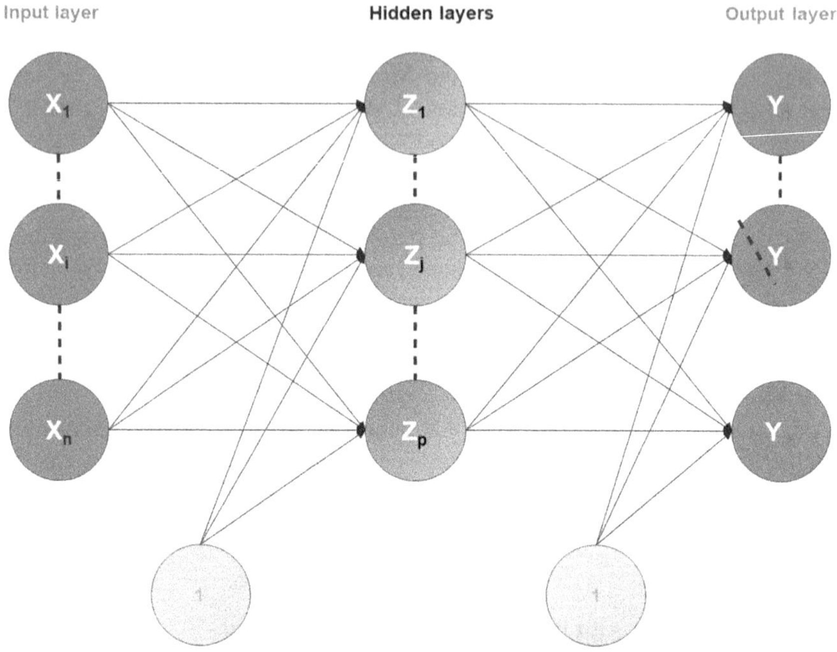

FIGURE 16.1 Back-propagation feed-forward of neural network.

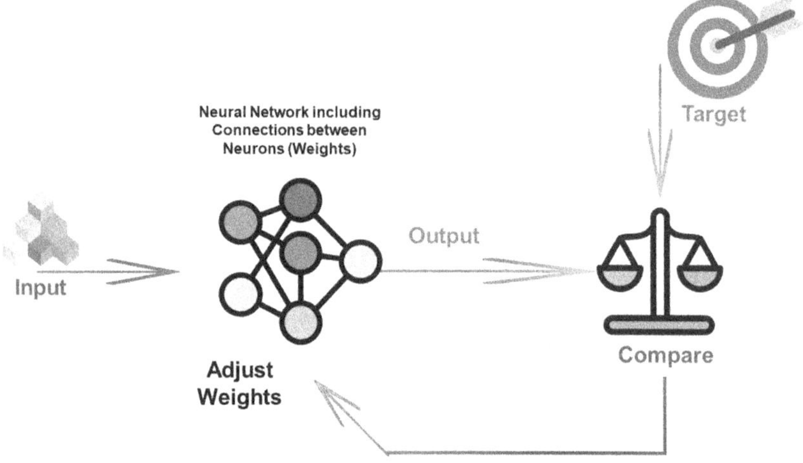

FIGURE 16.2 ANN's learning sequence.

The equation given in MSE for the network errors indicates:

$$\text{MAE} = \frac{1}{N}\sum_{i=1}^{N}\left|P_i - O_i\right| \tag{16.1}$$

Pearson's correlation coefficient is mentioned in the formula below:

$$R = \frac{\sum_{i=1}^{N}\left(O_i - \bar{O}\right)\left(P_i - \bar{P}\right)}{\sqrt{\sum_{i=1}^{N}\left(O_i - \bar{O}\right)^2}\sqrt{\sum_{i=1}^{N}\left(P_i - \bar{P}\right)^2}} \tag{16.2}$$

The equation to calculate the relationship between the target and actual output nodes is given by N, i, j, and t_i, ai for the observations; a_i, a_j for the targets (deceived and average target); and t_j, a_j for the averages (averages) of the target and actual output nodes.

16.4 EXPERIMENTAL PROGRAM

One hundred and fifty millimeter of concrete cubes were utilized in the experiment. For preserving strength variation, 344 cubes were cast in the laboratory. Before being tested, the cubes were immersed in a water tank for 7 and 28 days. Compression testing was performed on each cube to determine its compressive strength. The compressive strength was calculated using the results of the tests, as shown in Figure 16.3.

16.5 DATA SETS

The concrete mixtures are subjected to compressive strength testing before being used. Using these inputs and a single variable (compressive strength), the impact of numerous input variables as well as the total number of data is critical in determining the result of models. The data set includes 344 compressive strength test results, which will be used

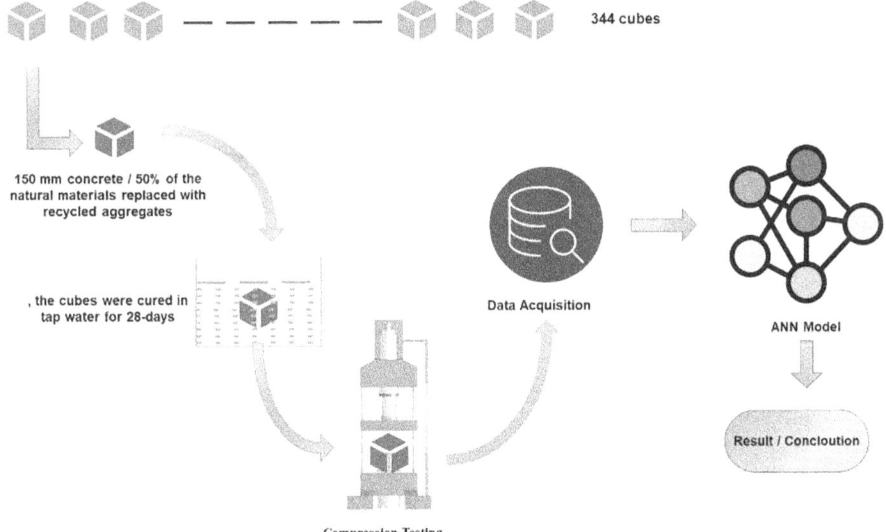

FIGURE 16.3 The methodology of work.

to construct neural network models. The ANN model was run in MATLAB® utilizing backpropagation in combination with the MATLAB® software. In Table 16.1, it can be observed that the statistical analysis description of all the parameters is included.

16.6 ANALYTIC STUDY USING

ANN is comprised of essential components that work in parallel and are inspired by the biological nervous system. The connections of components influence network function, just as they do in nature. An ANN may be trained for a specific purpose by observing changes in the values of the connections between components. Neural networks are often modified or trained for a specific output for a specified input (Demuth and Beale, 1993). Due to the reason that ANNs contain three levels: input layer, an output layer, and hidden layers that are linked with neurons, parallel distributed processing systems are also referred to as ANNs. Neurons receive weighted inputs from other neurons, and other neurons use the outputs of the neurons they receive in a process known as a function, which is a mathematical operation. The main objective of training the model for minimizing the difference between expected and actual outputs to the smallest possible value (Badagha et al., 2018).

The hidden layer's neurons form the relationship between inputs and outputs. The input components (such as cement, sand, coarse aggregate, superplasticizer, water–cement ratio, age, and so on) influence the ANN model's output. One or more output layers may compute the ANN model output. The computational propagation is carried out in a feed-forward way from the input layer to the output layer to compare the targets and the obtained discoveries. If the error is significant, it is propagated back to the network, where it is utilized to adjust the input components to decrease it. Figures 16.2 and 16.4 depict the ANN algorithm.

For the mix design, ANN assessed compressive strength using multiple network models. The ANN model is depicted schematically in Figure 16.4. After training

TABLE 16.1
The Description of Input Parameters*

	Water (kg/m³)	Cement (kg/m³)	FA (kg/m³)	NA (kg/m³)	RA (kg/m³)	SP (kg/m³)	SRA (mm)	DRA (kg/m³)	WRA (%)	Strength (MPa)
Count	344	344	344	344	344	344	344	344	344	344
Mean	184.45	386.86	681.75	398.88	649.86	1.32	19.75	2231.06	4.80	44.39
Std	30.23	82.16	205.22	370.14	378.41	2.05	4.02	580.95	2.26	15.62
Min	117.6	158	0	0	52	0	10	0	0	13.4
25%	165	340	642	0	297.75	0	19	2320	3.7	35.475
50%	180	380	698	471	552	0	20	2362.5	4.9	43.25
75%	205.05	413	811.78	733.5	982	2.59	20	2420	5.8	51.53
Max	452	600	1010	1448.25	1778	7.8	32	2661	10.9	108.5

* FA, fine aggregate; NA, natural aggregate; SP, superplasticizer, SRA, maximum size of recycled aggregates; DRA, density of recycled aggregates; WRA, water absorption of recycled aggregates.

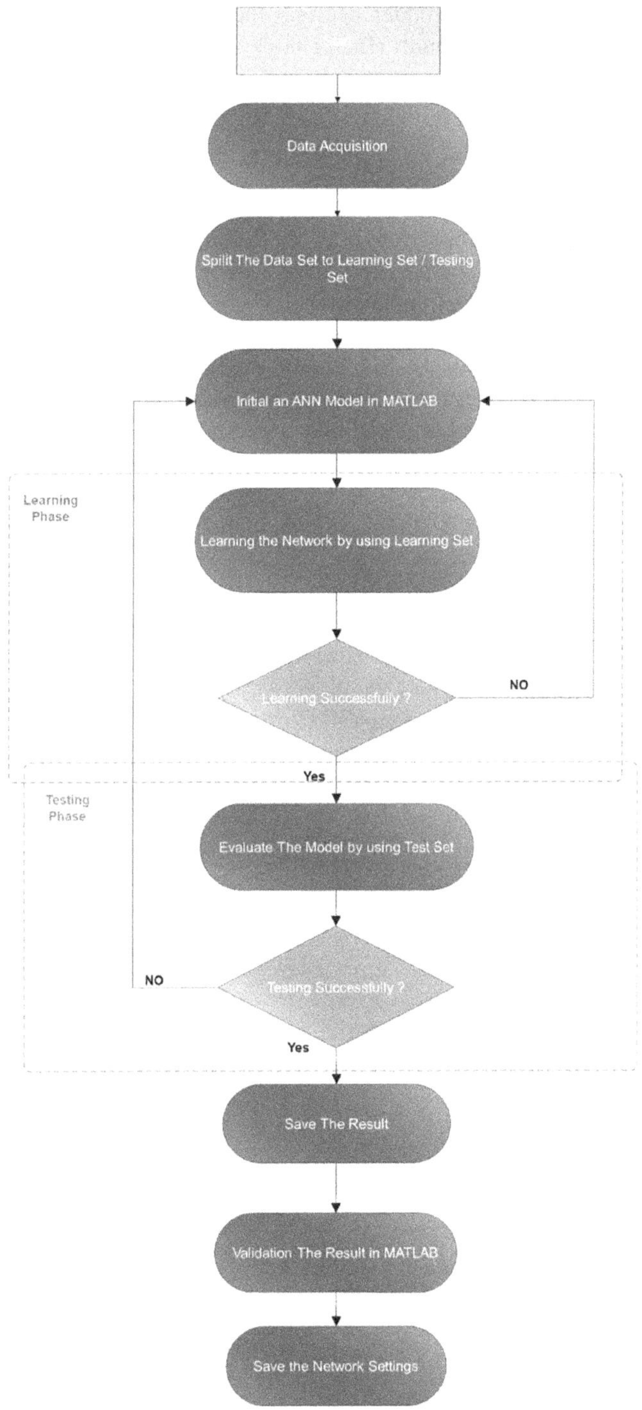

FIGURE 16.4 ANN development algorithmic framework.

multiple networks with varied hidden neurons, the predicted results were compared to the actual compressive strength (Badagha et al., 2018). A strong link has been established between actual compressive strength and expected results. Furthermore, a relation was found between actual and predicted values of ANNs.

16.7 RESULT AND DISCUSSION

It is necessary to conduct comparisons between the targets and the obtained discoveries so that the feed-forward computational propagation. As soon as the error becomes significant, it propagates back to the network, where it can be utilized to modify the input components in order to decrease the error. In this experiment, the ANN toolbox, which is integrated with the MATLAB® program was applied (Figure 16.5). Combining the backpropagation training methodology with the Levenberg–Marquardt technique resulted in developing a two-layer feed-forward network with a recurrent network structure. While dividing the data sets into training–testing sets is tricky, many researchers prefer using *k*-fold cross-validation to evaluate machine learning models. It can give accurate indicators about the model generalization by avoiding biased segregation of the data.

The *k*-fold cross-validation techniques divide data sets into *K* subsets where data cannot be overlapped between any data sets; therefore, each fold is used as a hold-out test set once while the other folds are used for training. After training the model for *K* times, the mean performance of the model evaluated on each fold is reported. In this study, five folds are used to ensure 20% testing data at each step.

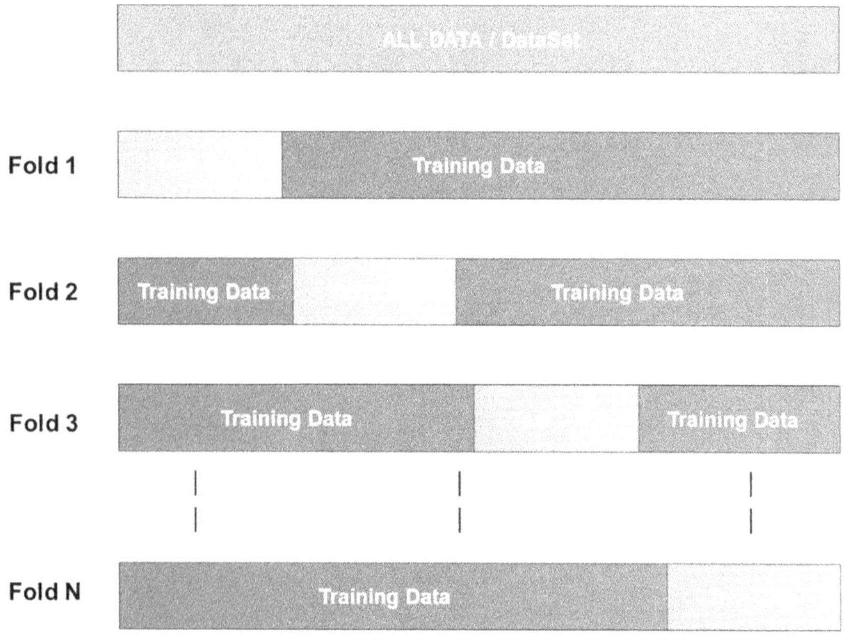

FIGURE 16.5 Integration of the MATLAB® program based on *k*-fold cross-validation.

In order to predict the strength of specific approaches, the final equations from Figure 16.6 have been established using the *k*-fold cross-validation ANNs for different numbers of nodes in the hidden layers which have been found that using is giving the best model among them all as shown in Table 16.2 With $R=0.93$ and NMSE=0.13. Based on Figure 16.9, the coefficient of determination for the curve is 0.95431, showing that the predicted and actual strengths are well correlated, and the training performance was 21.3366 at epoch 196, as shown in Figure 16.7 and A representation of each parameter used in the mixes, as well as their respective frequency distributions and frequencies, is shown in Figure 16.8. Therefore, the use of analysis by ANN is proven to be quite beneficial in forecasting the strength of single techniques.

The ANN models, as shown in Figure 16.9, provide a statistical analytical depiction of the actual and expected findings derived from the ANN models, as well as their

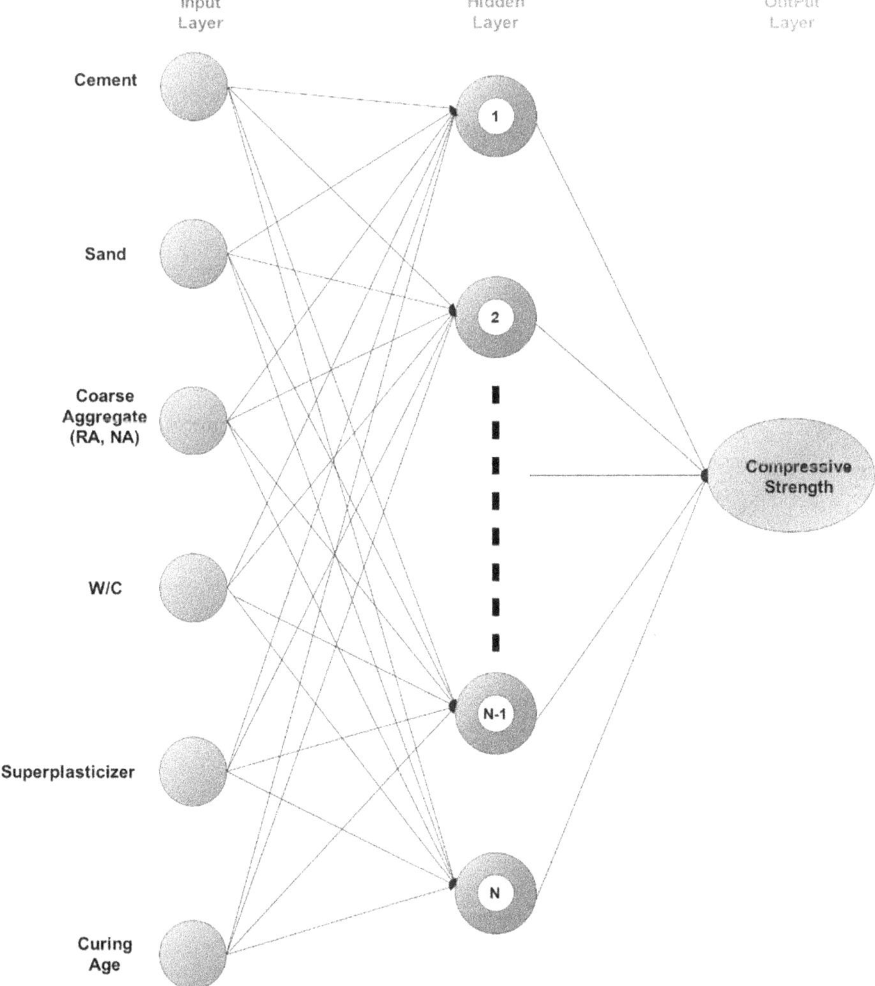

FIGURE 16.6 Structure of the suggested research work based on the ANN algorithm.

TABLE 16.2
The Results of the *k*-Fold Cross-Validation ANNs

Model	No. of Nodes	R	NMSE
1	3	0.92	0.16
2	5	0.86	0.27
3	**7**	**0.93**	**0.13**
4	9	0.90	0.19
5	11	0.93	0.14
6	13	0.91	0.18
7	15	0.88	0.24
8	17	0.87	0.24
9	19	0.88	0.27
10	21	0.92	0.15
11	23	0.92	0.16
12	50	0.91	0.17

Bold values indicate higher coefficient of correlation with lowers errors.

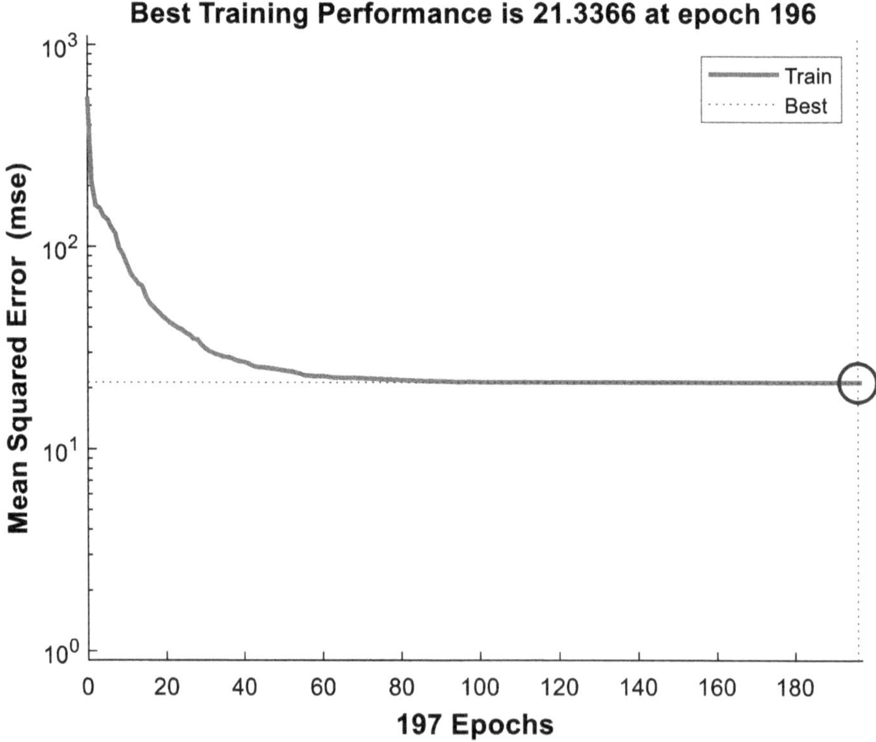

FIGURE 16.7 Training performance.

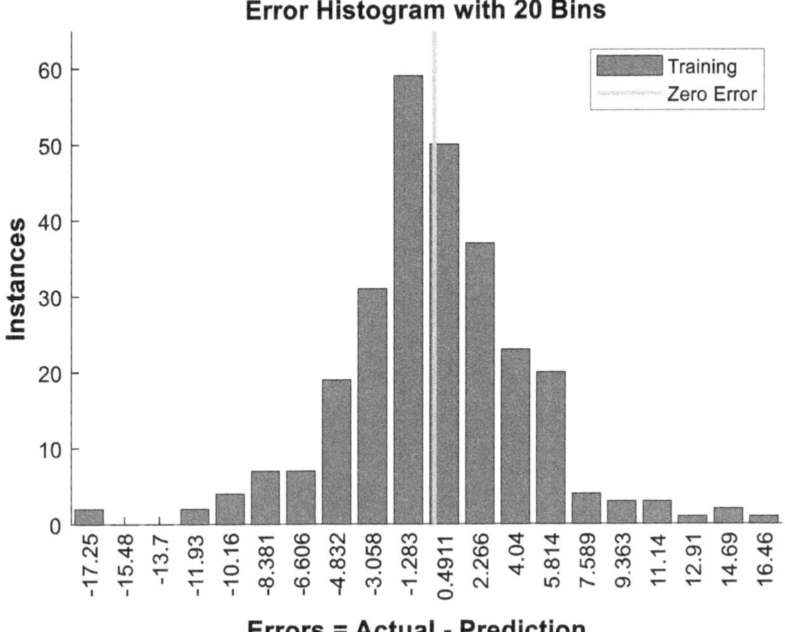

FIGURE 16.8 Error histogram.

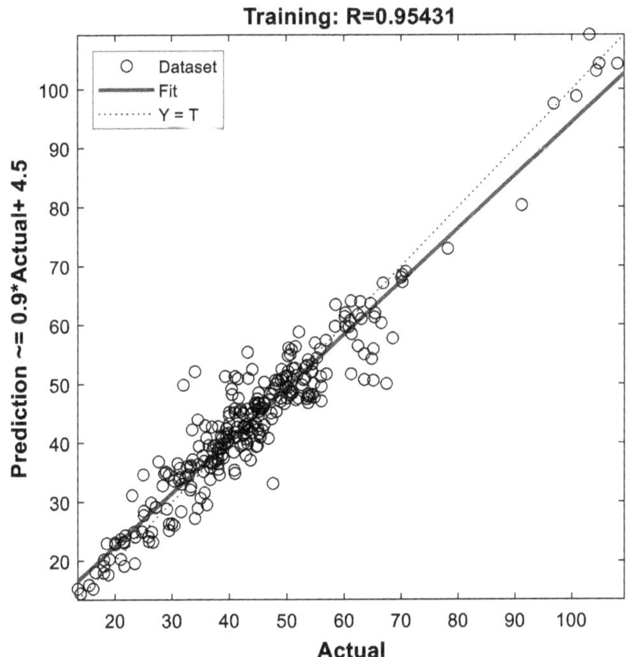

FIGURE 16.9 Compressive strength relation between predicted and actual values of the best model.

TABLE 16.3
The Relationship between Actual and ANN Strengths Values

NO.	Strength	Equation	Correlation (*R*)
1	Compressive Strength	$0.9\,f'c + 4.5$	0.95431

$f'c$, actual value of compressive strength.

TABLE 16.4
The Results of the *k*-Fold Cross-Validation ANNs after Deleting Features

Parameters	Total	Without Cement (kg/m³)	Without SRA (mm)	Without Cement (kg/m³) and SRA (mm)	With only Cement (kg/m³) and SRA (mm)
R best	0.95	0.95	0.90	0.85	0.78
NMSE	0.13	0.3092	0.19	0.28	0.4

error distributions. Figure 16.9 depicts the coefficient correlation (*R*) value, indicating that the model works very well in forecasting. Table 16.3 shows the connection between the actual and predicted strengths, as well as the interaction between the two.

As indicated in Table 16.4, this research investigates the effect of different variables on the compressive strength. The variables that have been provided have a significant impact on the predicted outcomes. The graph demonstrates that cement made the most significant contribution, while the lowest came from recycled coarse aggregates of the largest particle size. The researcher employed the following method to determine how much each variable contributed to the model's output. The first one includes all input parameters, whereas the second omitted cement and had correlation *R* and normalized mean square error (NMSE) of 0.95 and 0.3092, respectively, indicating that it is insufficient. The third did not include Spectro-Radiometric Calibration Assembly (SRCA). The correlation *R* and the NMSE were 0.9 and 0.19, respectively, which seems to be effective, and the second step of the study was carried out without both parameters, and the *R*-value and the NMSE were 0.85, 0.28, respectively. Only these two parameters were included in the final analysis, and the *R*-value and NMSE were both 0.78.

16.8 CONCLUSION

The book chapter shows that an ANN can predict a material's strength that is incredibly near to the real strength of the material. The prediction accuracy depends on the training of neural networks, and various mix designs are employed to achieve this. The ANN network's accuracy in predicting the compression and compressive strength has been shown to be acceptable. The neural network model 8 is determined to be the most optimum of all models, producing the least MSE and the highest *R*-value with the target outputs using a single hidden layer of seven nodes.

When the final chosen and adjusted ANN models were evaluated using the data sets, the *R*-value was 0.95431. This demonstrates the value of the created neural network models. Thus, dividing the data into five groups (folds) using the *k*-fold cross-validation helps create and choose the best ANN model and estimate its ultimate error, demonstrating a strong relationship between the predicted and actual strength of concrete cubes. As a result, ANN analysis has been proved valuable for forecasting the strength of single approaches.

Utilizing the *k*-fold cross-validation technique can further validate the back-propagation model's outstanding performance. There was a significant difference between predicting compressive strength and forecasting the maximum size of recycled aggregate depending on which cement was used and which SRCA size was utilized. When predicting the compressive strength, engineers may make use of ANNs, which are comprised of the following ingredients: cement, fly ash, silica sand, coarse aggregate, water, and a superplasticizer (giving suitable or desired compressive strength).

ACKNOWLEDGMENT

The authors would like to acknowledge the Department of Civil Engineering and the Centre of Excellence in Energy Science and Technology Department, Faculty of Engineering & Technology, Shoolini University, Department of Civil and Architectural Engineering, KTH Royal Institute of Technology, and The Computer Science and Engineering Department, Vel Tech University, for their assistance and support.

REFERENCES

Ahmadi-Nedushan, B., 2012. An optimized instance based learning algorithm for estimation of compressive strength of concrete. *Engineering Applications of Artificial Intelligence*, 25(5), pp. 1073–1081.

Akbari, M., Kabir, H.D., Khosravi, A. and Nasirzadeh, F., 2021. ANN-based LUBE model for interval prediction of compressive strength of concrete. *Iranian Journal of Science and Technology, Transactions of Civil Engineering*, pp. 1–11.

Badagha, D.G. and Modhera, C.D., 2017. M55 grade concrete using industrial waste to minimize cement content incorporating CO_2 emission concept: An experimental investigation. *Materials Today: Proceedings*, 4(9), pp. 9768–9772.

Badagha, D.G., Modhera, C.D. and Vasanwala, S.A., 2018. Mix proportioning and strength prediction of high performance concrete including waste using artificial neural network. *International Journal of Civil and Environmental Engineering*, 12(2), pp. 169–172.

Bandyopadhyay, G. and Chattopadhyay, S., 2007. Single hidden layer artificial neural network models versus multiple linear regression model in forecasting the time series of total ozone. *International Journal of Environmental Science & Technology*, 4(1), pp. 141–149.

Basma, A.A., Barakat, S.A. and Al-Oraimi, S., 1999. Prediction of cement degree of hydration using artificial neural networks. *ACI Materials Journal*, 96(2), pp. 167–172.

Demuth, H. B., and Beale, M. H., 2000. *Neural network toolbox: for use with MATLAB: user's guide*. MathWorks, Incorporated.

Etim, M.A., Babaremu, K., Lazarus, J. and Omole, D., 2021. Health risk and environmental assessment of cement production in Nigeria. *Atmosphere*, 12(9), p. 1111.

Fernando, S., Gunasekara, C., Law, D.W., Nasvi, M.C.M., Setunge, S. and Dissanayake, R., 2021. Life cycle assessment and cost analysis of fly ash–rice husk ash blended alkali-activated concrete. *Journal of Environmental Management*, 295, p. 113140.

Flood, I. and Kartam, N., 1994. Neural networks in civil engineering. I: Principles and understanding. *Journal of Computing in Civil Engineering*, 8(2), pp. 131–148.

Gencel, O., Erdugmus, E., Sutcu, M. and Oren, O.H., 2020. Effects of concrete waste on characteristics of structural fired clay bricks. *Construction and Building Materials*, 255, p. 119362.

Getahun, M.A., Shitote, S.M. and Gariy, Z.C.A., 2018. Artificial neural network based modelling approach for strength prediction of concrete incorporating agricultural and construction wastes. *Construction and Building Materials*, 190, pp. 517–525.

Ghezal, A. and Khayat, K.H., 2002. Optimizing self-consolidating concrete with limestone filler by using statistical factorial design methods. *Materials Journal*, 99(3), pp. 264–272.

Jain, A., Siddique, S., Gupta, T., Jain, S., Sharma, R.K. and Chaudhary, S., 2021. Evaluation of concrete containing waste plastic shredded fibers: Ductility properties. *Structural Concrete*, 22(1), pp. 566–575.

Kapadia, D. and Jariwala, N., 2021. Prediction of tropospheric ozone using artificial neural network (ANN) and feature selection techniques. *Modeling Earth Systems and Environment*, pp. 1–10.

Karakoç, M.B., Demirboğa, R., Türkmen, I. and Can, I., 2011. Modeling with ANN and effect of pumice aggregate and air entrainment on the freeze–thaw durabilities of HSC. *Construction and Building Materials*, 25(11), pp. 4241–4249.

Mukherjee, A. and Biswas, S.N., 1997. Artificial neural networks in prediction of mechanical behavior of concrete at high temperature. *Nuclear Engineering and Design*, 178(1), pp. 1–11.

Nazari, A., and Riahi, S., 2012. Retracted: Experimental investigations and ANFIS prediction of water absorption of geopolymers produced by waste ashes. *Journal of Non Crystalline Solids*. 358(1), pp. 40–46.

Okamura, H. and Ouchi, M., 2003. Self-compacting concrete. *Journal of Advanced Concrete Technology*, 1(1), pp. 5–15.

Ouchi, M., 2000. Self-compacting concrete-development, applications and investigations. *Nordic Concrete Research-Publications*, 23, pp. 29–34.

Plevris, V. and Asteris, P.G., 2014. Modeling of masonry failure surface under biaxial compressive stress using Neural Networks. *Construction and Building Materials*, 55, pp. 447–461.

Sarıdemir, M., 2009. Predicting the compressive strength of mortars containing metakaolin by artificial neural networks and fuzzy logic. *Advances in Engineering Software*, 40(9), pp. 920–927.

Schalkoff, R.J., 1997. *Artificial Neural Networks*. McGraw-Hill Higher Education.

Shahmansouri, A.A., Yazdani, M., Ghanbari, S., Bengar, H.A., Jafari, A. and Ghatte, H.F., 2021. Artificial neural network model to predict the compressive strength of eco-friendly geopolymer concrete incorporating silica fume and natural zeolite. *Journal of Cleaner Production*, 279, p. 123697.

Thakur, M.S., Pandhiani, S.M., Kashyap, V., Upadhya, A. and Sihag, P., 2021. Predicting bond strength of FRP bars in concrete using soft computing techniques. *Arabian Journal for Science and Engineering*, 46(5), pp. 4951–4969.

Upadhya, A., Thakur, M.S., Pandhian, S.M. and Tayal, S., 2021a. Estimation of marshall stability of asphalt concrete mix using neural network and M5P tree. In *Computational Technologies in Materials Science* (pp. 223–236). CRC Press.

Upadhya, A., Thakur, M.S., Sharma, N. and Sihag, P., 2021b. Assessment of soft computing-based techniques for the prediction of Marshall stability of asphalt concrete reinforced with glass fiber. *International Journal of Pavement Research and Technology*, pp. 1–20.

Wang, B., Man, T. and Jin, H., 2015. Prediction of expansion behavior of self-stressing concrete by artificial neural networks and fuzzy inference systems. *Construction and Building Materials*, 84, pp. 184–191.

Yadollahi, A., Nazemi, E., Zolfaghari, A. and Ajorloo, A.M., 2016. Optimization of thermal neutron shield concrete mixture using artificial neural network. *Nuclear Engineering and Design*, 305, pp. 146–155.

Yazdanbakhsh, A., Bank, L.C., Baez, T. and Wernick, I., 2018. Comparative LCA of concrete with natural and recycled coarse aggregate in the New York City area. *The International Journal of Life Cycle Assessment*, 23(6), pp. 1163–1173.

Yeh, I.C., 1998. Modeling of strength of high-performance concrete using artificial neural networks. *Cement and Concrete Research*, 28(12), pp. 1797–1808.

Zamora-Castro, S.A., Salgado-Estrada, R., Sandoval-Herazo, L.C., Melendez-Armenta, R.A., Manzano-Huerta, E., Yelmi-Carrillo, E. and Herrera-May, A.L., 2021. Sustainable development of concrete through aggregates and innovative materials: A review. *Applied Sciences,* 11(2), p. 629.

17 Compressive Strength Prediction and Analysis of Concrete Using Hybrid Artificial Neural Networks

Ahmad Alyaseen, Arunava Poddar, Fadi Almohammed, and Salwan Tajjour
Shoolini University

Karam Hammadeh
Vel Tech Rangarajan Dr. Sagunthala R&D Institute of Science and Technology

Hussain Alahmad
KTH Royal Institute of Technology

CONTENTS

DOI: 10.1201/9781003184331-17

17.1 INTRODUCTION

The interest in repurposing waste has increased considering recent environmental regulations such as fly ash (FA) and granulated blast furnace slag that regulate waste disposal. A portion of FA and granulated blast furnace slag is mixed with Portland cement to substitute cement in concrete (Yang et al., 2019). FA, a byproduct of coal-fired power plant emissions, has been utilized in sizeable concrete mass applications to decrease hydration heat and encourage early crack development. Concrete buildings made with FA demonstrate improved strength and longevity over time (Yang et al., 2019; ALABI and Mahachi, 2020). FA has recently been a commonly used component in the production of high-strength concrete. In order to evaluate FA concrete's mechanical and physical characteristics, such as workability, compressive strength, tensile strength, and elastic modulus, FA concrete structure's workability must be established. Model creation begins with the mechanical compressive strength test, which may be used to estimate other attributes (Yang et al., 2019; ALABI and Mahachi, 2020; Ravitheja et al., 2021).

Based on experimental data, analytical formulae in regression equations are often employed to determine the concrete properties. Depending on the design mix components used in concrete manufacture. These empirical correlations fail to provide the needed prediction accuracy when the number of factors impacting the tangible object is nonlinear, complex, or unknown neural networks, modeled after the human brain learns, are more straightforward when modeling material behavior issues than prior approaches. Modeling concrete compressive strength dependent on the design mix percent is an excellent example of an unstructured issue (Chandwani et al., 2015, Hodhod and Ahmed 2013). The compressive strength of FA and Greedy Best First Search (GBFS) concrete has been shown improvement (Chandwani et al., 2015; Contrafatto et al., 2017).

Additionally, concrete with excellent strength and energy absorption (Dias and Pooliyadda, 2001; Jain et al., 2006) has greater compressive strength. The back-propagation (BP) technique taught by artificial neural networks (ANNs) was employed in this work for simulating the material behavior of concrete (Sardemir, 2009; Diab et al., 2014; Kumar et al., 2020) which is in addition to the previous research stated above. Even though back-propagation neural networks (BPNNs) are extensively utilized and well-known, these techniques suffer from the inherent problem of being trapped at local minimum values even while a much deeper minimum value is close. Also have a slow convergence rate, which makes them less efficient in their operations. Although these techniques are extensively utilized and have a broad range of applications, this is the case. The system would not find acceptable responses for these limitations since the random initial weights lead the ANNs to act differently each time it is retrained, making it impossible to discover the responses. It is possible to search for answers to several problems simultaneously because GAs, which are modeled after evolutionary processes such as natural selection and genetic variation, utilize stochastic search (Kumar et al., 2020). Su et al., (2011) found that the hybrid ANN-GA training method surpasses conventional BP training in neural network performance. Instead of many hybridization techniques to improve GA applications' initial weighting and biases, no combination of GA and ANN has been used to model concrete compressive strength. Using a hybrid approach, GA, stochastic global search, and the local search BP algorithm are utilized to generate the initial weights and biases. In concrete mixes, the components comprised cement, FA, sand, coarse aggregate, superplasticizer, and water, and a hybrid ANN-GA method was used to simulate the strength of the mix design. For simulating the compressive strength of the concrete mix design components, computer simulation software was used. The relative significance of each RMC component on the compressive strength value must be estimated using the synaptic weights from the qualifying ANN-GA model.

The purpose of this chapter is to find the optimum mixing percentage of sintered FA aggregate concrete using GA produced from neural networks. The ultimate objective is to increase the quantity of FA utilized in the concrete. Using multilayer feedforward neural networks, it is possible to predict the compressive strength of concrete while accounting for the layered impact of the concrete.

17.2 MATERIAL

17.2.1 Training and Testing Neural Networks with Exemplar Data

According to the way the human brain processes information, ANN is a learning paradigm that was developed. Neural networks may also generate associations and learn connections between input and output data by drawing on information received via the network. Neural networks are data-intensive because the overall quality and amount of data are essential for their success. A local RMC company provided concrete compressive strength example patterns for use in neural network modeling of concrete. The proportions of the concrete design mix and the compressive strength test results were among the findings. Table 17.1 shows the dataset results that included 144 concrete design mixes with varying cement proportions, FA, sand, coarse aggregate, superplasticizer, water content, and compressive strength. In order to avoid bias, the datasets were randomly divided: research, validation, and test, with

TABLE 17.1
The Description of Input Parameters[a]

	Cement	Sand	Aggregate	W/C	Curing	Fly Ash	Superplasticizer (b%)	Compressive Strength (MPa)
Count	144	144	144	144	144	144	144	144
Mean	382.875	682.75	998.5	0.415	112.167	0.15	0.012	49.669
Std	95.186	80.113	38.294	0.121	128.952	0.112	0.0093	18.496
Min	231	541	936	0.27	3	0.00	0.00	7.2
25%	310.5	627	960.25	0.320	7	0.075	0.00	38.225
50%	371	681.5	1008.5	0.375	59	0.15	0.015	49.65
75%	441	769.25	1028.25	0.55	180	0.225	0.0185	62.925
Max	600	788	1056	0.6	365	0.3	0.03	87.4

[a] C = Cement, S = Sand, A = Aggregate, W/C = Water-Cement ratio, C = Curing, FA = Fly ash, SP = superplasticizer, CS = Compressive Strength.

the training dataset accounting for around 70% of the total data. The remaining 30% of the data was shared evenly between validation and testing, except validation datasets used for both.

17.3 METHODS

The BPNN and GA were implemented in the neural network structure to carry out the research, and the global optimization algorithm was implemented in the very competitive MATLAB® program.

17.3.1 Data Pre-processing

As stated earlier, illustrative data for modeling concrete compressive strength includes cement, FA, sand, coarse aggregate, superplasticizer, water, and compressive strength value. The data must be normalized in the range of −1 to +1 or 0 to +1 since it consists of several components with varying content characteristics and maximum–minimum ranges. This data pre-processing ensures that the network weighs all variables equally to remove any hidden bias towards any variable. Another advantage of this design is that, since the output values are most sensitive to changes in the input values in this region, learning is faster (Alshihri et al., 2009). Previous research utilized a linear scaling function from −1 to +1, which was applied again in this study.

$$x_{\text{norm}} = \frac{2 * (x - x_{\min})}{(x_{\max} - x_{\min})} - 1 \tag{17.1}$$

where x_{norm} is normalized value and $x_{\max}$ and $x_{\min}$ are minimum and maximum values.

17.3.2 Training Parameters and Neural Networks Structure

ANNs is the newest technique in artificial intelligence; it is the main element to develop deep learning and create an intelligent system with its control. It has stimulated the human brain, an intelligent approach to humans designed as a complex neuron cell network (Liu et al., 2017, Gevrey et al., 2003).

ANN architecture had a set of nodes which divided into layers (input layer, hidden layer can be multi-layers or zero, output layer), biases, weights between these nodes, and the ways of flow the information on the neural network as feedforward networks, the signal in this model transfer in one way from the input to output layer. Furthermore, in feedback networks, the signal can travel in both directions by the loops between the hidden layers, and this type of connection has a memory for an internal state to process a sequence of inputs, as shown in Figure 17.1 (Abraham, 2005). According to Benardos and Vosniakos (2007), there are four elements to clarify optimal ANN's architecture:

1. Number of layers (hidden layers).
2. Number of nodes.

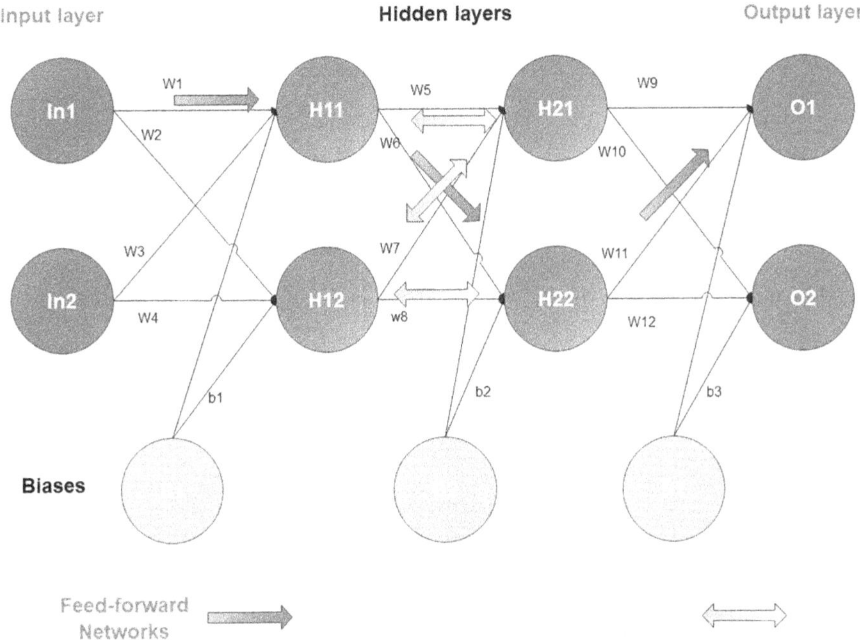

FIGURE 17.1 ANN architecture.

3. Type of active function.
4. The algorithm used in training plays an important role in determining the value of weight and biases of nodes.

In this method; the algorithm repeatedly updates the weights of nodes using the gradient descent or delta rule, referred to as gradient descent or delta rule (Zweiri et al., 2003; Ampazis et al., 1999).

Figure 17.2 shows that the backpropagation depends on two steps: forward propagation and backward propagation to update the weight and parameters of nodes to reach a minimum error and then the best prediction.

Synapses between nodes in ANNs may be weighted to imitate actual synapses in the human brain in a virtual environment. Nature's layers of nodes, also known as artificial neurons, known as the building elements of computer systems. A neural network can include several layers of connections. When it comes to prediction and forecasting, multilayer feedforward neural networks are commonly employed. An input layer, output, and some hidden layers are sandwiched between these three layers of information. The feedforward neural network is completely wired, with only forward connections between inter-layer neurons permitted between the layers of the network's architecture.

It comprises neurons responsible for receiving signals from the outside environment, referred to as "the input layer." For the purposes of this research, the "input layer" comprised of seven neurons, which were as follows: cement, sand (coarse aggregate), superplasticizer, and water content. In the "output layer," one neuron

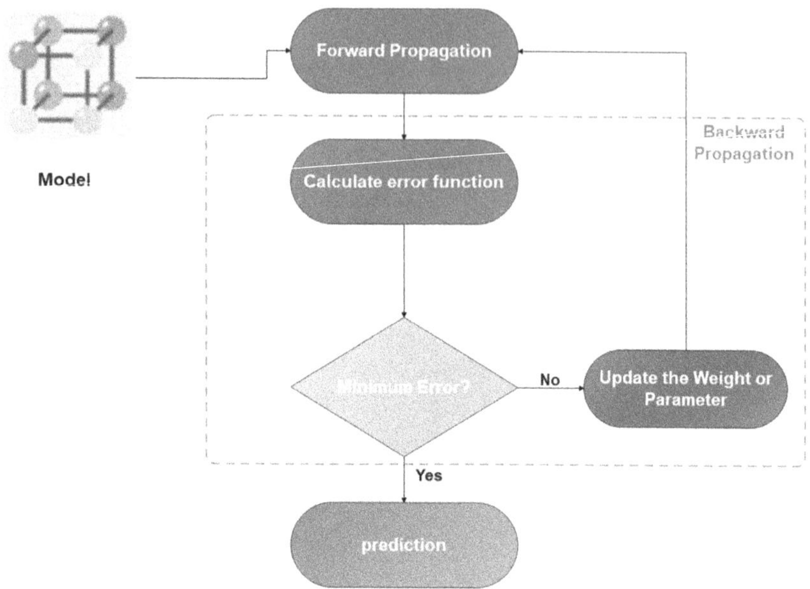

FIGURE 17.2 The backpropagation algorithm.

represented the compressive strength in millimeters, and this was the sole neuron in the system. The number of hidden layers, as well as neurons that must be utilized in the approximation technique, must be set per the complexity of the approximation method being used (Hornik et al., 1989). A hidden layer neural network may approximate any functional link, and it has the number of hidden layer neurons necessary to do the required task (Tamura and Tateishi, 1997; Hunter et al., 2012).

After a considerable amount of trial and error, it is eventually established that the number of hidden layers and neurons is adequate in this case. Applying a single hidden layer neural network architecture to a problem required seven different designs with hidden numbers ranging from five to eleven. Figure 17.3 shows the neural network design of five hidden layer neurons for modeling the concrete strength, which was developed to mimic the concrete strength and was shown in the previous paragraph.

Using a transfer mechanism to send a weighted total of all signals arriving at one neuron to all other neurons, input neurons get messages. Nonlinear connections between input–output data pairs are learned as they are received and processed. The hyperbolic transition characteristic is used when transferring input neurons to hidden neurons, as shown in this chapter. Using a linear transfer function, information was sent from the hidden to output layer, which was compared to the projected output of the ANN, to verify that the information was appropriately communicated.

The multilayer feedforward neural networks are trained using a BP method. BP networks are the most often used type of network in practice (BPNN). Gradient descent is a gradient-based approach used to find the values of weights and biases of neural networks that minimize the expected and the actual values. While the BP method may be speeded up if a significant learning rate is used, it is not possible to

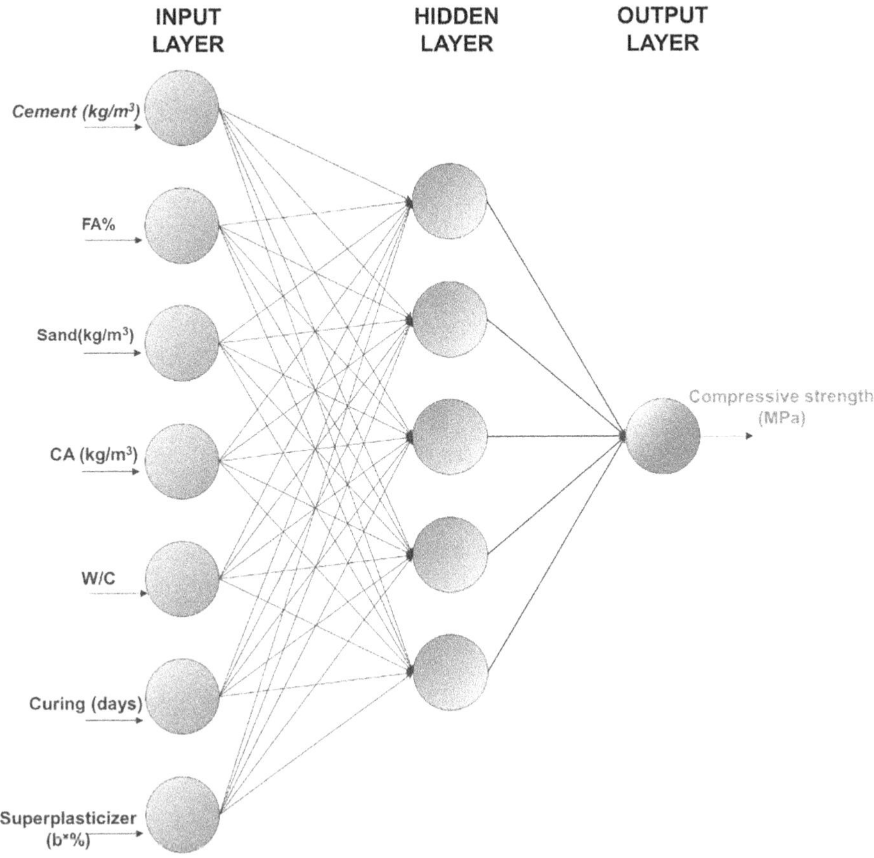

FIGURE 17.3 Neural network structure.

use them to converge the process in a short amount of time; ultimately, doing so would require re-evaluating the weights and biases regularly during each learning cycle. In addition to speeding up convergence, a high learning rate improves the network's chance to exceed the global minimum value. The momentum coefficient is used to adjust synaptic weights in order to maintain synaptic connectivity. Because of this, the learning rate may be increased, which speeds up convergence without causing weight oscillations. Due to the weight movements that have influenced the present course of travel in the weights, it effectively filters out high-frequency oscillations of error surfaces (Rajasekaran and Pai, 2003). When the BPNN's learning rate is proportional to the momentum coefficient, the BPNN's convergence is accelerated. As in the previous study, a learning rate of 0.45 and a momentum coefficient of 0.85 were used, as was the Lavenberg–Marquardt BP method, which also employed a learning rate of 0.45 a momentum coefficient of 0.85 in this study (Wilamowski et al., 1999). The Gauss–Newton technique is employed when the solution is near the local minimum.

17.3.3 ANN Training and Determination Using the BP Algorithm

Following the procedures described in the previous section, the Lavenberg–Marquardt BP style of training was used to train the different neural network topologies created by varying the number of hidden layer neurons, and the training parameters of neural networks were trained. The initial weights and biases were chosen at random and set to a range between −0.5 and +0.5 before training in order to prevent bias from being introduced into the analysis. Initially, a training dataset was sent to the neural network, and the amount of training error was determined by comparing the accuracy of the neural network's predictions with the actual results of the experiment. In order to prevent the testing dataset from overfitting, an early stopping method was developed. By distributing the validation dataset to certified neural networks and monitoring training errors at each training period, it is possible to compute and track a validation error as well as a tracking training error over time. When the validation error increases, the neural network training stops, although the training error continues to decrease. The neural network model is selected based on having the lowest validity error to model a concrete slump. Figure 17.4 illustrates a flowchart for the creation and assessment of a neural network using the Backpropagation method. The neural networks model (7-5-1) was shown to produce the least amount of validity error, and as a result, it was selected for use in this study.

17.3.4 Using GA to Evolve Neural Networks' Initial Weights and Biases and Then Training Them Using the BP Technique

The genetic algorithm (GA), one of the most intelligent and global search techniques, is used to optimize a dynamic problem with several difficult-to-obtain parameters or functions (Gopan et al., 2018; Ahmad et al., 2010; Kumar et al., 2020). This evolutionary optimization algorithm is comprised of four steps: population, selection, crossover, and mutation. The way the population is presented (chromosome), the health mechanism in the selection, and the mutation rate play a significant role in obtaining an optimum response.

In order to identify critical parameters of an ANN, such as the starting weights and the optimum number of hidden layers on the network, GA may be used. The primary goal of using a hybrid GA model is to minimize the objective function while increasing the power and speed of ANN by adjusting a set of weights and biases, as shown in Figure 17.5 This hybrid algorithm is ubiquitous in engineering, particularly civil engineering, for problems such as the Travel salesman problem, concrete, and anvil design (Gopan et al., 2018; Momeni et al., 2014; Tonnizam Mohamad et al., 2016; Saemi et al., 2007).

In Figure 17.6, according to (Tonnizam Mohamad et al., 2016; Goldberg and Holland, 1988), Note that after process dataset/feature on GA, the result will be input for ANN in training and testing, and the output will evaluate according to fitness and reprocessing with next generation of GA.

Both GA and ANN are utilized in the hybridization process, which is split into two phases: first, GA is used to evolve neural networks' start weights and biases, followed by the BP approach, which is a variant of the BP method. After processing the data, an ANN must be run and compare the output with the predicted values.

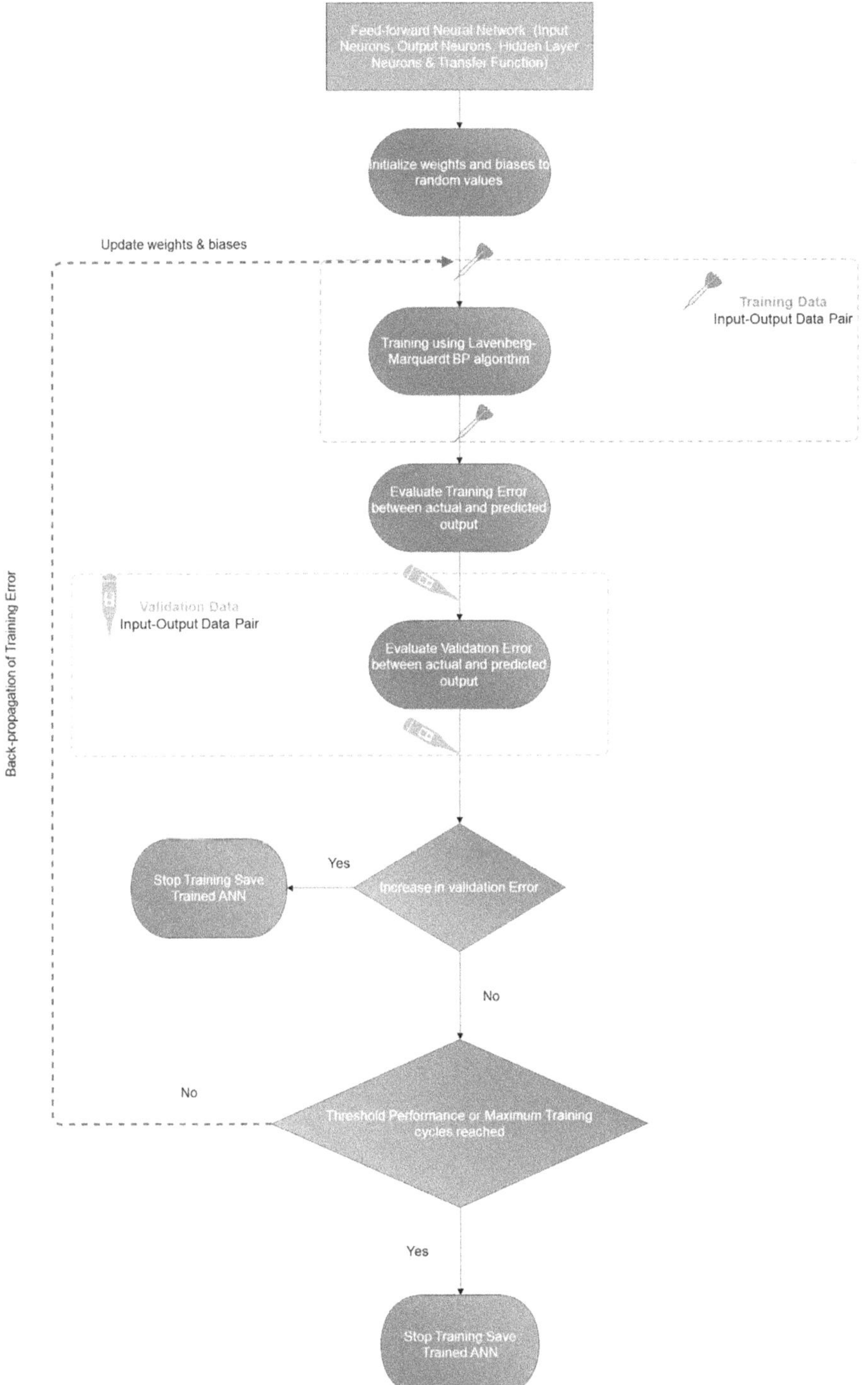

FIGURE 17.4 Training and validation of neural networks using the BP algorithm.

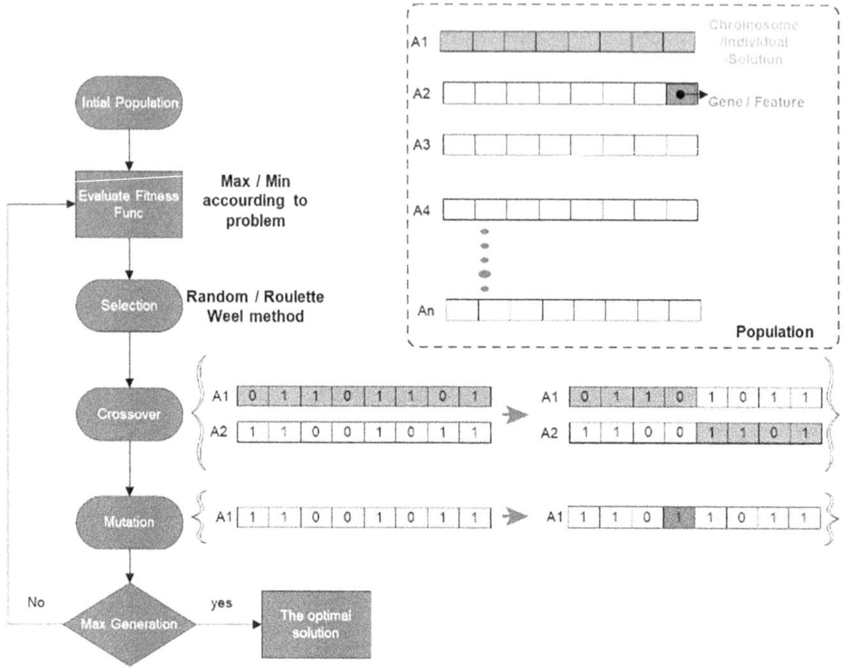

FIGURE 17.5 The component and mechanism of GA.

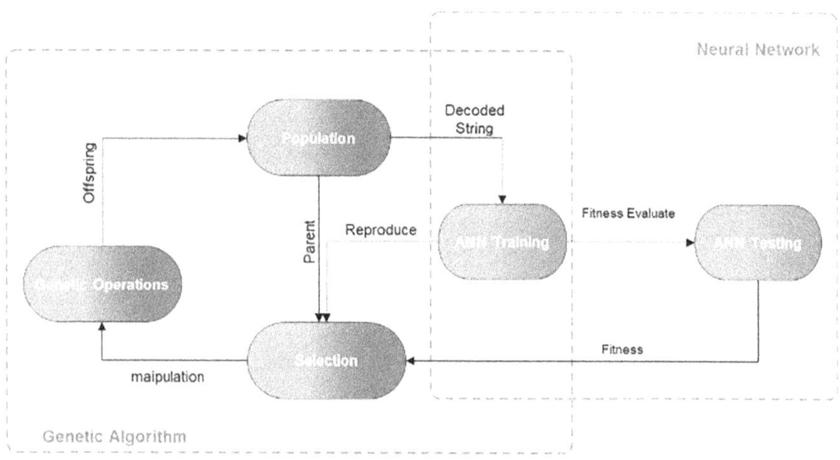

FIGURE 17.6 The hybrid algorithm of ANN and GA.

Then, the root means square error (RMSE) should be computed between the actual and anticipated outputs and repeat the process for each chromosome. The exercise machine provides a metric for differentiating between optimum and sub-optimal survival solutions, which compares potential survival solutions. Chromosome reproduction efficiency is measured by the number of chromosomes that reproduce.

GA can perform a wide range of stochastic chromosomal operations, including genetic operations such as crossover and mutation and evolutionary operations such as selection. The sorting operator separates the fitter chromosomes from the fewer fit chromosomes, which improves population fitness across generations.

When two or more chromosomes combine to form a new generation, this is known as crossover recombination; by combining the strengths of the two or more parent chromosomes, crossover recombination aids in the development of more competent individuals in the population. The cross-site between the two chromosomes is selected randomly, and the genes are exchanged between the parent and offspring chromosomes to give rise to better offspring; it is possible to replicate healthy chromosomes. By spontaneously changing the foundation blocks, mutation incorporates genetic variation into the new population. It encourages the algorithm to explore the whole solution space and keeps it stuck in local minima. The next wave of the population is generated by the crossover and mutation operators working together. The method is repeated until the number of generations or a specific value is achieved or the health characteristics plateau at a particular value has been reached. A roulette wheel selection approach was utilized in the current sample with an initial population size of 50 chromosomes, a scattered crossover operator with a crossover likelihood of 0.9, a standardized mutation with a mutation probability of 0.01, and a cumulative number of generations of 100.

The GA-evolved neural network weights and preferences are then used to train an ANN using the BP technique, with the ANN weights and preferences acquired via the GA being used in the training process. Using the BP technique, which is a variant of the B algorithm and is begun using GA-generated weights and biases, the ANN is trained until it is as accurate as possible. As shown in Figure 17.7, GA may be used to change the weights and biases of ANN models, which are then used to construct ANN models using the changed weights and biases.

17.4 STATISTICAL ANALYSIS

Five different statistical measures were used for all of the BPNN and hybrid ANN-GA models that qualified. For example, RMSE, correlation coefficient, Nash–Sutcliffe efficiency, MAPE, and Small and Medium Business Enterprise (SMBE) are some of the statistical output metrics that may be computed (Number algorithm, NMBE).

$$\text{RMSE} = \sqrt{\frac{1}{N}\sum_{i=1}^{N}(T_i - P_i)^2} \tag{17.2}$$

$$R = \frac{\sum_{i=1}^{N}\left((T_i - \bar{T})(P_i - \bar{P})\right)}{\sqrt{\sum_{i=1}^{N}(T_i - \bar{T})^2 \sum_{i=1}^{N}(P_i - \bar{P})^2}} \tag{17.3}$$

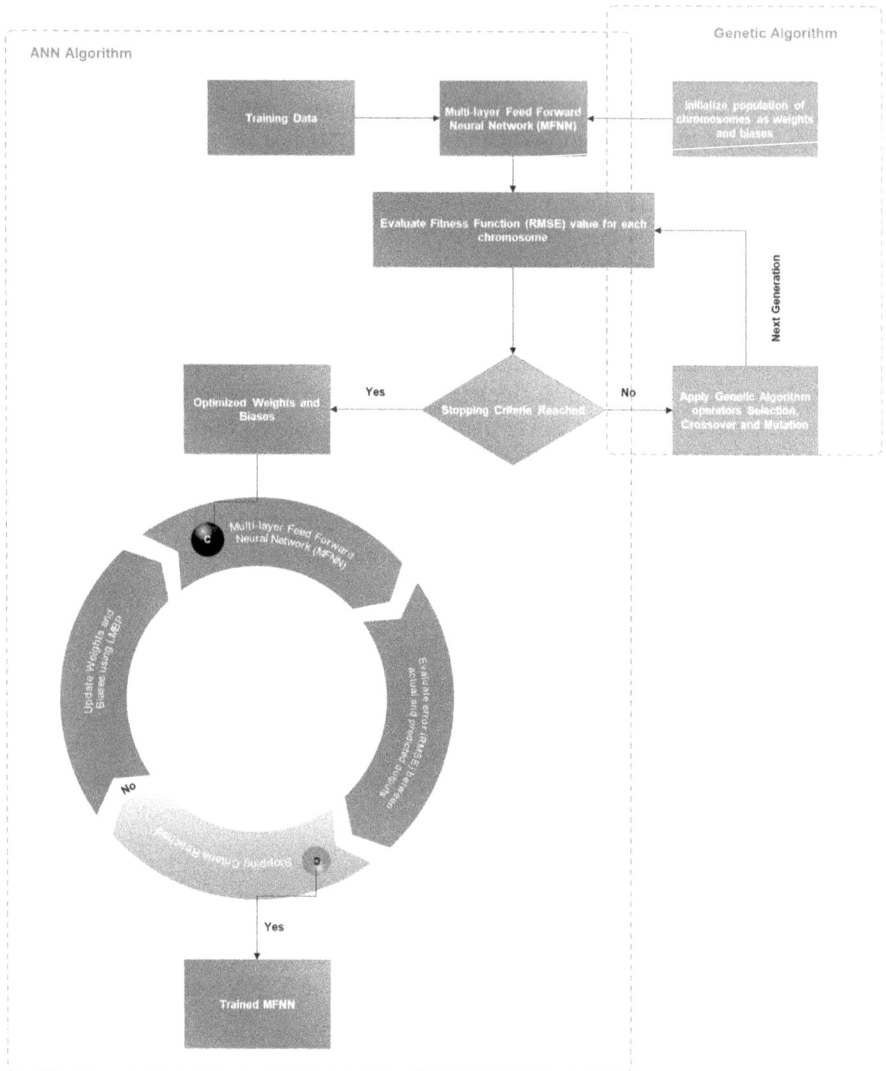

FIGURE 17.7 Incorporating GA with BP to evolve weights and biases.

$$E = 1 - \frac{\sum_{i=1}^{N} (T_i - P_i)^2}{\sum_{i=1}^{N} (T_i - \overline{T})^2} \tag{17.4}$$

$$\text{MAPE}(\%) = \frac{1}{N} \sum_{i=1}^{N} \frac{|T_i - P_i|}{T_i} \times 100 \tag{17.5}$$

$$\text{NMBE}(\%) = \frac{\frac{1}{N}\sum_{i=1}^{N}(P_i - T_i)}{\frac{1}{N}\sum_{i=1}^{N}T_i} \times 100 \qquad (17.6)$$

where T_i is the target values, P_i is the predicted values, $\overline{T}$ is the root mean square error of target values, and $\overline{P}$ is the root mean square error of predicted values.

A better prediction has a smaller RMSE; nevertheless, this statistic is biased towards forecasts with significant degrees of inaccuracy. In statistics, the correlation coefficient (R) reflects how strongly two variables are related to one another. This indicates that the connection between the actual and projected values is nearly perfect in this instance. In fact, a score of 1.0 was produced by the simulation. E is defined as the ratio of the residual error variance in actual outcomes to the anticipated variance in a regression model with a fixed efficiency coefficient (fixed coefficient of efficiency). In this instance, the model is correct, as shown by the value of one. The mean relative absolute error (MAPE) statistic is calculated by dividing the mean absolute error by the observed value, given in percentage. The lower the value of this statistic, the higher the accuracy of the forecasting procedure. The capacity of the model to estimate a value that varies from the mean value is evaluated using the NMBE. A high NMBE indicates that the model overpredicts, whereas a low NMBE indicates that the model underpredicts a given situation (Srinivasulu and Jain, 2006). As previously stated, when the performance utilization and prediction ability metrics are combined, they provide a reasonable approximation of the neural network models' prediction ability.

17.5 RESULTS

Following the pattern outlined in the previous section, the hybridization of GA and ANN occurred throughout two different periods of time. During the first phase, the GA was used to determine the optimum weight and bias settings for training ANNs using the BP approach, and during the second phase, the BP methodology was used. For this second training, the model was trained for 1000 generations on 200 populations using the identical ANN parameters used for the first training. A total of 1000 generations of runs was required for the GA performance to converge to optimum weights and biases, which took 342 seconds (Figure 17.8).

To forecast concrete compressive strength (7-5-1), the second step of the neural network architecture began with weights and biases developed by the Georgia Institute of Technology and concluded with the BP technique utilized in the first stage. The hybrid ANN-GA achieved the targeted performance goal of 0.003 in 43 epochs in 41.73 seconds, faster than the previous record of 41.73 seconds set by another hybrid ANN-GA (Figure 17.9). For achieving the required performance objective, it took 1374 epochs and 19 seconds (Figure 17.10) and the best validation of the BPNN model performance, which is shown in Figure 17.11. It took 1374 epochs and 19 seconds for the BPNN model to reach its desired validation performance. It took

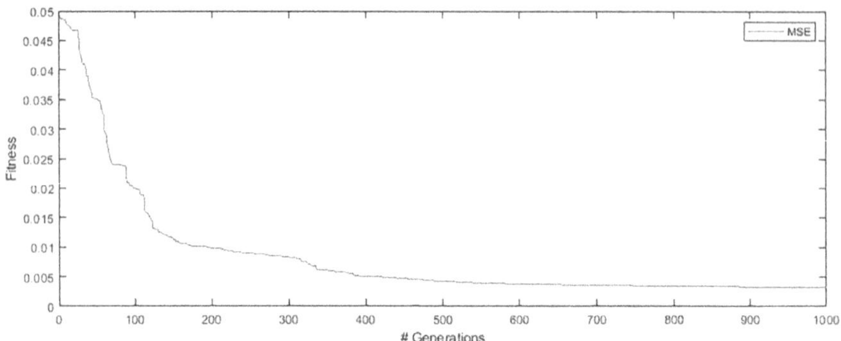

FIGURE 17.8 Generations versus fitness function.

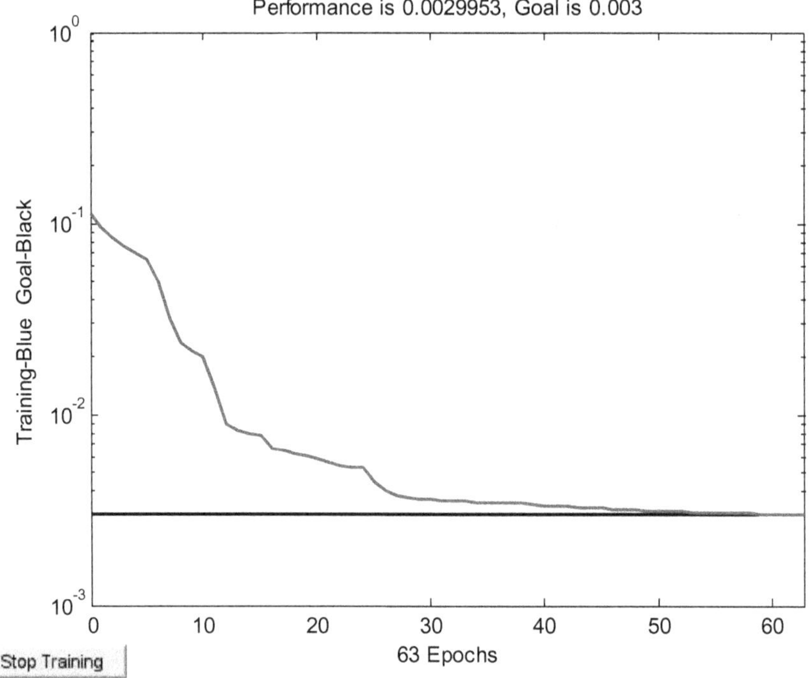

FIGURE 17.9 Training of ANN-GA model.

1374 epochs and 19 seconds to train the same neural networks design using the BP method. Neural networks trained using the BP method use the same set of weights has been trained.

The validated and tested BPNN and ANN-GA models were trained. Table 17.2 shows the findings in terms of statistical performance measures.

The trained models were also put to the test against all of the data from the concrete mix design. Figures 17.12 and 17.13 illustrate regression lines between actual and projected compressive strength values for BPNN and ANN-GA models. Table 17.3 shows a statistical performance of entire datasets.

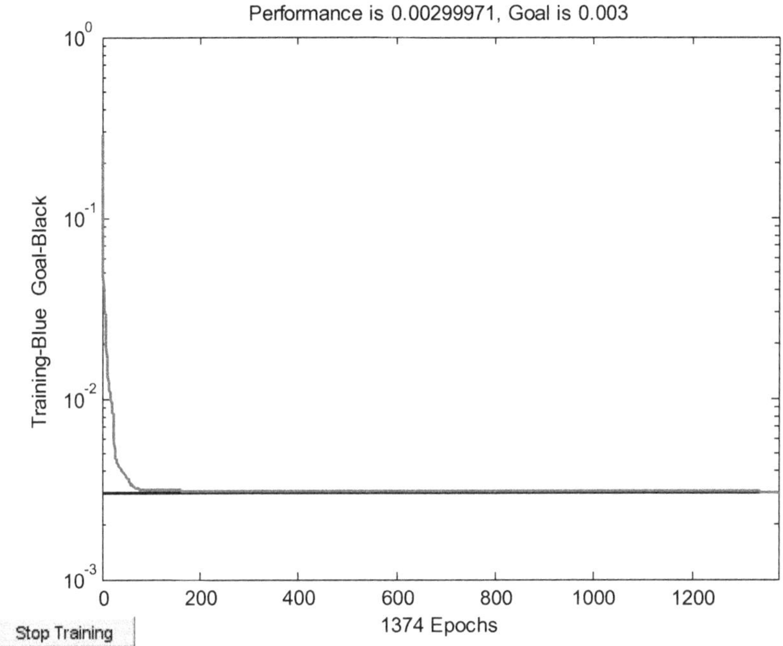

FIGURE 17.10 Training of BPNN model.

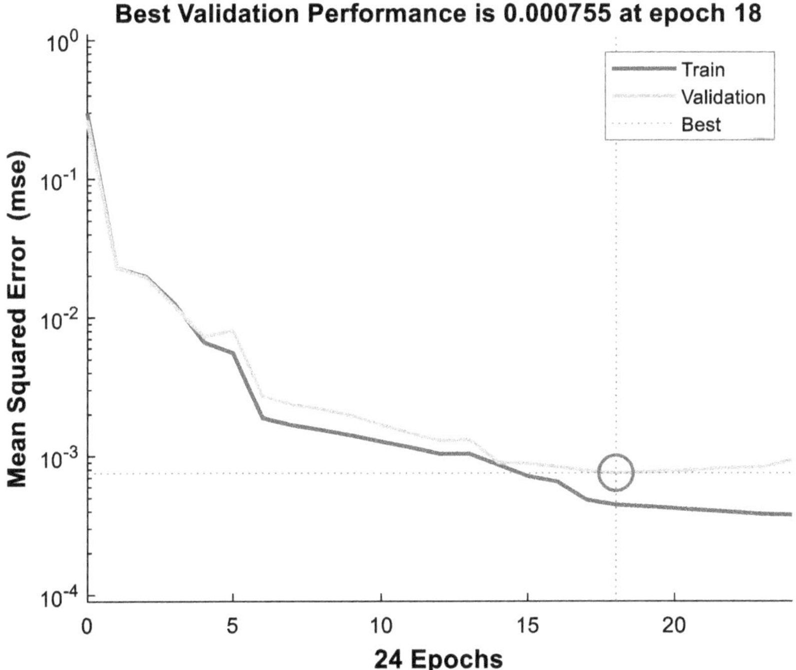

FIGURE 17.11 Best validation of BPNN model.

TABLE 17.2
BPNN ANN-GA Model Statistical Performance

Model	RMSE (mm)	R	E	MAPE (%)	NMBE (%)
			Training		
BPNN	0.0264	0.9981	0.9772	0.2028	0.07
ANN-GA	0.0482	0.9921	0.9625	0.2471	0.00062
			Testing		
BPNN	0.0358	0.9699	0.9370	0.2529	0.0012
ANN-GA	0.068	0.9694	0.8948	0.3218	0.08

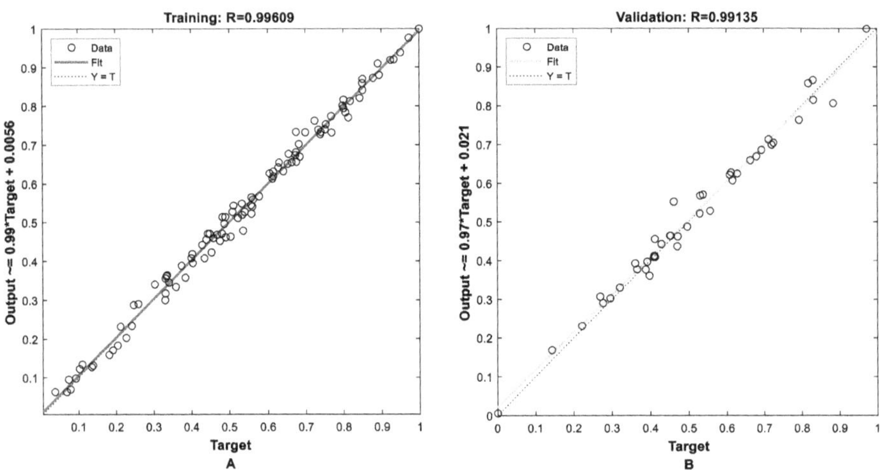

FIGURE 17.12 Regression plot for BPNN model.

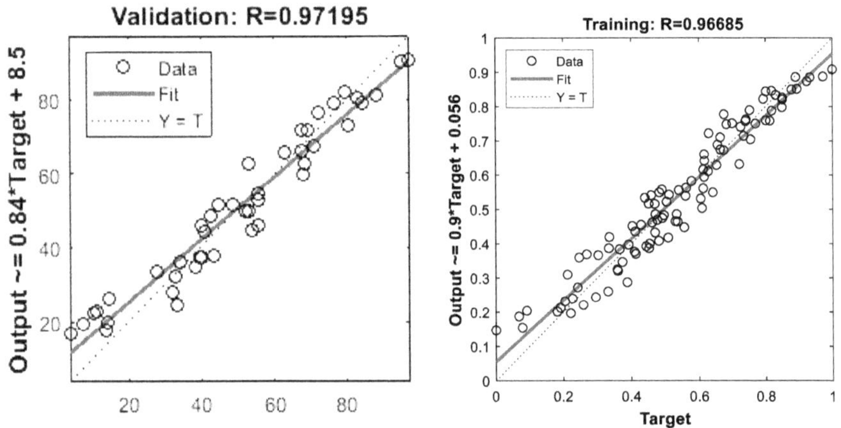

FIGURE 17.13 Regression plot for ANN-GA model.

TABLE 17.3
ANN Model Statistical Performance

Model	RMSE (mm)	*R*	*E*	MAPE (%)	NMBE (%)
BPNN	0.03	0.994	0.9648	0.22	0.006
ANN-GA	0.06	0.984	0.9646	0.28	0.02

17.6 DISCUSSION

It is fair to presume that GA and ANN can address the BP algorithm's issues of being stuck at local minima and experiencing delayed convergence due to the chapter findings. ANN training using the BP technique required 1374 trials and 19 seconds, while the ANN-GA needed 41.73 seconds to reach an acceptable level of performance.

This made accurate predictions during both the training and testing phases, demonstrating its superior accuracy. In fact, statistics *R* and *E*, as well as lower values of MAPE and RMSE, demonstrate that this is the case. While the ANN-GA model yielded values of 0.07% and 0.0012%, the BPNN model produced values of 0.00062% and 0.08%. This section uses numerical examples to demonstrate the predictability that the hybrid BPNN model provides.

The trained BPNN model was found to have lower RMSE and MAPE values (both at 0.03 millimeter and 0.22%) and higher values of *E* and *R* (both at 0.9648 and 0.994). As seen by the NMBE statistics for both the BPNN and ANN-GA models being 0.006 and 0.02, respectively, it is evident that the ANN-GA model underpredicts the compressive strength values in this experiment. However, the BPNN was able to predict compressive strength values that were within a few percent of their optimal values. As a result, thc statistical research demonstrates that BPNN models regularly outperform ANN-GA models regarding the accuracy of prediction predictions.

17.7 CONCLUSION

Modeling concrete compressive strength based on design mix proportions is a highly nonlinear issue that is difficult to solve using traditional mathematical methods. Because of that, the ANN with BP training method has historically been employed to simulate material behavior, and the initial weights and biases heavily affect its convergence probability since it relies on the gradient descent methodology. This method has a significant disadvantage that adversely affects the performance of ANN. It is feasible to hide the BP algorithm's tendency to converge at suboptimal locations by using ANN and GA. While combining two different methods increases the global search capacity of GA while simultaneously improving the searchability of the BP algorithm, it does it more effectively. According to the presented hybrid method, the GA algorithm was used to find the optimum initial set, then fine-tuned using the Levenberg-Marquardt (LMBP) technique for the initial set of neural network weights and biases.

Compared to the commonly used hybrid ANN-GA model, the research found that the BPNN method provided consistent predictions throughout the training and testing stages, demonstrating the resilience of the hybrid modeling approach. Furthermore, the BPNN model converged to the required performance level faster than the ANN-GA model. When used as a decision support tool, the proposed model may assist technical staff in estimating the compressive strength of a specific concrete mix design. With this technique, the amount of labor and time needed to produce a concrete mix with a particular compressive strength would be substantially reduced, and there would be no need for repeated experiments.

REFERENCES

Abraham, A., 2005. Artificial neural networks. *Handbook of Measuring System Design*, edited by Peter H. Sydenham and Richard Thorn. Oklahoma State University, Stillwater, OK, USA.

Ahmad, F., Mat-Isa, N. A., Hussain, Z., Boudville, R., & Osman, M. K., 2010, July. Genetic algorithm-artificial neural network (GA-ANN) hybrid intelligence for cancer diagnosis. In *2010 2nd international conference on computational intelligence, communication systems and networks* (pp. 78–83). IEEE.

Alshihri, M.M., Azmy, A.M. and El-Bisy, M.S., 2009. Neural networks for predicting compressive strength of structural light weight concrete. *Construction and Building Materials*, 23(6), pp. 2214–2219.

Alabi, A. S., and Mahachi, J., 2020. Behaviour of ground cupola furnace slag blended concrete at elevated temperature. *Journal of Materials and Engineering Structures (JMES)*, 7(1), pp. 35–46.

Ampazis, N., Perantonis, S. and Taylor, J., 1999. Dynamics of multilayer networks in the vicinity of temporary minima. *Neural Networks*, 12(1), pp. 43–58.

Benardos, P. and Vosniakos, G., 2007. Optimizing feedforward artificial neural network architecture. *Engineering Applications of Artificial Intelligence*, 20(3), pp. 365–382.

Chandwani, V., Agrawal, V., and Nagar, R., 2015. Modeling slump of ready mix concrete using genetic algorithms assisted training of Artificial Neural Networks. *Expert Systems with Applications*, 42(2), pp. 885–893.

Contrafatto, L., Cuomo, M., and Greco, L., 2017. Meso-scale simulation of concrete multiaxial behaviour. *European Journal of Environmental and Civil Engineering*, 21(7–8), pp. 896–911.

Diab, A.M., Elyamany, H.E., Abd Elmoaty, M. and Shalan, A.H., 2014. Prediction of concrete compressive strength due to long term sulfate attack using neural network. *Alexandria Engineering Journal*, 53(3), pp. 627–642.

Dias, W.P.S. and Pooliyadda, S.P., 2001. Neural networks for predicting properties of concretes with admixtures. *Construction and Building Materials*, 15(7), pp. 371–379.

Gevrey, M., Dimopoulos, I. and Lek, S., 2003. Review and comparison of methods to study the contribution of variables in artificial neural network models. *Ecological Modelling*, 160(3), pp. 249–264.

Gopan, V., Wins, K. and Surendran, A., 2018. Integrated ANN-GA approach for predictive modeling and optimization of grinding parameters with surface roughness as the response. *Materials Today: Proceedings*, 5(5), pp. 12133–12141.

Hodhod, O.A. and Ahmed, H.I., 2013. Developing an artificial neural network model to evaluate chloride diffusivity in high performance concrete. *HBRC Journal*, 9(1), pp. 15–21.

Hornik, K., Stinchcombe, M. and White, H., 1989. Multilayer feedforward networks are universal approximators. *Neural Networks*, 2(5), pp. 359–366.

Hunter, D., Yu, H., Pukish III, M.S., Kolbusz, J. and Wilamowski, B.M., 2012. Selection of proper neural network sizes and architectures—A comparative study. *IEEE Transactions on Industrial Informatics*, 8(2), pp. 228–240.

Jain, A., Jha, S.K. and Misra, S., 2008. Modeling and analysis of concrete slump using artificial neural networks. *Journal of Materials in Civil Engineering*, 20(9), pp. 628–633.

Kumar, N., Maharshi, S., Poddar, A. and Shankar, V., 2020, July. Evaluation of artificial neural networks for estimating reference evapotranspiration in western Himalayan region. In *2020 International Conference on Computational Performance Evaluation (ComPE)* (pp. 163–167). IEEE.

Liu, W., Wang, Z., Liu, X., Zeng, N., Liu, Y. and Alsaadi, F., 2017. A survey of deep neural network architectures and their applications. *Neurocomputing*, 234, pp. 11–26.

Momeni, E., Nazir, R., Jahed Armaghani, D. and Maizir, H., 2014. Prediction of pile bearing capacity using a hybrid genetic algorithm-based ANN. *Measurement*, 57, pp. 122–131.

Ravitheja, A., Kumar, G. P., and Anjaneyulu, C. M., 2021. Impact on cementitious materials on high strength concrete–A review. *Materials Today: Proceedings*, 46, pp. 21–23.

Saemi, M., Ahmadi, M. and Varjani, A., 2007. Design of neural networks using genetic algorithm for the permeability estimation of the reservoir. *Journal of Petroleum Science and Engineering*, 59(1–2), pp. 97–105.

Srinivasulu, S. and Jain, A., 2006. A comparative analysis of training methods for artificial neural network rainfall–runoff models. *Applied Soft Computing*, 6(3), pp. 295–306.

Su, C.L., Yang, S.M. and Huang, W.L., 2011. A two-stage algorithm integrating genetic algorithm and modified Newton method for neural network training in engineering systems. *Expert Systems with Applications*, 38(10), pp. 12189–12194.

Tabrizi, S. S., and Sancar, N., 2017. Prediction of Body Mass Index: A comparative study of multiple linear regression, ANN and ANFIS models. *Procedia computer science*, 120, pp. 394–401.

Tamura, S.I. and Tateishi, M., 1997. Capabilities of a four-layered feedforward neural network: four layers versus three. *IEEE Transactions on Neural Networks*, 8(2), pp. 251–255.

Wilamowski, B.M., Chen, Y. and Malinowski, A., 1999, July. Efficient algorithm for training neural networks with one hidden layer. In *IJCNN'99. International Joint Conference on Neural Networks. Proceedings (Cat. No. 99CH36339)* (Vol. 3, pp. 1725–1728). IEEE.

Yang, Z., Dai, Z., Yang, Y., Carbonell, J., Salakhutdinov, R. R., and Le, Q. V., 2019. Xlnet: Generalized autoregressive pretraining for language understanding. *Advances in Neural Information Processing Systems*, 32, pp. 1–11.

Yeh, I.C., 2007. Modeling slump flow of concrete using second-order regressions and artificial neural networks. *Cement and Concrete Composites*, 29(6), pp. 474–480.

Zweiri, Y., Whidborne, J. and Seneviratne, L., 2003. A three-term backpropagation algorithm. *Neurocomputing*, 50, pp. 305–318.

Index

For Product Safety Concerns and Information please contact our EU
representative GPSR@taylorandfrancis.com
Taylor & Francis Verlag GmbH, Kaufingerstraße 24, 80331 München, Germany

www.ingramcontent.com/pod-product-compliance
Lightning Source LLC
Chambersburg PA
CBHW060330220326
41598CB00023B/2663

* 9 7 8 1 0 3 2 0 2 6 3 5 0 *